Sustainable Environment and Carbon Balance Management

Sustainable Environment and Carbon Balance Management

Edited by Eugene Richards

☐ SYRAWOOD
PUBLISHING HOUSE

New York

Published by Syrawood Publishing House,
750 Third Avenue, 9th Floor,
New York, NY 10017, USA
www.syrawoodpublishinghouse.com

Sustainable Environment and Carbon Balance Management
Edited by Eugene Richards

International Standard Book Number: 978-1-68286-794-5 (Hardback)

Cataloging-in-Publication Data

Sustainable environment and carbon balance management / edited by Eugene Richards.
    p. cm.
Includes bibliographical references and index.
ISBN 978-1-68286-794-5
1. Sustainable development. 2. Environmental policy. 3. Carbon--Environmental aspects.
4. Global environmental change. I. Richards, Eugene.
GE105 .S87 2019
338.927--dc23

TABLE OF CONTENTS

PREFACE

Carbon dioxide undergoes exchange between the earth, biosphere, air and water. The processes of exchange between these reservoirs are mediated through the processes of transpiration, combustion, respiration and decomposition. The balance achieved between these is called the carbon balance. The widespread use of fossil fuels has led to the accumulation of carbon dioxide in the atmosphere resulting in adverse consequences, such as global warming. These effects can be managed by adhering to carbon cycle rebalancing strategies, which involve carbon capture and storage, promotion of carbon-offset mechanisms, adoption of sustainable energy, design and transport, etc. An example of a sustainable environmental practice is the burning of domestic waste to generate power. This book brings forth some of the most innovative concepts and elucidates the unexplored aspects of carbon balance management. Also included in this book is a detailed explanation of the strategies for sustainable environment. It will serve as a reference to a broad spectrum of readers.

This book unites the global concepts and researches in an organized manner for a comprehensive understanding of the subject. It is a ripe text for all researchers, students, scientists or anyone else who is interested in acquiring a better knowledge of this dynamic field.

I extend my sincere thanks to the contributors for such eloquent research chapters. Finally, I thank my family for being a source of support and help.

Editor

An assessment of the carbon stocks and sodicity tolerance of disturbed *Melaleuca* forests in Southern Vietnam

Da B Tran[1*], Tho V Hoang[1] and Paul Dargusch[2]

Abstract

Background: In the lower Mekong Basin and coastal zones of Southern Vietnam, forests dominated by the genus *Melaleuca* have two notable features: most have been substantially disturbed by human activity and can now be considered as degraded forests; and most are subject to acute pressures from climate change, particularly in regards to changes in the hydrological and sodicity properties of forest soil.

Results: Data was collected and analyzed from five typical *Melaleuca* stands including: (1) primary *Melaleuca* forests on sandy soil (VS1); (2) regenerating *Melaleuca* forests on sandy soil (VS2); (3) degraded secondary *Melaleuca* forests on clay soil with peat (VS3); (4) regenerating *Melaleuca* forests on clay soil with peat (VS4); and (5) regenerating *Melaleuca* forests on clay soil without peat (VS5). Carbon densities of VS1, VS2, VS3, VS4, and VS5 were found to be 275.98, 159.36, 784.68, 544.28, and 246.96 tC/ha, respectively. The exchangeable sodium percentage of *Melaleuca* forests on sandy soil showed high sodicity, while those on clay soil varied from low to moderate sodicity.

Conclusions: This paper presents the results of an assessment of the carbon stocks and sodicity tolerance of natural *Melaleuca cajuputi* communities in Southern Vietnam, in order to gather better information to support the improved management of forests in the region. The results provide important information for the future sustainable management of *Melaleuca* forests in Vietnam, particularly in regards to forest carbon conservation initiatives and the potential of *Melaleuca* species for reforestation initiatives on degraded sites with highly sodic soils.

Keywords: Carbon sequestration, Climate change, *Melaleuca*, REED+, Sodicity

Background

Numerous studies have shown that tropical wetlands typically contain large carbon stocks [1–7]. Protecting and restoring tropical coastal wetlands is considered a critical part of how society adapts to and mitigates global climate change [8].

Large areas of *Melaleuca* forests in Vietnam are disturbed ecosystems that experience extreme conditions, and are associated with floods and/or sodic soils. They mostly occur in the lower Mekong Basin, which has been severely impacted by climate change [9–12]. Little is known about the carbon sequestration potential of disturbed *Melaleuca* forests in Australasia and South-East Asia where the genus occurs. Carbon stocks of *Melaleuca* forests are generally considered to be low (i.e. about 27.8 tC/ha estimated by Australian Government Office [13]). However, Tran et al. [14] suggested that this has been grossly under-estimated and that *Melaleuca cajuputi* forests on peatland soils in Vietnam, Indonesia and Malaysia are likely to have a high potential for carbon sequestration.

Sea level rise has significant impacts on the coastal zone, where soils will become saline and/or highly sodic [15]. Sodic soils are distinguished by an excessively high concentration of Sodium (Na) in their cation exchange complex. High sodicity causes soil instability due to poor physical and chemical properties, which affects plant growth and can have a more significant impact than excessive salinity growth [16, 17]. Sodicity impacts plant

*Correspondence: tranbinhda@gmail.com
[1] The Vietnam Forestry University, Hanoi, Vietnam
Full list of author information is available at the end of the article

growth in three ways, including: soil dispersion, specific ion effects, and nutritional imbalance in plants [18, 19]. Excessive sodium concentrations cause clay dispersion which is the primary physical effect of the sodic soil. Sodium-induced dispersion can reduce water infiltration, decrease hydraulic conductivity, and increase soil surface crusting that strongly affect roots such as root penetration, root development, and blocking plant uptake of moisture and nutrients [19].

Except for those containing mangroves and other halophytes, most ecosystems are severely affected by salinity and/or sodicity. A few studies have examined saline-sodic soils in shrimp farming areas in the coastal regions of Vietnam (i.e. ECe = 29.25 dS/m and exchangeable sodium percentage ranged from 9.63 to 72.07%, which had a big impact on plant cultivation systems [20]).

Several studies (such as Dunn et al. [21], Niknam and McComb [22], van der Moezel et al. [23, 24]) have examined the tolerance of woody species such as *Acacia*, *Eucalyptus*, *Melaleuca*, and *Casuarina* species to salinity and/or sodicity, but more research is required. This paper examines the carbon stocks of disturbed *Melaleuca* forests and the sodicity tolerance of *M. cajuputi* forests in Southern Vietnam.

Results and discussion

Characteristics of the typical *Melaleuca* forests in the study areas

The major characteristics of five *Melaleuca* forests types examined include standing trees, an understory, and saturated conditions (Table 1). The variation in these characteristics not only distinguishes the different stands but also improves understanding of their carbon stocks.

The stand densities of the five typical *Melaleuca* forest types varied considerably: they were 2,330, 10,950, 980, 9,833, and 6,867 trees/ha for VS1, VS2, VS3, VS4, and VS5, respectively (Table 1). Within each study site, the tree densities of regenerating forests (VS2, VS4, and VS5) were significantly higher than primary forests (VS1) and secondary forests (VS2) (Figure 1a). The increased stand densities of types VS2, VS4, and VS5 were mostly comprised of trees with a diameter at breast height (DBH) <10 cm. In contrast, VS1 was dominated by trees with DBH < 20 cm (accounting for 84.3%), with the balance of trees having a DBH $\geq$ 20 cm (including 4.2% of trees with DBH $\geq$ 30), while VS3 was mostly dominated by trees with a 5 cm $\leq$ DBH < 20 cm (accounting for 96%), with the balance having a 20 cm $\leq$ DBH < 40 cm (accounting for 4%) (Table 1).

Average DBH of all stand classes were 16.71, 5.36, 12.93, 5.88, and 6.20 for VS1, VS2, VS3, VS4, and VS5, respectively (Figure 1b). There was a significant difference in DBH in the five *Melaleuca* forest types (χ^2 = 446.86,

$p = 2.2e^{-16}$). However, post hoc test shows that there is no significant difference in tree DBH between VS1 and VS3, and between VS2, VS4, and VS5 (Additional file 1: 2b).

Average total height of all stand classes were 14.69, 7.11, 9.69, 5.68, and 7.50 m for VS1, VS2, VS3, VS4, and VS5, respectively (Figure 1c). There was a significant difference in the total height of the five *Melaleuca* forest types (χ^2 = 11.616, p = 0.0088) (Additional file 1: 2c). Furthermore, the tree density of the five forest types was generally very high, especially of VS2, VS4 and VS5 (over 2,000 individuals/ha), which can contribute to a large biomass. The basal areas shown in Figure 1d further confirm the potential high biomass of VS2, VS4 and VS5 (BA = 28.41, 30.14, and 23.14 m²/ha, respectively). Furthermore, the basal area of VS1 is significantly greater than VS3, accounting for 41.45 and 10.29 m²/ha, respectively (F = 3.341, p = 0.0423) (Additional file 1: 2d).

Different species were found in the understorey of the various *Melaleuca* forest types. Key species for VS1 and VS2 include *Leptocarpus* sp., *Lepironia* sp., *Hanguana* sp., *Eleocharis* sp., *Euriocaulon* sp., *Xyris* sp., *Stenochlaena* sp., *Melastoma* sp., and *Imperata cylindrica*. For VS3, VS4, VS5, the following species dominate the understorey: *Stenochlaenapalustris* sp., *Phragmitesvallatoria* sp., *Melastomadodecandrum* sp., *Diplaziumesculentum* sp., *Lygodiumscandens* sp., *Aspleniumnidus* sp., *Scleriasumatrensis*, *Cassia tora*, *Paederiafoetida* sp., *Flagellariaindica* sp., and *Cayratiatrifolia* sp. (Table 1).

Carbon stocks of *Melaleuca* forests

The carbon densities of five typical *Melaleuca* forests in Southern Vietnam were 275.98, 159.36, 784.68, 544.28, and 246.96 tC/ha, respectively, for primary *Melaleuca* forests on sandy soil (VS1), regenerating *Melaleuca* forests on sandy soil (VS2), degraded secondary *Melaleuca* forests on clay soil with peat (VS3), regenerating *Melaleuca* forests on clay soil with peat (VS4), and regenerating *Melaleuca* forests on clay soil without peat (VS5) (Figure 2a). There is significant difference in carbon densities between the forest types (χ^2 = 10.419, p = 0.0339) (Additional file 1: 2e). On sandy soils, the carbon density of VS1 was significantly greater (1.7 times) than VS2. The carbon density of *Melaleuca* forests on clay soil with peat was still high after disturbance (VS3 was 1.4 times higher than VS4). The carbon density of VS5 was lower than VS3 and VS4 because there was no peat layer.

On sandy soil, the stands and soil layers were the highest contributors to carbon density of VS1 (accounting for 41.34 and 29.11%, respectively), while VS2 has a high contribution from the soil layer, then stands (soil and stand categories contribute for carbon density of 56.15 and 28.53%, respectively) (Figure 2b). However, in

Table 1 Major characteristics of five typical _Melaleuca_ forests in the study areas

Forest types	Tree classes	Code	Stand trees Density (trees/ha) Mean	SE	DBH (cm) Mean	SE	BA (m²/ha) Mean	SE	Height (m) Mean	SE	Understory	Saturation levels
Primary _Melaleuca_ on sandy soil	DBH < 5 cm	VS1C0	800	248.3	3.87	0.11	na	na	6.00	0.28	_Leptocarpus sp._	Including non-inundated, seasonal, and permanent inundation
	5 cm ≤ DBH < 10 cm	VS1C1	400	100.0	7.18	0.36	na	na	9.81	0.68	_Lepironia sp._	
	10 cm ≤ DBH < 20 cm	VS1C2	750	273.8	14.63	0.22	na	na	14.80	0.26	_Hanguana sp._	
	20 cm ≤ DBH < 30 cm	VS1C3	285	34.0	24.33	0.49	na	na	18.44	0.40	_Eleocharis sp._	
	30 cm ≤ DBH < 40 cm	VS1C4	80	28.3	34.37	0.90	na	na	20.17	0.97	_Euriocaulon sp._	
	DBH ≥ 40 cm	VS1C5	20	10.0	48.73	3.75	na	na	22.20	1.77	_Xyris sp._	
	All classes	VS1	2,330	558.0	16.71	0.55	41.54	6.16	14.69	0.30	_Stenochlaena sp._ _Melastoma sp._ _Imperata sp._	
Regenerating _Melaleuca_ on sandy soil	DBH < 5 cm	VS2C0	5,450	2,850.0	3.63	0.07	na	na	6.13	0.16		Including non-inundated and seasonal inundation
	5 cm ≤ DBH < 10 cm	VS2C1	5,500	700.0	7.07	0.14	na	na	8.08	0.15		
	DBH ≥ 10 cm	na	na	na	na	na	na	na	na	na		
	All classes	VS2	10,950	3,550.0	5.36	0.14	28.41	3.14	7.11	0.13		
Degraded secondary _Melaleuca_ on clay soil with peat	DBH < 5 cm	VS3C0	150	na	4.41	0.23	na	na	5.00	0.29	_Stenochlaenapalustris_	Including seasonal and permanent inundation
	5 cm ≤ DBH < 10 cm	VS3C1	350	na	7.12	0.68	na	na	4.57	0.38	_Phragmitesvallatoria_	
	10 cm ≤ DBH < 20 cm	VS3C2	440	20.0	13.11	0.36	na	na	10.44	0.41	_Melastomadodecandrum_	
	20 cm ≤ DBH < 30 cm	VS3C3	30	na	25.00	1.20	na	na	14.33	0.17	_Diplaziumesculentum_	
	30 cm ≤ DBH < 40 cm	VS3C4	10	na	35.35	na	na	na	12.50	na	_Lygodiumscandens_ _Aspleniumnidus_	
	DBH ≥ 40 cm	VS3C5	na	na	na	na	na	na	na	na	_Scleriasumatrensis_ _Cassia tora_	
	All classes	VS3	980	560.0	12.93	0.71	10.29	4.74	9.69	0.45	_Paederiafoetida_ _Flagellariaindica_ _Cayratiatrifolia_	
Regenerating _Melaleuca_ on clay soil with peat	DBH < 5 cm	VS4C0	3,867	2,258.6	3.84	0.06	na	na	4.15	0.11		Including seasonal and permanent inundation
	5 cm ≤ DBH < 10 cm	VS4C1	5,967	176.4	7.20	0.12	na	na	6.68	0.17		
	DBH ≥ 10 cm	na	na	na	na	na	na	na	na	na		
	All classes	VS4	9,833	2,265.9	5.88	0.12	30.14	1.46	5.68	0.13		
Regenerating _Melaleuca_ on clay soil without peat	DBH < 5 cm	VS5C0	2,133	592.6	3.82	0.09	na	na	4.95	0.17		Including seasonal and permanent inundation
	5 cm ≤ DBH < 10 cm	VS5C1	4,733	1,560.3	7.27	0.13	na	na	8.65	0.31		
	DBH ≥ 10 cm	na	na	na	na	na	na	na	na	na		
	All classes	VS5	6,867	1,970.1	6.20	0.14	23.02	8.53	7.50	0.25		

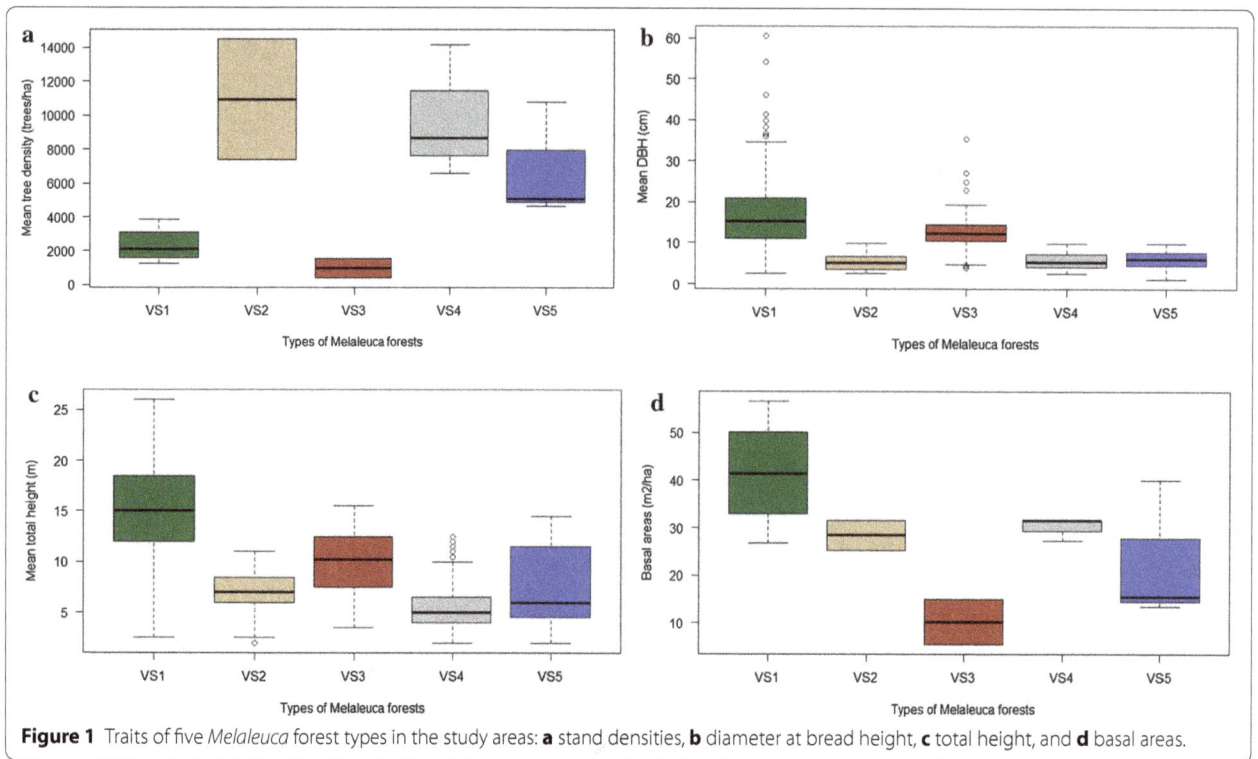

Figure 1 Traits of five *Melaleuca* forest types in the study areas: **a** stand densities, **b** diameter at bread height, **c** total height, and **d** basal areas.

the peat land, the greatest contribution of carbon densities for VS3 and VS4 are the peat and soil categories (accounting for 61.41%, 22.10% of VS3, and 57.66, and 16.72% of VS4, respectively). Separately, carbon density of VS5 is mostly linked to the soil, deadwood, and stand categories (accounting for 33.54, 32.16, and 14.66%, respectively) (Figure 2b).

Variability of carbon stocks in different types of *Melaleuca* forests

This study investigated the carbon stocks of six categories: stands, understory, deadwood, litter, root, and soil for five types of *Melaleuca* forests in Southern Vietnam (Figure 3).

The carbon densities of stands of the various forest types were 110.67, 44.27, 22.79, 48.25, and 37.20 tC/ha for VS1, VS2, VS3, VS4, and VS5, respectively (Figure 3a). There was a significant difference in stand carbon density between the forest types ($\chi^2 = 48.3184$, $p = 8.1e^{-10}$) (Additional file 1: 2f). The carbon density of the stand VS1 is the highest and is 2.5, 4.9, 2.3, and 3.0 times higher than VS2, VS3, VS4, and VS5. Surprisingly, there is no statistical difference in stand carbon densities between secondary forests (VS3) and regenerating forests (VS2, VS4 and VS5) (Additional file 1: 2f). These carbon stocks were lower those from other studies of different forests (e.g. 144 tC/ha for Asian tropical forests [25]; 200.23 tC/ha and 92.34 tC/ha of primary and secondary

swamp forests in Indonesia (involving *Melaleuca* vegetation), respectively [26]).

The carbon densities of the understory in the *Melaleuca* forests of Vietnam were 2.45, 2.48, 6.23, 1.65, and 5.27 tC/ha for VS1, VS2, VS3, VS4, and VS5, respectively (Figure 3b). There was a statistically significant difference in understory carbon density between the forest types ($\chi^2 = 30.7189$, $p = 3.49e^{-6}$) (Additional file 1: 2g). However, there was no significant difference in understory carbon density between *Melaleuca* forest types on sandy soils (VS1 and VS2). On clay soils, the understory carbon densities of VS3 and VS5 were significantly higher than VS4.

The carbon densities of deadwood in the forest types were 30.47, 0, 67.90, 45.06, and 74.59 tC/ha for VS1, VS2, VS3, VS4, and VS5, respectively (Figure 3c). There was a statistically significant difference in deadwood carbon density between the *Melaleuca* forest types ($\chi^2 = 3.0978$, $p = 0.5416$), but pairwise comparisons show no significant differences (Additional file 1: 2 h). Surprisingly, deadwood was not present in regenerating forests in the study sites on Phu Quoc Island. This is probably due to frequent forests fires and/or fuelwood collection by people associated crop cultivation.

Some of the carbon stock of *Melaleuca* forests is contributed by layers of coarse and fine litter. The carbon densities of the total litter layer of the forest types were

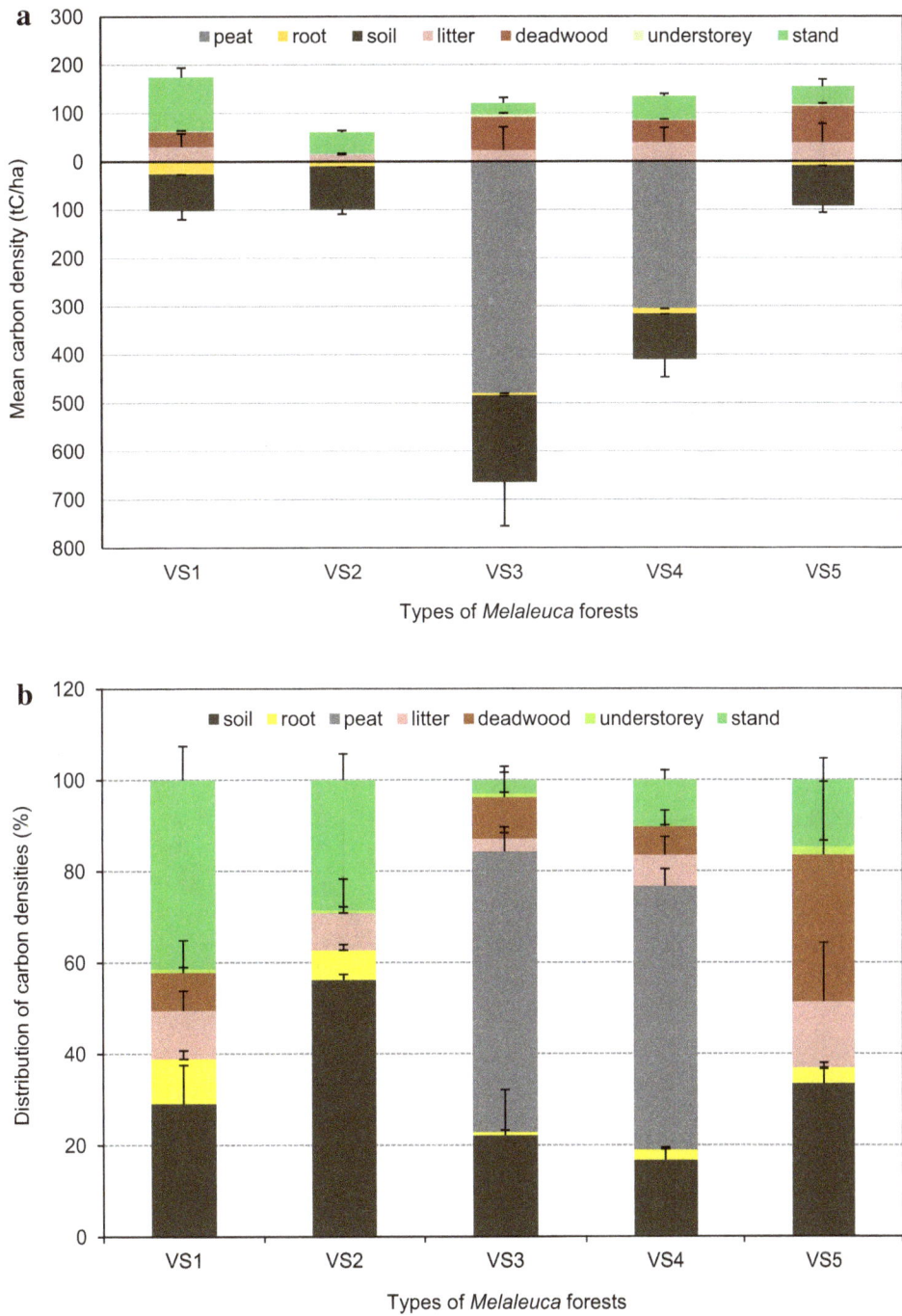

Figure 2 Carbon densities of five typical *Melaleuca* forests in the study areas: **a** mean carbon density, and **b** distribution of carbon densities.

31.03, 14.45, 23.76, 57.35, and 39.23 tC/ha for VS1, VS2, VS3, VS4, and VS, respectively (Figure 3d). There was a statistically significant difference in overall litter carbon density between these forest types ($\chi^2 = 1.5619$, $p = 0.08156$), but pairwise comparisons show no significant differences (Additional file 1: 2i).

The carbon densities from peat of the *Melaleuca* forests were 479.62 and 294.57 tC/ha for secondary forests (VS3) and regenerating forests (VS4), respectively (Figure 3e). The carbon density from peat of VS3 is significantly greater than that of VS4 ($\chi^2 = 5.2359$, $p = 0.0221$) (Additional file 1: 2j). This is almost certainly due to

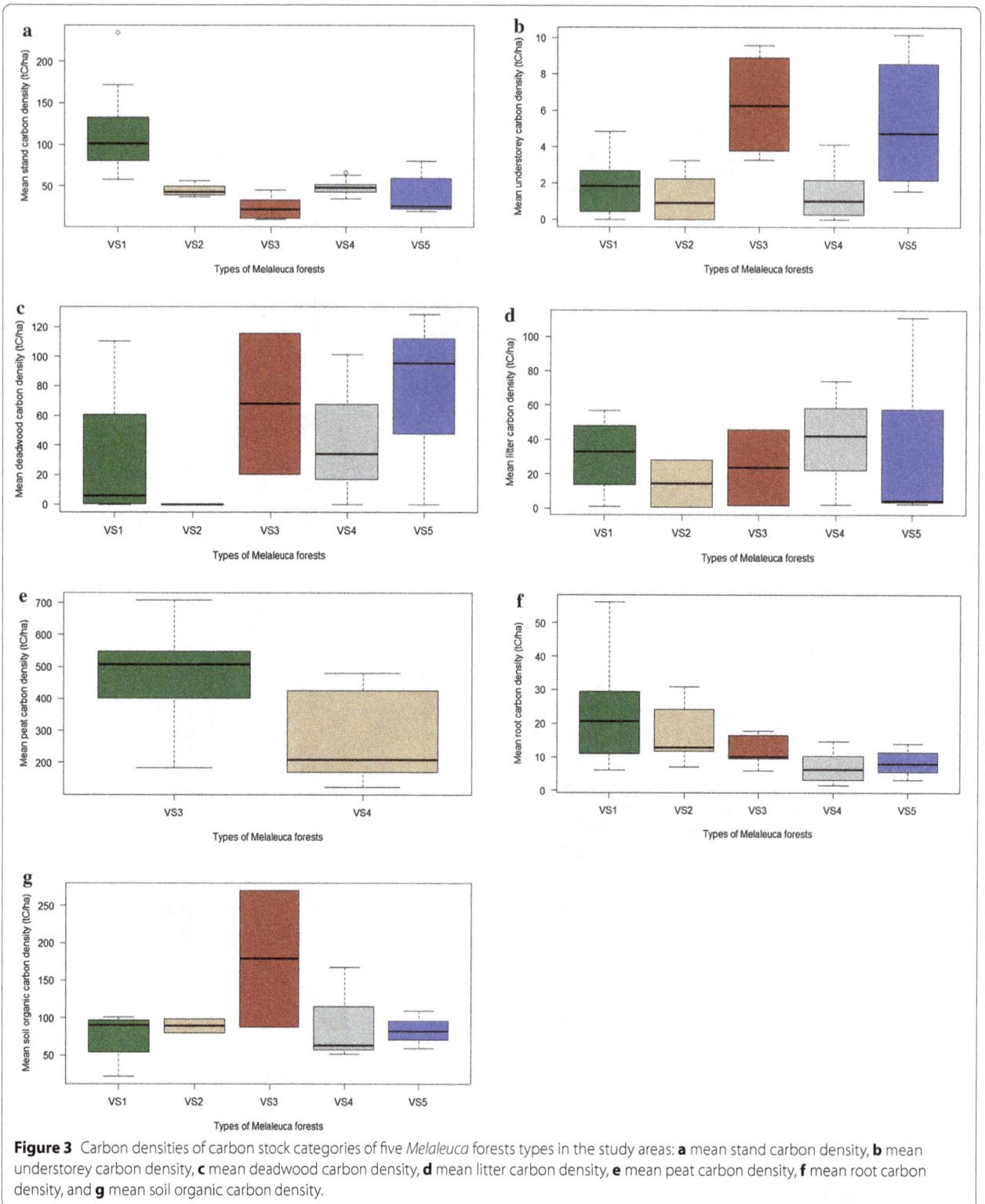

Figure 3 Carbon densities of carbon stock categories of five *Melaleuca* forests types in the study areas: **a** mean stand carbon density, **b** mean understorey carbon density, **c** mean deadwood carbon density, **d** mean litter carbon density, **e** mean peat carbon density, **f** mean root carbon density, and **g** mean soil organic carbon density.

peat being partly burned in the regenerating forest by the severe fire of 2002. In U Minh Thuong National Park, peat comprises the top soil layer, with a deep layer of clay below. The depth of the peat layer ranged from 15 to 62 cm in 18 soil cores, and the peat bulk density ranged from 0.19 to 0.3. The depths of the peat layer in this study were much thinner than in other forests (i.e. primary peat layer in U Minh Thuong was over 90 cm depth [27], and the thick peat layer in U Minh Ha was over 120 cm depth [28]).

The carbon densities of roots in the *Melaleuca* forests were 22.75, 16.97, 11.97, 6.99, and 8.35 tC/ha for VS1, VS2, VS3, VS4, and VS5, respectively (Figure 3f). There was a statistically significant difference in root carbon density between the forest types ($\chi^2 = 22.437$, $p = 0.00016$). The carbon densities of roots in *Melaleuca* forests in sandy soil were higher than those in clay soil, in particular, the root carbon density of VS2 was significant higher than that of VS4 (Additional file 1: 2k).

Organic soil carbon densities to a 30 cm depth in the study areas were 75.81, 89.22, 178.93, 93.94, and 83.58 tC/ha for VS1, VS2, VS3, VS4, and VS5, respectively (Figure 3 g). There was a statistically significant difference in organic soil carbon density between the forest types ($\chi^2 = 1.7333$, $p = 0.230$), but pairwise comparisons showed no significant differences (Additional file 1: 2k). These results are consistent with those of other studies of soil carbon stocks in wetlands (e.g. organic soil carbon stocks in swamp forests in Indonesia (with *Melaleuca* vegetation) were 106.00 and 135.63 tC/ha in the top 30 cm of soil of primary and secondary forests, respectively [29]).

Overall, the carbon density of *Melaleuca* forests on sandy soil in Southern Vietnam ranged from 159.36 tC/ha for regenerating forests to 275.98 tC/ha for primary forests. The carbon densities of forests on clay soil ranged from 246.96 tC/ha for regenerating forests without peat to 784.68 tC/ha of secondary forests with peat. Compared with the carbon stocks of other forests on peatland (e.g. the carbon density of mangrove forests in the Indo-Pacific region was 1,030 tC/ha [30]), the carbon density of disturbed *Melaleuca* forests on the peatland of Southern Vietnam is about one half, but the results are consistent with other studies on peat swamp forests (e.g. the carbon density of undisturbed swamp forests in South-East Asia ranged from 182 to 306 tC/ha [31]). Despite this, *Melaleuca* forests in the peatlands of Vietnam still have high

potential as carbon stores. The case of U Minh Thuong National Park is an example. The total carbon stock of 8,038 ha of *Melaleuca* forests in the park is about 2.69 M tC (Table 2), which is equivalent 9.43 M tCO$_2$e. Furthermore, there were 8,576 hectares of *Melaleuca* forested peatland in U Minh Ha National Park that have peat layers ranging from 40 cm to over 120 cm deep [32], which provides an even higher potential carbon store.

Sodicity tolerance of *Melaleuca cajuputi* forests toward the adaptation to global climate change

Sea-level rise is a consequence of global climate change that will severely affect coastal and wetland ecosystems. *Melaleuca* forests are largely located in coastal and wetland areas that may be affected by climate change [33], so the risk of salinization of the region will increase. Salinity in soils can damage woody plant species by stunting buds, reducing leaf size and causing necroses in buds, roots, leaf margins and shoot tips [34]. Salinity can also inhibit seed germination, and can even kill non-halophytic species [35]. Both vegetative and reproductive growth of woody species are also reduced by high concentrations of sodium chloride in soil [35, 36]. The combination of flooding and salinity can create a more pronounced effect on growth and survival of plants than either stress alone [35]. High concentrations of sodium can affect the structure of sodic soils [37–39]. In contrast, low sodium concentration, soil structure is not affected by salinity in saline soil [40]. Sodicity and salinity always occur together and coming to have negative impacts on soil properties and plants [38, 41], but sodic soils may be either non-saline or saline [17].

The lower Mekong Basin and coastal regions of southern Vietnam are highly vulnerable to global climate change impacts [9, 33, 42, 43]. Most of Vietnam's *Melaleuca* forests occur in these areas and will be affected projected sea-level rise. Fortunately, this study has shown that *M. cajuputi* has the ability to tolerant increase in sodic soils.

About 28 soil samples collected from *Melaleuca* forests in Southern Vietnam were examined and all were

Table 2 Potential carbon storage in *Melaleuca* peat-swamp forests: case in U Minh Thuong National Park

Land cover type	Area (ha)	Carbon density (tC/ha)	Carbon storage tC
Mature *Melaleuca* forests on clay soil without peat	1,765	305.06	538,431
Mature *Melaleuca* forests on clay soil with peat	601	784.68	471,593
Regenerating *Melaleuca* on clay soil with peat	2,106	544.28	1,146,254
Regenerating *Melaleuca* on clay soil without peat	1,106	246.96	273,138
Others (open water, reeds and grasses)	2,460	107.91	265,459
Total	8,038		2,694,874

The areas of *Melaleuca* forests in U Minh Thuong National Park are taken from a Vietnam Environment Protection Agency report [48].

Table 3 Chemical element concentration and sodicity levels of the *Melaleuca* forest soils in the study areas

Forest types	Soil layers (cm)	pH$_{(KCl)}$		Ca^{2+} (meq/100 g)		Mg^{2+} (meq/100 g)		Na^{+} (meq/100 g)		K^{+} (meq/100 g)		Al^{3+} (meq/100 g)		Fe^{3+} (mg/100 g)		ESP (%)		Sodicity
		Mean	SE	Mean	SE	Mean	SE	Mean	SE	Mean	SE	Mean	SE	Mean	SE	Mean	SE	
Primary *Melaleuca* on sandy soil (VS1)	0–10	3.97	0.15	1.413	0.75	1.783	1.58	1.790	1.56	0.600	0.47	0.910	0.29	3.303	1.20	32.05	4.28	High
	10–30	4.12	0.17	1.065	0.41	1.138	1.02	1.708	1.58	0.383	0.33	0.660	0.24	7.310	2.23	39.78	7.90	High
Regenerating *Melaleuca* on sandy soil (VS2)	0–10	3.68	0.03	0.690	0.10	0.310	0.10	0.310	0.15	0.155	0.02	1.860	0.14	1.615	0.36	21.16	7.82	High
	10–30	3.86	0.04	0.645	0.03	0.175	0.00	0.150	0.02	0.065	0.02	1.280	0.24	1.810	0.56	14.49	2.28	Moderate
Degraded secondary *Melaleuca* on clay soil with peat (VS3)	0–10	4.12	0.25	7.585	1.82	6.320	1.81	1.705	0.81	0.455	0.19	0.100	0.10	37.155	17.63	10.61	2.37	Moderate
	10–30	4.07	0.32	7.585	2.96	5.795	2.59	1.470	0.41	0.705	0.07	1.680	1.64	48.245	7.19	9.45	1.09	Low
Regenerating *Melaleuca* on clay soil with peat (VS4)	0–10	4.67	0.19	8.845	0.55	6.685	0.58	1.760	0.51	0.585	0.27	0.00	0.00	47.550	16.06	9.85	3.05	Low
	10–30	5.00	0.04	5.855	0.33	4.860	0.42	1.320	0.19	0.575	0.18	0.00	0.00	54.825	36.49	10.47	1.27	Moderate
Regenerating *Melaleuca* on clay soil without peat (VS5)	0–10	4.16	0.26	11.580	4.19	5.557	0.55	1.330	0.24	0.663	0.07	5.533	5.53	67.433	9.03	6.95	0.49	Low
	10–30	3.91	0.40	8.603	1.65	5.170	0.08	1.507	0.24	0.717	0.09	8.000	7.02	78.440	10.37	9.42	0.37	Low

shown to be sodic (Table 3). While the exchangeable sodium percentage (ESP) of soil layers of *Melaleuca* forests on clay soil (VS3, VS4, and VS5) ranges from low to moderate sodicity, those of *Melaleuca* forests on sandy soil (VS1 and VS2) were significantly higher, particularly VS1, which had an ESP of up to 39.78% in soil taken from depths of 10–30 cm (Table 3). This indicates that both mature and young *M. cajuputi* forests have a high tolerance of sodic soils. Furthermore, *M. cajuputi* seeds can germinate and grow in highly sodic soil [e.g. *M. cajuputi* in forest type VS2 was able to grow in highly sodic soil with ESP up to 21.16% in the top 0–10 cm (Table 3)].

With the exception of mangroves, few woody species can tolerate saline and/or sodic soils. Many woody species have been examined for their tolerance of salinity and/or sodicity. For example, *Eucalyptus*, *Melaleuca*, *Acacia*, *Casuarina* [21–24], *Grevillea robusta*, *Lophostemon confertus* and *Pinus caribea* [44], and *Moringa olifera* [45] have been examined and their tolerance to salinity assessed in the field and in glasshouses. In extremely saline soils in Australia, Niknam and McComb [22] suggested that the land care benefit of establishing species such as *Melaleuca* or *Casuarina* is more important than their commercial value. As well as the land care value, this study has shown that *M. cajuputi* forests in Vietnam can adapt to climate change through their tolerance to sodicity, and other harsh conditions [33], and can help to mitigate climate change through their carbon storage abilities.

Conclusion

By undertaking original field data, this study examined the carbon sequestration potential of five types of *Melaleuca* forests including 'Primary *Melaleuca* forests on sandy soil' (VS1), 'Regenerating *Melaleuca* forests on sandy soil' (VS2), 'Degraded secondary *Melaleuca* forests on clay soil with peat' (VS3), 'Regenerating *Melaleuca* forests on clay soil with peat' (VS4), and 'Regenerating *Melaleuca* forests on clay soil without peat' (VS5). The study also assessed the sodicity tolerance of *M. cajuputi* forests in coastal and wetland regions of Vietnam.

The carbon densities of VS1, VS2, VS3, VS4, and VS5 were 275.98 ($\pm$38.62) tC/ha, 159.36 ($\pm$21.01) tC/ha, 784.68 ($\pm$54.72) tC/ha, 544.28 ($\pm$56.26) tC/ha, and 246.96 ($\pm$27.56) tC/ha, respectively. Most carbon stocks were contributed from the soil (including peat) and stands.

The exchangeable sodium percentage (ESP) of soil from *Melaleuca* forests on clay soil (VS3, VS4, and VS5) ranged from low to moderate sodicity, but those from *Melaleuca* forests on sandy soil (VS1 and VS2) were highly sodic.

The results provide important information for the future sustainable management of *Melaleuca* forests in Vietnam, particularly in regards to forest carbon conservation initiatives and the potential of *Melaleuca* species for reforestation initiatives on degraded sites with highly sodic soils. In Vietnam, forest carbon conservation initiatives such as REDD+ have hereto, in our view, not placed appropriate priority or consideration on the protection of carbon stocks of *Melaleuca* forests. The results presented in this paper suggest that *Melaleuca* forests in Vietnam, particularly those on peatland areas, hold globally significant carbon stocks—arguably greater than those found in upland rainforest ecosystems, which have so far been given higher priority in REDD+ planning in Vietnam. Furthermore, the results presented in this paper suggest that some *Melaleuca* forest species in Vietnam, particularly those on sandy soils, exhibit a tolerance for highly sodic soils. This suggests that those species might be useful in reforestation initiatives on degraded sites with highly sodic soils. As degradation pressures including climate change continue to alter the hydrological features of soil systems in areas such as the Mekong Delta in Vietnam, and the sodicity of soils in some areas increases, *Melaleuca* species could offer a useful option for reforestation and rehabilitation initiatives.

The results in this research provide further scientific information to support better *Melaleuca* ecosystem management. The results should help policy makers make better decisions in an era of global change. The results have particular relevance for the application of REED+ in the Southeast Asia.

Methods
Study sites and disturbance context

Melaleuca cajuputi is naturally distributed as scattered shrub populations along the coastal regions in the middle Provinces and up to the Northern hilly regions, and as tall forests in the Mekong Delta of Vietnam [46]. Thus, the study focussed on the sites in Southern Vietnam (involving Mekong Delta). The study investigated two sites: the Phu Quoc National Park and U Minh Thuong National Park, which both contain extensive *Melaleuca* forests in coastal wetlands (Figure 4). A total of 14 plots were randomly selected for carbon storage assessment, covering five types of *Melaleuca* stands: 'Primary *Melaleuca* forests on sandy soil'(VS1), 4 plots; 'Regenerating *Melaleuca* forests on sandy soil' (VS2), 2 plots; 'Degraded secondary *Melaleuca* forests on clay soil with peat' (VS3), 2 plots; 'Regenerating *Melaleuca* forests on clay soil with peat' (VS4), 3 plots; and 'Regenerating *Melaleuca* forests on clay soil without peat' (VS5), 3 plots.

Phu Quoc National Park is located on the northern Phu Quoc Island of Vietnam (at N 10°12′07″–N 10°27′02″, E 103°50′04″–E 104°04′40″) (Figure 4). *Melaleuca* forest areas cover 1,667.50 ha out of the total area

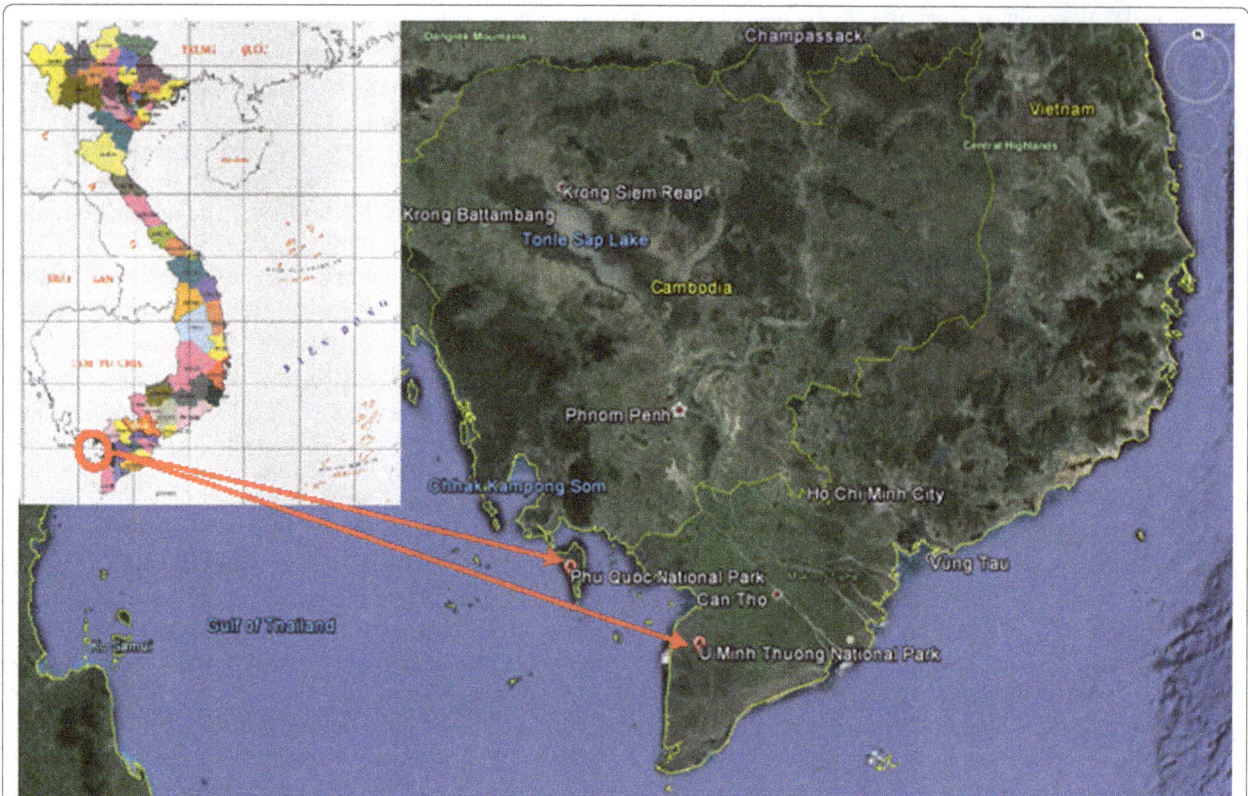

Figure 4 The study locations in Southern Vietnam: Phu Quoc National Park and U Minh Thuong National Park. Source: map from Department of Information Technology, Vietnam. Image Landsat from Google Earth (free version).

of 28,496.90 ha. These *Melaleuca* forests naturally occur on lowland regions of the island where they are seasonally inundated and/or permanent saturated, and also on permanent sand bars where no inundation occurs [47]. The rest areas of the park are hilly and mountainous forests. Two *Melaleuca* forest types were found in the park: primary *Melaleuca* forest (VS1); and regenerating *Melaleuca* forest (VS2). Before the park was established in 2001, key disturbance included forest fires and human intrusion for crop cultivation. The regenerating *Melaleuca* forests were up to 10–12 years of age at the time this study was conducted.

U Minh Thuong National Park is located in the Kien Giang Province (at N 9° 31′–N 9° 39′, E 105° 03′–E 105° 07′) (Figure 4). *Melaleuca* forest on swamp peatland is an endemic ecosystem in the lower Mekong Basin of Vietnam. The core area of the park is 8,038 ha, which is surrounded by a buffer zone of 13,069 ha. Here, the key disturbance is fire, with the last major fire occurring in April 2002, which burnt the primary vegetation as well as the peat soil. The Vietnamese Environment Protection Agency [48] reported that 3,212 hectares of *Melaleuca* forests was almost destroyed, so a canal system was built as a key management solution

to increase water inundation of the forest to prevent fires. Currently, there are three *Melaleuca* forest types in U Minh Thuong National Park: VS3, VS4, and VS5. At the time of this study, the VS4 and VS5 areas were up to 10 years old.

Field sampling and data collection

The major plots were set out as 500 m² quadrats (20 m × 25 m), and all trees with a DBH ≥ 10 cm were measured and recorded. Sub-plots also were set out as 100 m² quadrats (20 m × 5 m) within the major plots to measure all trees with DBH < 10 cm and a total height of >1.3 m (modified from Van et al. [49]). Data on DBH, alive or dead, and height were recorded for all standing trees.

Deadwood (dead fallen trees) with a diameter ≥10 cm were measured within the major plots (500 m²), while deadwood with 5 cm ≤ diameter < 10 cm were measured within the sub-plots (100 m²). Diameters at both ends of the trunk (D1 and D2), length (if ≥50 cm length), and the decay classes (involved sound, intermediate, and rotten [50, 51]) were recorded for all deadwood.

Seventy random quadrats (1 m × 1 m) were located in the main plots to collect and record the 'fresh weight' of the understory. Samples of all species from the understory were collected in each major plot and taken back to the Vietnam Forestry University laboratory for drying.

Seventy random coarse litter samples and seventy random fine litter samples were collected in the major plots. The fresh weight of each litter sample was recorded. Each litter type (coarse litter and fine litter) collected in every major plot were well mixed and taken to the laboratory for drying.

Two soil samples, one from the upper (0–10 cm) soil layer and one from the lower (10–30 cm) soil layer, were taken from each of 14 plots, giving a total of 28 soil samples. The 28 soil samples were taken back to the National Institute of Agricultural Planning and Projection laboratory for further analysis. Various soil chemical properties of the 28 samples were tested including: pH_{KCl}, total C, total N, Ca^{2+}, Mg^{2+}, Na^+, K^+, Al^{3+}, and Fe^{3+}. Twenty-eight duplicate soil samples were collected and analyzed for bulk density.

Sample analysis

Each understory and litter sample was divided into three sub-samples and dried in a drying oven at 60°C to measure the moisture content, based on the Eq. (1) below:

$$R_{moist} = \frac{\sum_{i=1}^{n} \frac{W_{fi} - W_{di}}{W_{fi}}}{n}. \tag{1}$$

where R_{moist} = moist ratio [0:1], W_{fi} = fresh weight of sub-sample i, W_{di} = dry weight of sub-sample i, n = number of sub-samples. The scales used to weight sub-samples were accurate to ±0.01 g.

Total organic carbon (C%) was measured using the Walkley–Black method, which is commonly used to examine soil organic carbon via oxidation with $K_2Cr_2O_7$ [52, 53]. Total nitrogen was measured using the Kjeldahl method, which is the standard way to determine the total organic nitrogen content of soil [54]. A standard bulk density test was used to analyze all soil bulk samples in a dryven. Bulk density was calculated using Eq. (2):

$$BD = \frac{Ms}{V}. \tag{2}$$

where BD = the bulk density of the oven-dry soil sample (g/cm³), Ms = the oven dry-mass of the soil sample (gram), V = the volume of the ring sample (cm³).

Exchangeable sodium percentage (ESP) was calculated using Eq. (3) [55–57], and classified with four sodic levels as non-sodic soil (ESP < 6), low sodic soil (ESP = 6–10), moderately sodic soil (ESP = 10–15), and highly sodic soil (ESP > 15) [55–57].

$$ESP = \frac{Na^+}{\Sigma \left[Na^+ \right] \left[K^+ \right] \left[Mg^{2+} \right] \left[Ca^{2+} \right]} \times 100. \tag{3}$$

Basal area (BA) was calculated with Eq. (4) (modified from Jonson and Freudenberger [58]):

$$BA = \frac{\sum_{n}^{1} \left[\pi \times (DBH_i/200)^2 \right]}{S_{plot}} \times 10,000 \tag{4}$$

where BA = basal area (m²/ha), DBH_i = diameter at bread height of tree i (cm), i = stand individual (i = [1:n]), n = number of trees of sample plot, S_{plot} = area of the sample plot (m²).

Biomass allometric computation

Nine allometric equations, which are most common way to measure forest carbon stocks, were applied to calculate the above-ground and root biomass of the stands (Table 4). The selected allometric equations were tested for statistical significance using the R Statistic Program (Additional file 1: 1). Using these equations, the average biomass was analyzed for five typical Melaleuca stands (VS1, VS2, VS3, VS4, and VS5). To convert from fresh to dry biomass, a moisture rate of 0.5 was applied as suggested by Van et al. [49] for the allometric equation of Finlayson et al. [59]. According to the Global Wood Density Database, the density of M. cajuputi timber ranges from 0.6 to 0.87 g/cm³ [60], so 0.6 g/cm³ was applied for the above-ground biomass allometric equation of Chave et al. [61].

The fallen deadwood biomass were calculated using Eq. (5) ([62], p 12):

$$B = \pi \times r^2 \times L \times \delta \tag{5}$$

where B = biomass (kg), r = ½ diameter (cm), L = length (m), and δ = wood density (= 0.6 g/cm³).

Then, the biomass of the fallen deadwood was determined using the IPCC [50, 51] density reduction factors (sound = 1, intermediate = 0.6, and rotten = 0.45). The biomass of standing dead trees was measured using the same criteria as live trees, but a reduction factor of 0.975 is applied to dead trees that have lost leaves and twigs, and 0.8 for dead trees that have lost leaves, twigs, and small branches (diameter <10 cm) ([51], p 4.105).

To convert biomass to carbon mass for all categories (stands, roots, deadwood, understory, and litter), a factor of 0.45 was applied.

Soil organic carbon (SOC) was calculated using Eq. (6) [50, 51]:

$$SOC = Dep \times BD \times C_{sample} \times 100 \tag{6}$$

where SOC = Soil organic carbon, Dep = depth of soil layer (m), BD = bulk density (g/cm³), C_{sample} = organic

Table 4 List of allometric equations applied to examine stand biomass of the *Melaleuca* forests

Allometric equations	R^2	Vegetation	Sites	References
$\log_{10}(FW) = 2.266\log_{10}(D) - 0.502$ where FW = fresh above-ground biomass (kg/tree), D = diameter at breast height (cm)	0.98	*Melaleuca* spp.	Northern Territory	Finlayson et al. [59]
$y = 0.124 \times DBH^{2.247}$ where y = above-ground biomass (kg/tree), DBH = diameter at breast height (cm)	0.97	*Melaleuca cajuputi*	Vietnam	Le [63]
$y = \exp[-2.134 + 2.53\ln(D)]$ where y = above-ground biomass (kg/tree), D = diameter at breast height (cm)	0.97	Mixed species	Tropical, moist forest	IPCC [51] or Brown [64]
$\ln(y) = 2.4855\ln(x) - 2.3267$ where y = above-ground biomass (kg/tree), x = diameter at breast height (cm)	0.96	Native sclerophyll forest	NSW, ACT, VIC, TAS, and SA	Keith et al. [65]
$\ln(AGB) = -1{,}554 + 2.420\ln(D) + \ln(\rho)$ where AGB = above-ground biomass (kg/tree), D = diameter at breast height (cm), ρ = wood density (g/cm^3)	0.99	Tropical forests	America, Asian and Oceania	Chave et al. [61]
$\ln(RBD) = -1{,}085 + 0.926\ln(ABD)$ where RBD = root biomass density (tons/ha), ABD = above-ground biomass density (tons/ha)	0.83	Upland forests	Worldwide	IPCC [51] or Cairn et al. [66]
$y = 0.27x$ where y = total root biomass (tons/ha), x = total shoot biomass (tons/ha)	0.81	Natural forests	Worldwide	Mokany et al. [67]
$Wr = 0.0214 \times D^{2.33}$ where W_r = coarse root biomass (kg/tree), D = diameter at breast height (cm)	0.94	Tropical secondary forests	Sarawak, Malaysia	Kenzo et al. [68]
$W_r = 0.023 \times D^{2.59}$ where W_r = coarse root biomass (kg/tree), D = diameter at breast height (cm)	0.97	Tropical secondary forests	Sarawak, Malaysia	Niiyama et al. [69]

NSW New South Wales, *ACT* Australian Capital Territory, *VIC* Victoria, *TAS* Tasmania, *SA* South Australia.

carbon content of soil sample (%), and 100 is the default unit conversion factor.

Statistical analysis

One-way ANOVA tests were applied to compare stand densities, DBH, height classes, basal areas, and six categories of carbon stocks of the five *Melaleuca* forest types. LSD post hoc tests were also used for all pairwise comparisons between group means. Statistical analysis was undertaken using Microsoft Excel 2010 and the R Statistic Program.

Authors' contributions

DBT conducted design of the study, field data collection, carried out all analyses and drafted the manuscript. TVH and PD helped field data collection, guided the research, and assisted with the writing. All authors read and approved the final manuscript.

Author details

[1] The Vietnam Forestry University, Hanoi, Vietnam. [2] School of Geography, Planning and Environmental Management, The University of Queensland, Brisbane, QLD, Australia.

Acknowledgements

This study was authorized to access and collect vegetation and soil samples by the director boards of two national parks including the Phu Quoc National Park and U Minh Thuong National Park. All work was approved by the Vietnam Forestry University. We would like to thank the staffs of Phu Quoc National Park; U Minh Thuong National Park; the National Institute of Agricultural Planning and Projection; and the Vietnam Forestry University for their association of doing fieldwork and laboratory work. We also specially thank the anonymous reviewers for their excellent comments on the earlier version of this manuscript. We gratefully thank International Foundation for Science (IFS) for research funds.

Compliance with ethical guidelines

Competing interests

The authors declare that they have no competing interests.

References

1. Mitsch W, Bernal B, Nahlik A, Mander Ü, Zhang L, Anderson C et al (2013) Wetlands, carbon, and climate change. Landsce Ecol 28(4):583–597. doi:10.1007/s10980-012-9758-8
2. Bernal B, Mitsch WJ (2012) Comparing carbon sequestration in temperate freshwater wetland communities. Glob Change Biol 18(5):1636–1647. doi:10.1111/j.1365-2486.2011.02619.x
3. Mitsch W, Nahlik A, Wolski P, Bernal B, Zhang L, Ramberg L (2010) Tropical wetlands: seasonal hydrologic pulsing, carbon sequestration, and methane emissions. Wetlands Ecol Manag 18(5):573–586. doi:10.1007/s11273-009-9164-4
4. Bernal B, Wolski P, Nahlik A, Ramberg L, Zhang L, Mitsch WJ (2010) Tropical wetlands: seasonal hydrologic pulsing, carbon sequestration, and methane emissions. Wetlands Ecol Manag 18(5):573–586
5. Mitsch WJ, Tejada J, Nahlik A, Kohlmann B, Bernal B, Hernández CE (2008) Tropical wetlands for climate change research, water quality management and conservation education on a university campus in Costa Rica. Ecol Eng 34(4):276–288. doi:10.1016/j.ecoleng.2008.07.012
6. Bernal BS (2008) Carbon pools and profiles in wetland soils: the effect of climate and wetland type. The Ohio State University, Ohio
7. Bernal B, Mitsch WJ (2008) A comparison of soil carbon pools and profiles in wetlands in Costa Rica and Ohio. Ecol Eng 34(4):311–323. doi:10.1016/j.ecoleng.2008.09.005

8. Irving AD, Connell SD, Russell BD (2011) Restoring coastal plants to improve global carbon storage: reaping what we sow. PLoS One 6(3):e18311

9. Erwin K (2009) Wetlands and global climate change: the role of wetland restoration in a changing world. Wetlands Ecol Manag 17(1):71–84. doi:10.1007/s11273-008-9119-1

10. Renaud FG, Kuenzer C (2012) Climate and environmental change in River Deltas globally: expected impacts, resilience, and adaptation. In: Renaud FG, Kuenzer C (eds) Mekong delta system: interdisciplinary analyses of a River Delta, vol Book. Springer Netherlands, Whole

11. Bastakoti RC, Gupta J, Babel MS, van Dijk MP (2014) Climate risks and adaptation strategies in the Lower Mekong River basin. Reg Environ Change 14(1):207–219. doi:10.1007/s10113-013-0485-8

12. Le TVH, Nguyen HN, Wolanski E, Tran TC, Haruyama S (2007) The combined impact on the flooding in Vietnam's Mekong River delta of local man-made structures, sea level rise, and dams upstream in the river catchment. Estuar Coast Shelf Sci 71(1):110–116. doi:10.1016/j.ecss.2006.08.021

13. MIG (2008) Australia's State of the forests report: five-yearly report 2008. Montreal process implementation group for Australia, Bureau of Rural Sciences, Canberra

14. Tran DB, Dargusch P, Herbohn J, Moss P (2013) Interventions to better manage the carbon stocks in Australian Melaleuca forests. Land Use Policy 2013(35):417–420. doi:10.1016/j.landusepol.2013.04.018

15. Renaud FG, Le T, Lindener C, Guong V, Sebesvari Z (2014) Resilience and shifts in agro-ecosystems facing increasing sea-level rise and salinity intrusion in Ben Tre Province, Mekong Delta. Clim Change 1–16. doi:10.1007/s10584-014-1113-4

16. Rengasamy P, Olsson K (1991) Sodicity and soil structure. Soil Res 29(6):935–952. doi:10.1071/SR9910935

17. Bernstein L (1975) Effects of salinity and sodicity on plant growth. Annu Rev Phytopathol 13:295–312

18. Mahmood K (2007) Salinity, sodicity tolerance of Acacia ampliceps and identification of techniques useful to avoid early stage salt stress. Kassel Univ. Press, Kassel

19. Warrence NJ, Bauder JW, Pearson KE (2002) Basics of salinity and sodicity effects on soil physical properties. Land Resources and Environmental Sciences Department, Montana State University, Bozeman

20. Tho N, Vromant N, Hung NT, Hens L (2008) Soil salinity and sodicity in a shrimp farming coastal area of the Mekong Delta, Vietnam. Environ Geol 54(8):1739–1746. doi:10.1007/s00254-007-0951-z

21. Dunn GM, Taylor DW, Nester MR, Beetson TB (1994) Performance of twelve selected Australian tree species on a saline site in southeast Queensland. For Ecol Manag 70(1–3):255–264. doi:10.1016/0378-1127(94)90091-4

22. Niknam SR, McComb J (2000) Salt tolerance screening of selected Australian woody species: a review. For Ecol Manag 139(1–3):1–19. doi:10.1016/S0378-1127(99)00334-5

23. van der Moezel PG, Pearce-Pinto GVN, Bell DT (1991) Screening for salt and waterlogging tolerance in Eucalyptus and Melaleuca species. For Ecol Manag 40(1–2):27–37. doi:10.1016/0378-1127(91)90089-E

24. van der Moezel P, Watson L, Pearce-Pinto G, Bell D (1988) The response of six Eucalyptus species and Casuarina obesa to the combined effect of salinity and waterlogging. Austr J Plant Physiol 15(3):465–474

25. VCS (2011) Methodology for sustainable grassland management (SGM). Verified Carbon Standard-A global Benchmark for Carbon

26. Taiyab N (2006) Exploring the market for voluntary carbon offsets. International Institute for Environment and Development (IIED), London

27. Polglase PJ, Reeson A, Hawkins CS, Paul KI, Siggins AW, Turner J et al (2013) Potential for forest carbon plantings to offset greenhouse emissions in Australia: economics and constraints to implementation. Clim Change 121(2):161–175. doi:10.1007/s10584-013-0882-5

28. Singh A, Nigam PS, Murphy JD (2011) Renewable fuels from algae: an answer to debatable land based fuels. Bioresour Technol 102(1):10–16. doi:10.1016/j.biortech.2010.06.032

29. Rahayu S, Harja D (2012) A study of rapid carbon stock appraisal: average carbon stock of various land cover in Merauke, Papua Province. World Agroforestry Centre (ICRAF-SEA)

30. Donato DC, Kauffman JB, Murdiyarso D, Kurnianto S, Stidham M, Kanninen M (2011) Mangroves among the most carbon-rich forests in the tropics. Nat Geosci 4:293–297

31. Verwer CC, Meer PJVD (2010) Carbon pool in tropical peat forest: toward a reference value for forest biomass carbon in relatively undisturbed peat swamp forests in Southeast Asia. Wageningen, Allterra Wageningen UR

32. Le PQ (2010) Inventory of peatlands in U Minh Ha Region, Ca Mau Province, Vietnam. Institute for Environment and Natural Resources, National University, HCM City

33. Tran DB, Dargusch P, Moss P, Hoang TV (2013) An assessment of potential responses of Melaleuca genus to global climate change. Mitig Adapt Strat Glob Change 18(6):851–867. doi:10.1007/s11027-012-9394-2

34. Larcher W (1980) Physiological plant ecology. vol Book, Whole. Springer, Berlin

35. Kozlowski TT (1997) Responses of woody plants to flooding and salinity. Tree Physiol 17(7):490. doi:10.1093/treephys/17.7.490

36. Greenway H, Munns R (1980) Mechanisms of salt tolerance in nonhalophytes. Annu Rev Plant Physiol 31(1):149–190. doi:10.1146/annurev.pp.31.060180.001053

37. Wong VL, Dalal R, Greene RB (2008) Salinity and sodicity effects on respiration and microbial biomass of soil. Biol Fertil Soils 44(7):943–953. doi:10.1007/s00374-008-0279-1

38. Department of Primary Industries (2008) Identifying, understanding and managing hostile subsoils for cropping. University of Adelaide-South Australian Research and Development Institute

39. Mavi MS, Marschner P, Chittleborough DJ, Cox JW, Sanderman J (2012) Salinity and sodicity affect soil respiration and dissolved organic matter dynamics differentially in soils varying in texture. Soil Biol Biochem 45:8–13. doi:10.1016/j.soilbio.2011.10.003

40. Howat D (2000) Acceptable salinity, sodicity and pH values for Boreal forest reclamation: Alberta Environment, Environmental Sciences Division, Edmonton Alberta. Report # ESD/LM/00-2. ISBN 0-7785-1173-1 (printed edition) or ISBN 0-7785-1174-X (on-line edition)

41. Nuttall JG, Armstrong RD, Connor DJ, Matassa VJ (2003) Interrelationships between edaphic factors potentially limiting cereal growth on alkaline soils in north-western Victoria. Soil Res 41(2):277–292. doi:10.1071/SR02022

42. Nicholls RJ, Wong PP, Burkett VR, Codignotto JO, Hay JE, McLean RF et al (2007) Coastal systems and low-lying areas. In: Parry ML, Canziani OF, Palutikof JP, Linden PJVD, Hanson CE (eds) Climate change 2007: impacts, adaptation and vulnerability. Contribution of working group II to the fourth assessment report of the intergovernmental panel on climate change (IPCC). Cambridge University Press, Cambridge, pp 315–356

43. Toan TL (2009) Impacts of climate change and human activities on environment in the Mekong Delta, Vietnam. Centre d'Etudes Spatiales de la Biosphère (CESBIO), Toulouse

44. Sun D, Dickinson G (1993) Responses to salt stress of 16 Eucalyptus species, Grevillea robusta, Lophostemon confertus and Pinus caribea var. hondurensis. For Ecol Manag 60(1–2):1–14. doi:10.1016/0378-1127(93)90019-J

45. Paul KI, Roxburgh SH, England JR, Ritson P, Hobbs T, Brooksbank K et al (2013) Development and testing of allometric equations for estimating above-ground biomass of mixed-species environmental plantings. For Ecol Manag 310:483–494. doi:10.1016/j.foreco.2013.08.054

46. Cuong NV, Quat HX, Chuong H (2004) Some comments on indigenous Melaleuca of Vietnam. Sci Technol J Agric Rural Dev (Vietnam). (11/2004)

47. Hoover CM, Smith JE (2012) Site productivity and forest carbon stocks in the United States: analysis and implications for forest offset project planning. Forests 3(4):283–299. doi:10.3390/f3020283

48. Vietnam Environment Protection Agency (2003) Report on peatland management in Vietnam. Ministry of Natural Resources and Environment

49. Van TK, Rayachetry MB, Center TD (2000) Estimating above-ground biomass of Melaleuca quinquenervia in Florida, USA. J Aquat Plant Manag 38:62–67

50. IPCC (2006) Good practice guidance for land use, land-use change and forestry. Institute for Global Environmental Strategies (IGES) for the IPCC, Kanagawa

51. IPCC (2003) Good practice guidance for land use, land-use change and forestry. Institute for Global Environmental Strategies (IGES) for the IPCC, Kanagawa

52. Walkley A (1947) A critical examination of a rapid method for determination of organic carbon in soils—effect of variations in digestion conditions and of inorganic soil constituents. Soil Sci 63:251–257

53. Schumacher BA (2002) Methods for the determination of total organic carbon (TOC) in soils and sediments. Ecological Risk Assessment Support Center Office of Research and Development, US Environmental Protection Agency

54. LABCONCO (1998) A guide to Kjeldahl nitrogen determination methods and apparatus. An Industry Service Publication, Houston

55. Rengasamy P, Olsson KA (1991) Sodicity and soil structure. Aust J Soil Res 29(6):935–952. doi:10.1071/SR9910935

56. Ford G, Martin J, Rengasamy P, Boucher S, Ellington A (1993) Soil sodicity in Victoria. Soil Res 31(6):869–909. doi:10.1071/SR9930869

57. Gj C (1999) Cation exchange capacity, exchangeable cations and sodicity. vol Book, Whole

58. Jonson JH, Freudenberger D (2011) Restore and sequester: estimating biomass in native Australian woodland ecosystems for their carbon-funded restoration. Aust J Bot 59(7):640–653. doi:10.1071/BT11018

59. Finlayson CM, Cowie ID, Bailey BJ (1993) Biomass and litter dynamics in a *Melaleuca* forest on a seasonally inundated floodplain in tropical, Northern Australia. Wetlands Ecol Manag 2(4):177–188

60. Thomas S, Hoegh-Guldberg OOHG, Griffiths A, Dargusch P, Bruno J (2010) The true colours of carbon. Nat Preced. http://precedings.nature.com/documents/5099/version/1

61. Chave J, Andalo C, Brown S, Cairns MA, Chambers JQ, Eamus D et al (2005) Tree allometry and improved estimation of carbon stocks and balance in tropical forests. Oecologia 145(1):87–99

62. Hairiah K, Sitompul S, Noordwijk MV, Palm C (eds) (2001) Methods of sampling carbon stocks above and below ground. ASB Lecture Note. International Centre for Research in Agroforestry (ICRAF)-Southeast Asian Regional Research Program, Bogor

63. Le ML (2005) Phương pháp đánh giá nhanh sinh khối và Ảnh hưởng của độ sâu ngập lên sinh khối rừng Tràm (*Melaleuca cajuputi*) trên đất than bùn và đất phèn khu vực U Minh Hạ tỉnh Cà Mau (Evaluation biomass and Effect of submergence depth on growth of *Melaleuca* planting on peat soil and acid sulfate Soil in U Minh Ha area—Ca Mau Province). Nong Lam University, Ho Chi Minh City

64. Brown S (1997) Estimating biomass and biomass change of tropical forests: a primer. FAO, Quebec City

65. Keith H, Barrett D, Keenan R (2000) Review of allometric relationships for estimating woody biomass for New South Wales, the Australian Capital Territory, Victoria, Tasmania and South Australia

66. Cairns MA, Brown S, Helmer EH, Baumgardner GA (1997) Root biomass allocation in the world's upland forests. Oecologia 111(1):1–11

67. Mokany K, Raison RJ, Prokushkin AS (2006) Critical analysis of root:shoot ratios in terrestrial biomes. Glob Change Biol 12:84–96

68. Kenzo T, Ichie T, Hattori D, Itioka T, Handa C, Ohkubo T et al (2009) Development of allometric relationships for accurate estimation of above- and below-ground biomass in tropical secondary forests in Sarawak, Malaysia. J Tropic Ecol 25(4):371–386

69. Niiyama K, Kajimoto T, Matsuura Y, Yamashita T, Matsuo N, Yashiro Y et al (2010) Estimation of root biomass based on excavation of individual root systems in a primary dipterocarp forest in Pasoh Forest Reserve, Peninsular Malaysia. J Tropic Ecol 26(3):271–284

2

Community assessment of tropical tree biomass: challenges and opportunities for REDD+

Ida Theilade[1], Ervan Rutishauser[2]* (iD) and Michael K Poulsen[3]

Abstract

Background: REDD+ programs rely on accurate forest carbon monitoring. Several REDD+ projects have recently shown that local communities can monitor above ground biomass as well as external professionals, but at lower costs. However, the precision and accuracy of carbon monitoring conducted by local communities have rarely been assessed in the tropics. The aim of this study was to investigate different sources of error in tree biomass measurements conducted by community monitors and determine the effect on biomass estimates. Furthermore, we explored the potential of local ecological knowledge to assess wood density and botanical identification of trees.

Results: Community monitors were able to measure tree DBH accurately, but some large errors were found in girth measurements of large and odd-shaped trees. Monitors with experience from the logging industry performed better than monitors without previous experience. Indeed, only experienced monitors were able to discriminate trees with low wood densities. Local ecological knowledge did not allow consistent tree identification across monitors.

Conclusion: Future REDD+ programmes may benefit from the systematic training of local monitors in tree DBH measurement, with special attention given to large and odd-shaped trees. A better understanding of traditional classification systems and concepts is required for local tree identifications and wood density estimates to become useful in monitoring of biomass and tree diversity.

Keywords: Community monitoring, Tree biomass, Indonesia, Wood density, Species identification, MRV, REDD+

Background

Programs aiming at curbing deforestation and forest degradation in tropical regions (REDD+) rely upon cost-efficient techniques to monitor, report and verify forest carbon stocks. A complete enumeration of all living plants in a given landscape is impossible, and most studies rely upon a "sample plot" approach in which all trees are measured. However, the representativeness of a plot network for an entire landscape remains challenging to ascertain [1], but recommendations on the shape, size or number of sample plots have recently been proposed (e.g. [2–4]).

While professional foresters or scientists are generally in charge of establishing such sample plots, several

REDD+ projects have recently shown how local communities might represent a cheap and efficient alternative to external professionals [5–7]. In South East Asia, community monitoring was able to measure forest carbon stocks with similar accuracy as that of professional foresters [5]. Error in plot-level biomass estimates carried out by non-professional ranged between ±10% [5, 8]. At plot-level, error in biomass estimates can be divided into: (1) model error, such as the choice of a particular allometric model, prediction errors or error on the model parameters [9, 10], and (2) measurement error on the tree growth variables (e.g. tree diameter or height) or omission of trees. To mitigate these errors, standardized protocols and practices have been developed [11, 12] and generic allometric models to estimate tree biomass are now widely applied.

*Correspondence: er.rutishauser@gmail.com
[2] CarboForExpert, 1248 Hermance, Switzerland
Full list of author information is available at the end of the article

However, a significant difference in community vs forester's estimates of biomass (381 vs 449 Mg ha^{-1} respectively) was found in Indonesia by Danielsen and colleagues [5]. This discrepancy is exclusively due to measurement errors, as tree biomass was computed using the same model for both observers. In dense tropical forests, errors of measurement may be due to the presence of buttresses, irregular-shaped trunks, misplacement of the tape measure on the trunk, misreading of the actual measure or error of transcription on the tally sheet. Most REDD+ pilot programs use temporary sample plots to assess carbon stocks. The lack of repeated measurements prevents the assessment of measurements' accuracy and precision. Indeed, tree diameter could be measured accurately (mean of replicates close to the true value), but imprecisely (high variance among replicates), or precisely (low variance of replicates) but inaccurately (e.g. measured with an instrument calibrated with an incorrect standard) [13]. As a consequence, both imprecision or inaccuracy may inflate the uncertainty surrounding tree biomass estimates.

Large tropical trees are known to be more challenging to measure due to large buttresses or odd-shape stems [14], while they account for a large fraction of aboveground biomass [15]. Hence, forests with numerous large trees are more prone to be affected by errors of measurement and to large uncertainties in their biomass estimates. Due to lack of time, data precision and accuracy are barely assessed and reported in forest carbon monitoring. However, assessing main sources of error will help identifying areas where more investment in explanations and training are needed.

Another source of uncertainty relates to tree wood density (WD) that may vary at tree, species and landscape scales [16, 17]. In low accuracy estimation of carbon stocks (Tier 1), WD are approximated by an average regional default value [18, 19]. More sophisticated tree biomass estimates (Tiers 2 and 3) rely upon allometric models based on WD, tree height and diameter at breast height (DBH) [20]. Hence, botanical identification of trees is an important investment for REDD+ activities to accurately estimate tree biomass and monitor biodiversity. Due to the low number of tropical tree taxonomy experts, it has been proposed that para-taxonomists (people who lack formal education, but who are trained to undertake taxonomic tasks) can provide information at a greater rate and at a lower cost compared to expert botanists and conventional approaches [21]. Even though some communities seems to name trees consistently [22, 23], a previous study from Central Kalimantan, Indonesia resulted in poor matching between vernacular names and actual taxa, possibly due to the variety of dialects encountered [24]. On the other hand, wood densities have been found to be relatively homogeneous within Indonesian tree genera [25], and a congruent identification of the common genera by local monitors could replace the use of average WD with genus-specific values and reduce uncertainties in corresponding forest carbon stock estimates.

The present study addresses the following questions:

1. How accurate and precise are tree diameter measurements carried out by community monitors?
2. Does prior experience from logging inventories reduce measurement errors?
3. Is local ecological knowledge useful for tree identifications?
4. How do different sources of error propagate into tree biomass estimates?

Results

Source of errors in tree diameter measurements

Tree girth of 103 trees were measured by eleven local monitors, with 95% of all measurements comprised between −5.73 and 5.83 cm around the actual DBH value. Only 86 measurements out of 1,749 felt out of this confidence interval, designated hereafter to as "large errors". Large errors were more frequent and of greater magnitude (i.e. larger SD) among trees with large DBH (Figure 1). Errors were biased positively, and stand-level biomass was generally overestimated (range −4 to +20%; mean +7%). Half (52.3%) of this errors ($|DBH_{mes} − DBH_{mean}| > 6$ cm) were found among trees designated as having "odd shape" by local monitors, while these trees made up only 16% of the sample. A fifth of the measurements done on trees with odd shape was affected by large errors, significantly more than those carried on more regular stems (16 vs 3% respectively, $\chi^2 = 81.3$, df = 1, P < 10^{-5}).

Prior experience in measuring trees did not significantly decrease the likelihood of doing a large error ($\chi^2 = 2.5$, df = 1, P = 0.11). But when the repeatability of measurement was investigated, experienced monitors performed better. Difference in paired DBH measurements significantly differed (Pairwise Student test: t = −2.34, df = 146.4, P = 0.02) among experienced and inexperienced monitors, averaging 0.9 and 2.4 cm respectively (Figure 2).

Estimating wood hardness

For each tree, local monitors were also asked to estimate the wood density on a 3-classes scale (i.e. very light, light and heavy). While this simple classification returned generaly poor results (Figure 3), experienced monitors were able to discriminate trees with low wood densities (Figure 3, ANOVA: $F_{2,613} = 11.76$, P < 10^{-4}) while inexperienced monitors could not (ANOVA: $F_{2,511} = 0.424$, P = 0.655).

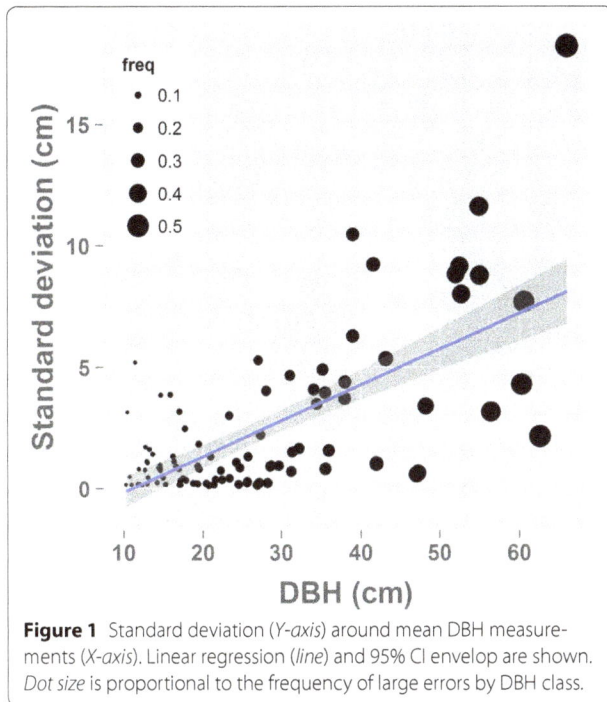

Figure 1 Standard deviation (*Y-axis*) around mean DBH measurements (*X-axis*). Linear regression (*line*) and 95% CI envelop are shown. *Dot size* is proportional to the frequency of large errors by DBH class.

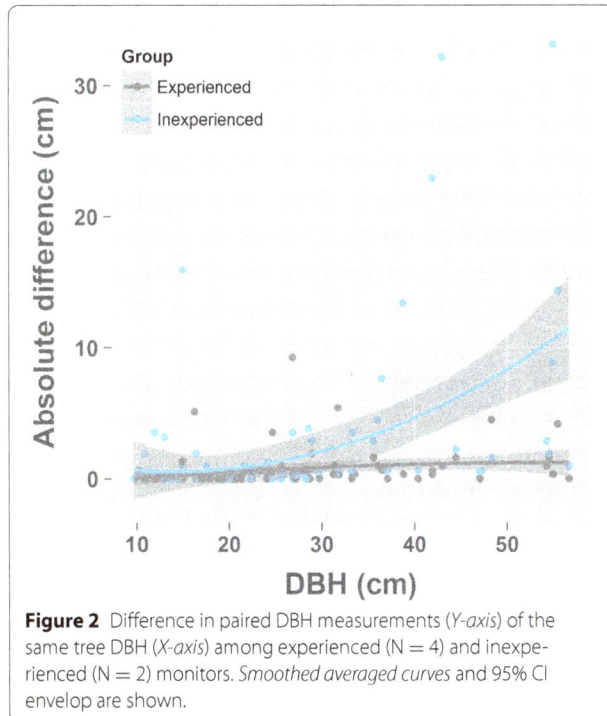

Figure 2 Difference in paired DBH measurements (*Y-axis*) of the same tree DBH (*X-axis*) among experienced (N = 4) and inexperienced (N = 2) monitors. *Smoothed averaged curves* and 95% CI envelop are shown.

Vernacular identification

The third information collected in the field was the vernacular name of each tree. Overall, there was very little agreement among observers in naming trees (Figure 4).

For instance, the number of vernacular names averaged nine per taxa. More consistency was found among Dipterocarp trees, which were better identified by experienced monitors than inexperienced ones (ANOVA: $F_{1,42} = 10.55$, $P = 0.002$).

Propagating error of DBH measurement and wood hardness into tree biomass estimates

For both experienced and inexperienced monitors, the bias increased with tree biomass (Figure 5). When accounting for DBH measurements and average wood density per wood hardness class, experienced monitors performed better and generated lower bias compared to their inexperienced counterparts (Figure 5, Estimates 1). When all trees were assigned the same wood density, biases lowered but remained high for large trees (Figure 5, Estimates 2).

Discussion
Tree diameter measurements
Overall, local monitors had good ability to measure trees, with 95% of the measurements found within 6 cm around the actual DBH. Large errors were not randomly distributed, but increased in frequency (i.e. number of occurence) and magnitude (i.e. breath of SD) with DBH (Figure 1). Half of these errors were found among odd-shaped trees, while these trees made up only 16% of the sample. A fifth of the repeated measures done on odd-shaped trees was affected by at least one large error, significantly more than among regular stems (16 vs 3%). When averaged out at stand level, we found a significant bias towards larger DBH measurements that resulted in an stand-level biomass overestimation of 7%. This error remain low and of similar magnitude as that reported in other studies [25, 26]. We have decided to use the most recent allometric models to calculate tree biomass, as generic models were shown to perform better at our site [27]. However, we acknowledge that the choice of a particular allometric model may result in greater inaccuracies than the physical measurements described above [9].

Beyond tree measurements
As botanical identification is mandatory to determine specific WD and calculate tree biomass, two methods were tested to see whether local knowledge could help towards this task. The introduction of a simple 3-scales wood hardness classification returned unconvincing results (Figure 3), as inexperienced monitors were not able to distinguish between hardwood classes while experienced monitors were able to distinguish very light wood only. Likewise, more consistency was found among experienced monitors to name Dipterocarp trees (Figure 4), i.e. the main commercial timber family

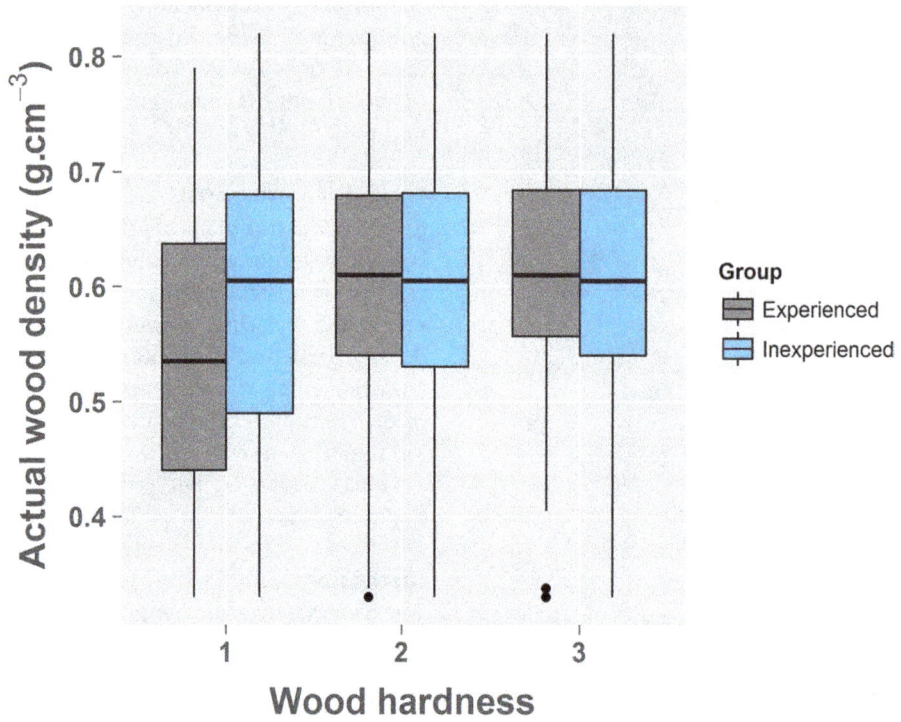

Figure 3 Boxplot of wood densities by wood hardness class estimated by experienced (*grey*) and inexperienced (*blue*) observers.

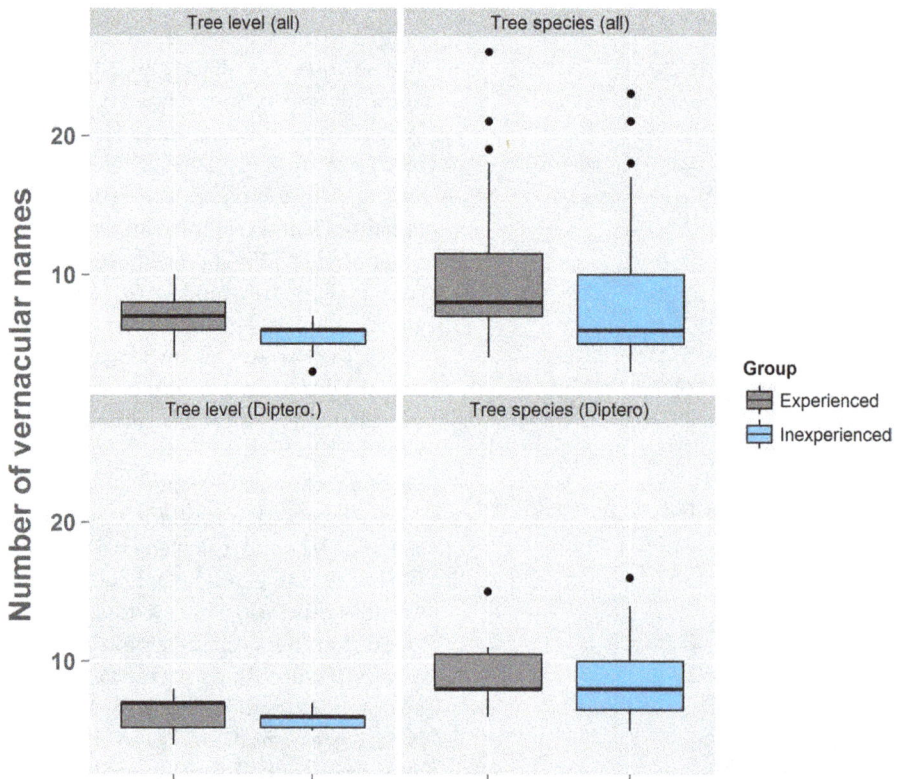

Figure 4 Number of vernacular names (*boxplots*) at tree and species by experienced and inexperienced monitors for all trees (*top*) and Dipterocarps only (*bottom*).

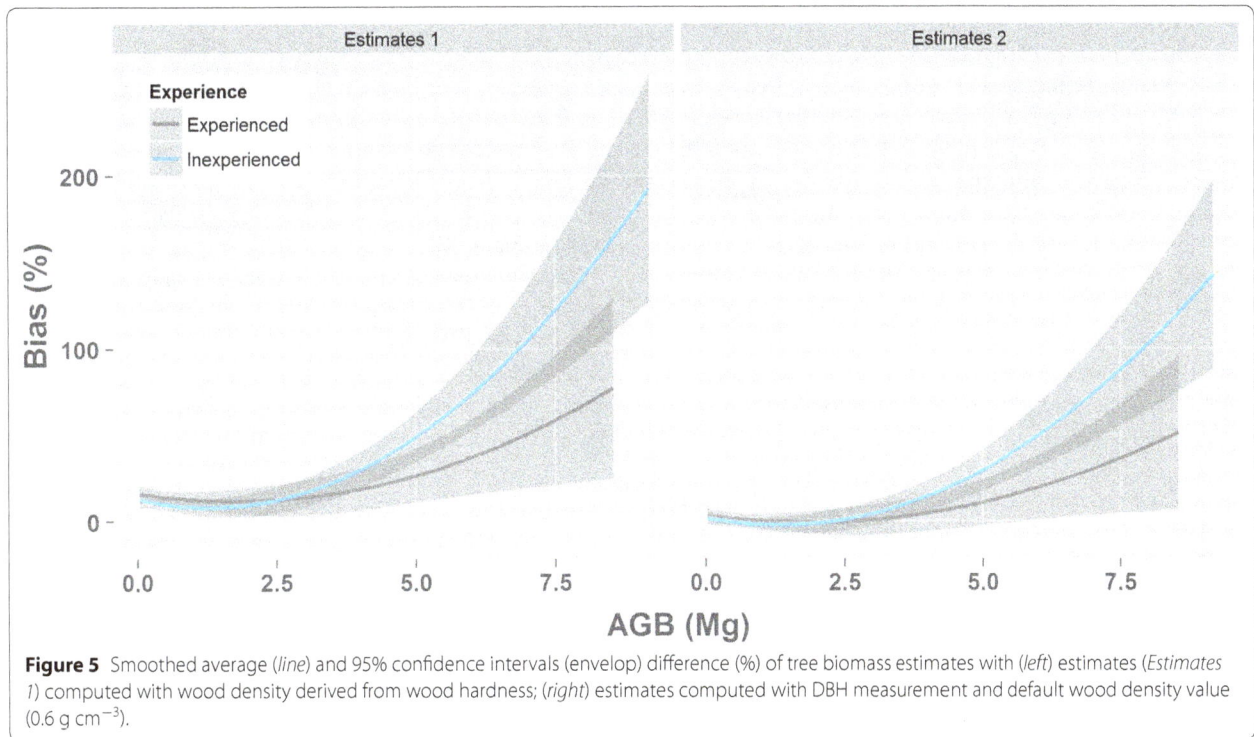

Figure 5 Smoothed average (*line*) and 95% confidence intervals (envelop) difference (%) of tree biomass estimates with (*left*) estimates (*Estimates 1*) computed with wood density derived from wood hardness; (*right*) estimates computed with DBH measurement and default wood density value (0.6 g cm^{-3}).

in the region. This is not surprising as their experience consists mainly in identifying commercial hard wood species, including Dipterocarps, during pre-logging inventories [28]. The overall inability of monitors to classify trees based on coarse wood hardness categories may arise from a misunderstanding of this peculiar concept. Local people usually possess sound knowledge on what different species can be used for, including wood properties such as workability, termite resistance, suitability for tools, firewood or boat-making. A possible explanation may lie in their inability to 'translate' this knowledge into this simple wood hardness scale. We suggest that future studies take point of departure in emic categories, i.e. categories defined by local people. Overall, there was little agreement among observers in naming trees. This result corroborate a previous study carried in Borneo, where only 10–20% of the vernacular names employed by Dayak para-taxonomists could be related to a given taxa [24]. The great variability in vernacular names in the region is a result of the numerous ethnic groups and dialects encountered in Borneo. Locally, trees are named based on local or traditional usage and names might be restricted to a community or even a group of villagers. Different species or genera having similar properties or usage are often given the same vernacular name. For instance, at our site, some trees were given names that can be translated as "big tree". Hence collection and interpretation of vernacular

names remains challenging. However, vernacular names remain employed in the logging industry and timber trade, but with little consistency with scientific taxonomy [23]. Refining the list of commonly used vernacular names of Bornean trees, and the corresponding botanical identification at species or genus level would improve forest inventories based on vernacular names.

Improvement of community monitoring in a REDD+ scheme

The discrepancy in forest biomass stocks measured by community monitors and foresters reported in a previous study at our site [5, 28], is likely to be due to the difficulty to accurately measure large trees in dense tropical forests. Measurement errors among odd-shaped trees is recurrent in carbon accounting studies. As tree biomass allometries relate dry mass with a theoretical taper or cylindrical bole diameter, biomass estimation requires tree measurements above any major irregularities of the trunk. Due to the polynomial form of current generic allometric models, a linear relationship between error and DBH (Figure 1) results mechanically in an exponential inflation of uncertainty when expressed in biomass (Figure 5). We have shown that error in biomass estimates inflates with tree biomass and inexperience. For instance, the biomass of a typical tree of 7.5 ton might be over/underestimated by 47 or 80% by an experienced or inexperienced observer respectively (Figure 5, Estimates

2). This difference goes up to 55 and 120% respectively, when estimated WD are included into biomass computation (Figure 5, Estimates 1).

This issue becomes more acute when monitoring forest biomass over time, as rapid radial increments of buttresses will compound the overestimation of biomass increase [29]. While local monitors accurately measured DBH of most trees, much attention and training should be paid on large trees (>60 cm DBH). Prior experience in measuring trees did not lower the likelihood of doing large errors, but increased accuracy of repeated measurements. Thereby, trained monitors are less prone to systematic bias, a key feature in terrestrial carbon monitoring where true biomass value is sought. Accuracy will also be requested to estimate changes in forest carbon stocks over repeated censuses. Indeed, error of measurements and data correction might prevent the detection of any directional change in biomass stock [30].

In a multi-country comparison of the efficiency (i.e. costs and accuracy) of local communities to monitor tree biomass stocks, Brofeldt and collaborators [28] relied at the second census upon a few community members trained initially, while the rest of team received a brief training only. Based on this study, we recommend that all community monitors involved in REDD+ programmes receive a complete training on tree measurement with special attention on dealing with large and odd-shaped trees. When multi-census has to be carried out, points of measurements should be clearly marked in the field (i.e. paint mark on the trunk). Technical improvements to increase accuracy of community-based measurements of carbon stock will likely facilitate the uptake and scaling up of local information as part of the national forest monitoring system (NFMS) and the associated monitoring, reporting, and verification (MRV) system for REDD+ [31]. This is in line with current United Nations Framework Convention on Climate Change (UNFCCC) texts and guidance documents on the technical aspects of REDD+ which outline explicit roles for indigenous people and local communities in implementing REDD+ [32–34].

Conclusion

Several REDD+ studies have recently shown how community monitors represent a cost-efficient and reliable alternative to external professionals. In this study, we have investigated different sources of error in tree diameter measurements conducted by community monitors and propagated those at both tree and stand levels biomass estimates.

Local monitors had good ability to measure tree DBH with 95% of all measurements found within a confidence interval of 6 cm around the actual DBH. Large errors were more frequent and of greater magnitude among trees with a large DBH (>60 cm DBH) and odd-shaped trunks. Monitors with experience from logging inventories performed better and generated lower bias compared to inexperienced monitors although the likelihood of large errors was identical among both groups. Overall, we found a directional bias towards overestimated DBH among monitors that led to a slight inflation of stand-level biomass (7%).

We suggest that future REDD+ programmes may benefit from the systematic training of local monitors in measuring tree DBH with special attention given to large and odd-shaped trees. A better understanding of traditional classification systems and concepts, possibly combined with a basic training of local monitors in taxonomy, is required for tree identifications to become useful in monitoring either forest biomass, or tree diversity.

Methods

Study site and community monitors

The study area is located in the district of Kutai Barat District, East Kalimantan, Indonesia. Monitoring plots were established in the customary forest surrounding the Dayak village of Batu Majang. The tropical lowland rainforest at 300 m.a.s.l. is characterised by species of the Dipterocarp family such as *Shorea* sp., *Dipterocarpus* sp., *Anisoptera* sp., and *Hopea* sp. among other high quality timber species. Despite the customary harvest of a few trees and other non-wood forest products, the forest structure is similar to that of a primary forest. The local community is committed to conserve the forest for various reasons, such as protecting the watershed and hunting/harvesting resources. Several permanent forest plots were established in 2012, in which all trees >10 cm DBH were tagged, measured and identified to species level [27].

Representatives of the local Dayak community helped select eleven participants (referred hereafter to as community monitors) based on their interest and experience with forest resources, to measure the girth, estimate wood density, and identify trees in the permanent plots. All community monitors were male, had attended primary school, and received 3 h of specific training on tree measurement in the field. Six monitors had a prior employment in timber companies, doing surveys (i.e. mapping harvestable stems) for logging operations. This group is referred to as "experienced", while others (n = 5) with no previous experience are referred to as "inexperienced".

Data collected

In 2014, 103 trees were randomly chosen among two permanent monitoring plots and measured by local monitors. While creating a tree-walk and numbering the trees, the community monitors were trained at measuring tree girth and estimate wood hardness. Girth measurement was done at 130 cm height using classical tapes with centimeter units. Monitors were instructed carefully to avoid common mistakes such as a twisted or lax tape, a thumb placed under the tape, and measuring below breast height.

When measurement was hampered by the presence of buttresses, lianas, or trunk deformities, i.e. extra efforts had to be made to measure tree, monitors were asked to record the tree as "odd shaped". Wood properties of common tree species is often known by local communities. To test whether such information could be used to refine tree biomass estimates, each monitor was asked to assess wood hardness using a simple classification: "1" for very light wood, "2" for floater (light wood) and "3" for sinker (heavy wood). These categories are used in the logging industry and are well-known to local people. Finally, monitors were asked to name each tree using Dayak common names. Community members worked in teams of two people, monitor A measuring the girth, assessing wood density and naming trees along the full tree-walk, and monitor B writing down information on a pre-prepared form.

Statistical analysis
Overall precision
We investigated the distribution of error measurements on a per-tree basis. As each tree was measured at least once by the different community members, we computed the differences between each measurement and the average DBH for each tree. We further used the 5th and 95th percentiles of these differences to identify large errors. For each tree, we defined the actual DBH (DBH_{mean}), as the average of all measurements comprised within the 5th and 95th percentiles. The minimum number of measurements used to compute the actual DBH is 12 (max = 17).

The precision of measurements of a given tree diameter refers to the variance of the different measurements. We used the standard deviation to estimate how the different measures spread out from the mean value. The bigger the error, the larger the standard deviation.

$$SD(\sigma) = \sqrt{\frac{1}{n} \sum \left(DBH_{mes,i} - DBH_{mean,i} \right)^2}$$

Repeatability of measurement
102 trees were measured twice by six observers. We estimated the repeatability of girth measurements among those observers, by calculating the absolute difference among both measurements.

Comparison of wood hardness and botanical estimation
In 2012, all trees were identified at species level by a professional botanist [27]. Trees were identified directly in the field to the lowest taxonomical level. Among the

102 trees accounted for in the present study, 70% were identified at species level and 30% at genus level (Additional file 1). From these identifications, wood densities were extracted from the Global Wood Density Database [35] and considered as actual wood densities (WD). The capacity of local observers to group trees in three classes of wood hardness was further assessed with a one-way ANOVA by wood hardness classes and observers experience.

Error propagation in tree biomass estimates
We integrated information gathered in the field by local monitors (i.e. wood hardness and DBH measurements) into biomass estimates. Wood hardness was associated to the 25, 50 and 75th percentile of actual wood densities respectively (1 = 0.55, 2 = 0.63, 3 = 0.73 g cm^{-3}). Tree biomass (Estimates 1) was computed using a generic allometric model [20], as follow:

$$AGB_{est} = \exp[-1.803 - 0.976 \times E + 0.976 \times \ln(WD)$$
$$+ 2.673 \times \ln(DBH) - 0.0299 \times \ln(DBH)^2]$$

where E is a synthetic index of temperature seasonality, maximum climatological water deficit, and precipitation seasonality (E = -0.09162301 at our site), WD is the wood density (g cm^{-3}), and DBH, the diameter at breath height (cm).

Alternatively, tree biomass (Estimates 2) was computed using a default WD value for Bornean forests (WD = 0.6, 37) to estimate a "Tier 1" level of uncertainty. Both estimates were further compared to the best tree biomass estimate (AGB_0), computed with actual WD and DBH (DBH_{mean}) as recommended by Tier 3 standard [19]. Differences in tree biomass are expressed as bias (e.g. [estimate$_1$ $-$ AGB_0]/AGB_0). To check if errors could cancel each other at stand level (i.e. no directional bias), tree biomass were summed for each monitor and the relative bias (%) per monitor was computed as follow:

$$bias_j(\%) = \frac{\sum AGB_{ij} - \sum AGB_0}{\sum AGB_0},$$

where i = the ith tree, j = the jth monitor and AGB_0 = best tree biomass estimate.

Authors' contributions
IT, ER, MKP equally contributed in designing the protocol and writing the manuscript; MKP conducted data collection and ER did the statistical analysis. All authors read and approved the final manuscript.

Author details
[1] Faculty of Science, Institute of Food and Resource Economics, University of Copenhagen, Rolighedsvej 25, 1958 Frederiksberg C, Denmark. [2] Carbo-ForExpert, 1248 Hermance, Switzerland. [3] Nordic Agency for Development and Ecology (NORDECO), Skindergade 23, 1159 Copenhagen K, Denmark.

Acknowledgements
We are most grateful to Pak Yosep, head of village of Batu Majang, Agus, Anse Latus, Lusang, Prin, Samuel Ajang, Simon, Sius, Syahdan, Vincen Idum, Vincensius Yen and Sarjuni for conducting field work. We thanks Yuyun Karniawan and Itong Sarjuni for facilitating liaison and transportation, and Pak Yosep for housing Itong Sarjuni during the field visit. We also thank Kristell Hergoualch (CIFOR) and Andreas de Neergaard (University of Copenhagen) for providing tree botanical identifications.

Compliance with ethical guidelines

Competing interests
The authors declare that they have no competing interests.

References
1. Chave J, Condit R, Aguilar S, Hernandez A, Lao S, Perez R (2004) Error propagation and scaling for tropical forest biomass estimates. Philos Trans R Soc B Biol Sci 359:409–420
2. Baraloto C, Molto Q, Rabaud S, Hérault B, Valencia R, Blanc L et al (2013) Rapid simultaneous estimation of aboveground biomass and tree diversity across neotropical forests: a comparison of field inventory methods. Biotropica 45:288–298
3. Wagner F, Rutishauser E, Blanc L, Herault B (2010) Assessing effects of plot size and census interval on estimates of tropical forest structure and dynamics. Biotropica 42:664–671
4. Walker SM, Pearson T, Casarim FM, Harris H, Petrova S, Grais A et al (2012) Standard operating procedures for terrestrial carbon measurement. Winrock International, USA
5. Danielsen F, Adrian T, Brofeldt S, van Noordwijk M, Poulsen MK, Rahayu S et al (2013) Community monitoring for REDD+: international promises and field realities. Ecol Soc 18:41
6. Larrazábal A, McCall MK, Mwampamba TH, Skutsch M (2012) The role of community carbon monitoring for REDD+: a review of experiences. Curr Opin Environ Sustain 4:707–716
7. Butt N, Slade E, Thompson J, Malhi Y, Riutta T (2013) Quantifying the sampling error in tree census measurements by volunteers and its effect on carbon stock estimates. Ecol Appl 23:936–943
8. Molto Q, Rossi V, Blanc L (2013) Error propagation in biomass estimation in tropical forests. Methods Ecol Evol 4:175–183
9. Picard N, Boyemba Bosela F, Rossi V (2014) Reducing the error in biomass estimates strongly depends on model selection. Ann For Sci. doi:10.1007/s13595-014-0434-9
10. GOFC-GOLD (2012) A sourcebook of methods and procedures for monitoring and reporting anthropogenic greenhouse gas emissions and removals caused by deforestation, gains and losses of carbon stocks in forests remaining forests, and forestation. Wageningen, Global Observation of Forest Cover and Land Dynamic (GOFC-GOLD)
11. IPCC (2014) 2013 Revised supplementary methods and good practice guidance arising from the kyoto protocol
12. Clark DB, Kellner JR (2012) Tropical forest biomass estimation and the fallacy of misplaced concreteness. J Veg Sci 23:1191–1196
13. Clark DA (2002) Are tropical forests an important carbon sink? Reanalysis of the long-term plot data. Ecol Appl 12:3–7
14. Slik J, Paoli G, McGuire K, Amaral I, Barroso J, Bastian M et al (2013) Large trees drive forest aboveground biomass variation in moist lowland forests across the tropics. Glob Ecol Biogeogr 22:1261–1271
15. Chave J, Muller-Landau HC, Baker TR, Easdale TA, Ter Steege H, Webb CO (2006) Regional and phylogenetic variation of wood density across 2,456 neotropical tree species. Ecol Appl 16:2356–2367
16. Henry M, Besnard A, Asante W, Eshun J, Adu-Bredu S, Valentini R et al (2010) Wood density, phytomass variations within and among trees, and allometric equations in a tropical rainforest of Africa. For Ecol Manag 260:1375–1388
17. Brown S (1997) Estimating biomass and biomass change of tropical forests: A primer. FAO Forestry Paper, vol 134. UN FAO, Rome **(FAO [series editor]: Forestry Paper)**
18. IPCC (2006) Guidelines for national greenhouse gas inventories, vol 4. Institute for Global Environmental Strategies (IGES), Hayama **(Eggelstons S, Buendia L, Miwa K, Todd N, Tanabe K [series editors])**
19. Chave J, Réjou-Méchain M, Búrquez A, Chidumayo E, Colgan MS, Delitti WB et al (2014) Improved allometric models to estimate the above-ground biomass of tropical trees. Glob Chang Biol 20:3177–3190
20. Sheil D, Lawrence A (2004) Tropical biologists, local people and conservation: new opportunities for collaboration. Trends Ecol Evol 19:634–638
21. Jinxiu W, Hongmao L, Huabin H, Lei G (2004) Participatory approach for rapid assessment of plant diversity through a folk classification system in a tropical rainforest: case study in Xishuangbanna, China. Conserv Biol 18:1139–1142
22. de Lacerda AEB, Nimmo ER (2010) Can we really manage tropical forests without knowing the species within? Getting back to the basics of forest management through taxonomy. For Ecol Manag 259:995–1002
23. Wilkie P, Saridan A (1999) The limitations of vernacular names in an inventory study, Central Kalimantan, Indonesia. Biodivers Conserv 8:1457–1467
24. Slik JWF (2006) Estimating species-specific wood density from the genus average in Indonesian trees. J Trop Ecol 22:481
25. Venter M, Venter O, Edwards W, Bird MI (2015) Validating community-led forest biomass assessments. PLoS One 10:e0130529
26. Butt N, Epps K, Overman H, Iwamura T, Fragoso JMV (2015) Assessing carbon stocks using indigenous peoples' field measurements in Amazonian Guyana. For Ecol Manag 338:191–199
27. Rutishauser E, Noor'an F, Laumonier Y, Halperin J, Rufi'ie, Hergoualch K, Verchot L (2013) Generic allometric models including height best estimate forest biomass and carbon stocks in Indonesia. For Ecol Manag 307:219–225
28. Brofeldt S, Theilade I, Burgess ND, Danielsen F, Poulsen MK, Adrian T et al (2014) Community monitoring of carbon stocks for REDD+: does accuracy and cost change over time? Forests 5:1834–1854
29. Sheil D (1995) A critique of permanent plot methods and analysis with examples from Budongo Forest, Uganda. For Ecol Manag 77:11–34
30. Muller-Landau HC, Detto M, Chisholm RA, Hubbell SP, Condit R (2014) Detecting and projecting changes in forest biomass from plot data. In: Coomes DA, Burslem DFRP, Simonsen WD (eds) Forests and global change. Cambridge University Press, Cambridge, pp 381–416
31. Torres A (2014) Potential for integrating community-based monitoring into REDD+. Forests 5:1815–1833
32. UNFCCC (2011) Framework convention on climate change, subsidiary body for scientific and technological advice (SBSTA), methodological guidance for activities relating to reducing emissions from deforestation and forest degradation and the role of conservation, sustainable management of forests and enhancement of forest carbon stocks in developing countries. Draft conclusions proposed by the Chair, Thirty-fifth session Durban, 28 November to 3 December 2011. UNFCCC, Bonn
33. UNFCCC (2011) Outcome of the work of the ad hoc working group on long-term cooperative action under the convention. Draft decision [-/CP.17]. UNFCCC, Bonn
34. UNFCCC (2009) Methodological guidance for activities relating to reducing emissions from deforestation and forest degradation and the role of conservation, sustainable management of forests and enhancement of forest carbon stocks in developing countries. Decision 4/CP.15, FCCC/CP/2009/11/Add.1. United Nations Framework Convention on Climate Change, Copenhagen
35. Zanne AE, Lopez-Gonzalez G, Coomes DA, Ilic J, Jansen S, Lewis SL et al (2009) Global wood density database. http://datadryad.org/repo/handle/10255/dryad.235

3

Spatial distribution of temporal dynamics in anthropogenic fires in miombo savanna woodlands of Tanzania

Beatrice Tarimo[1,2*], Øystein B Dick[3], Terje Gobakken[1] and Ørjan Totland[1]

Abstract

Background: Anthropogenic uses of fire play a key role in regulating fire regimes in African savannas. These fires contribute the highest proportion of the globally burned area, substantial biomass burning emissions and threaten maintenance and enhancement of carbon stocks. An understanding of fire regimes at local scales is required for the estimation and prediction of the contribution of these fires to the global carbon cycle and for fire management. We assessed the spatio-temporal distribution of fires in miombo woodlands of Tanzania, utilizing the MODIS active fire product and Landsat satellite images for the past ~40 years.

Results: Our results show that up to 50.6% of the woodland area is affected by fire each year. An early and a late dry season peak in wetter and drier miombo, respectively, characterize the annual fire season. Wetter miombo areas have higher fire activity within a shorter annual fire season and have shorter return intervals. The fire regime is characterized by small-sized fires, with a higher ratio of small than large burned areas in the frequency-size distribution ($\beta = 2.16 \pm 0.04$). Large-sized fires are rare, and occur more frequently in drier than in wetter miombo. Both fire prevalence and burned extents have decreased in the past decade. At a large scale, more than half of the woodland area has less than 2 years of fire return intervals, which prevent the occurrence of large intense fires.

Conclusion: The sizes of fires, season of burning and spatial extent of occurrence are generally consistent across time, at the scale of the current analysis. Where traditional use of fire is restricted, a reassessment of fire management strategies may be required, if sustainability of tree cover is a priority. In such cases, there is a need to combine traditional and contemporary fire management practices.

Keywords: Burned area, Carbon stocks, Fire history, Frequency-size distribution, Landsat, Miombo woodland, MODIS, Surface fires

Background

Anthropogenic fires are historically an integral component of African savannas. They strongly influence the composition, structure and distribution of mesic savannas in particular, where tree cover is not constrained by climatic conditions [1–4]. Fire regimes in African savannas, including the frequency and season of burning, are mainly human regulated [5]. The variability of fire regimes in African savannas is more dependent on

human drivers than on climate and thus human drivers may regulate the future of savanna fire regimes under changing climate conditions [6–9]. Human activities associated with fire ignitions and fragmentation of the landscape play a key role in determining the occurrence of fire and resulting spatial extents of burned areas [10–14]. Tropical savannas, predominantly in Africa, contributes the highest proportion of the global burned area [15], and their contribution to biomass burning emissions is substantial [16, 17]. The role of these fires as a management tool or as a threat to woody cover, and in the global carbon cycle, vary within savannas and is dependent on the fire regime. Efforts to change fire regimes in favor of

*Correspondence: beatrice.tarimo@nmbu.no
[1] Department of Ecology and Natural Resource Management, Norwegian University of Life Sciences, P.O. Box 5003, 1432 Ås, Norway
Full list of author information is available at the end of the article

management priorities, such as carbon sequestration, are being challenged in the light of traditional fire regimes that are more suited for the sustainability of savannas [10, 18–21]. Viable fire management plans aiming at maintenance of stored carbon requires an understanding of historical fire regimes at local scales, which is generally lacking for many parts of African savannas. This understanding is required for precise estimates of the contribution of savanna fires in the global carbon dynamics. Characterization of current fire regimes at local scales is required in order to set references against which assessment of changes in burning practices and their contribution to the carbon cycle will be made [22]. This is of particular importance in fire-adapted ecosystems, such as miombo woodlands, that also support a wide range of human subsistence activities.

Fire is regarded essential to the structure and stability of miombo woodlands [23, 24]. Intense fires suppress tree biomass when their frequency is higher than the rate of tree regeneration and growth [23, 25, 26]. Frequent and intense fires threatens the maintenance of stored carbon stocks, and consequently undermines the potential benefits of activities that comprise the reducing emissions from deforestation and forest degradation (REDD+) policy instrument [27]. Fire contributes to long-term degradation that, although significant, has proven difficult to quantify and monitor, and thus receive less attention in REDD+ negotiations compared to deforestation [28–30]. In addition, they impede the enhancement of carbon stocks for REDD+ payments and sustainability of tree cover at large. Tree recruitment and succession are constrained by recurrent fires [23, 31], which instead facilitates grass encroachment and colonization that may fuel more intense and frequent fires [32, 33]. Exclusion of fire on the other hand facilitates tree dominance of the ecosystem [34], which limits the growth of light demanding grasses and consequently fuel loading. The timing of burning further regulates fire effects, such that late dry season fires have adverse effects on both vegetation and soils, whereas prescribed early dry season fires may be a beneficial management tool [23, 26]. Fire management is of crucial importance for successful forest management [23]. However, it is impaired by the limited understanding on which controlled burning treatments are beneficial for respective components of woodland savannas, coupled with the socio-economic dependency from their surroundings, which play a major role in shaping fire regimes. Characterization of the long-term fire regime will contribute to the ongoing efforts to quantify carbon stocks and fluxes for the purposes of monitoring and verification in the context of REDD+ policy framework and for better fire management practices in general.

A key challenge to both the estimation of carbon fluxes from fires and fire management efforts in African savannas is lack of complete and consistent fire records. In Tanzania, the vast majority of fire events stem from anthropogenic ignitions for different purposes, including farm preparation, pasture management, hunting, honey harvesting, charcoal production, arsons, and for security around settlements and roads [35]. Fire records are limited to a few isolated areas that implement fire management plans. In the absence of long term systematic ground fire records, satellite data forms a unique source of the recent fire history [e.g. 36, 37]. Since tropical savanna fires are fueled mainly by grasses and litter, they sweep the ground surface and leave tree crowns and soil sub-surface unaffected. The resulting burned scars persist for a few weeks only [38–41]. Therefore, frequent observations are required to capture most of the area burned in the course of a fire season. Monthly composites of observation of fire events may be representative of the spatial and temporal distribution of African savanna fires [42–44]. Although the use of different satellite systems provides multiple acquisitions every month, data availability is constrained by cloud cover and other limitations.

Datasets on active fires and burned areas derived from along track scanning radiometer (ATSR), SPOT-VEGETATION and moderate resolution imaging spectroradiometer (MODIS), among other satellite sensors, are available in the public domain. They provide fire patterns at a coarse spatial resolution and at very short temporal coverages. However, comparisons of burned areas derived from coarse resolution (1 km) with those derived from finer resolutions (e.g. 30 m) satellites, show that the majority (up to 90%) of small burned areas characteristic of fragmented fires in tropical savannas, are not detected by coarse resolution burned area products [41, 42, 44, 45]. The low detectability of small-burned areas by coarse spatial resolution products limit the efficacy of these products at smaller spatial scales when detailed information is required. There is thus a need to quantify spatial and temporal fire patterns and resulting burned extents at finer resolutions than those available in the public domain.

The availability of Landsat satellite images in the public domain provides the opportunity to extract burned area records since the early 1970s. Methods are being developed for (semi)automatic burned area mapping at finer spatial resolution e.g. [46, 47], which facilitate frequent and complete mapping at local and regional scales. However, few studies have employed the utility of these methods in African savannas. Thus, burned area records are still missing despite the availability of satellite images. We aim at assessing the fire history during the past ~40 years and respective spatial patterns from satellite based data.

Burned areas are mapped by fuzzy classification using spectral indices that include infrared wavelengths, since they are more sensitive to fire induced changes than other spectral combinations [48–50]. We discuss the derived fire return intervals, seasonality and burned extents in Tanzanian miombo relative to those from other African savannas, and the observed frequency-size statistics relative to those reported from other ecosystems. We highlight the consistency in the fire regime across spatial and temporal scales and point out priority areas requiring further analyses and reassessment of management practices.

Results

Validation of detected burned areas

Table 1 summarizes classification accuracy analysis of detected burned areas. Omissions of burned pixels are mainly in the partially burned areas, which are not included in Table 1.

Based on an independent validation, the overall performance of the fuzzy classification when including partially burned areas was 57%, which is not as good as that of the completely burned areas (Table 1). It should however be noted that the definition of fuzzy membership scores to distinguish burned from partially burned areas (see "Validation of detected burned areas" in the "Methods" section) on one hand, and the subjective element of the result of the visual interpretation on the other hand, might have had an impact on the quantification of the performance of the fuzzy classification.

Spatial and temporal patterns of burned areas

Burned patch sizes

For each particular year, the majority (up to the third quartile) of burned patches are less than five hectares in size (Fig. 1). The annual median of burned patch size ranged from 0.8 to 1.4 ha. Small burned patches are more common in wet miombo than in dry miombo areas, with annual median ranging from 0.8 to 1.4 ha and 0.7 to 1.8 ha, respectively. Relatively few and very occasional big fires may reach sizes of up to ~60,000 ha. These account for a large proportion of the total area burned but they tend to decrease in frequency during the 1972–2011

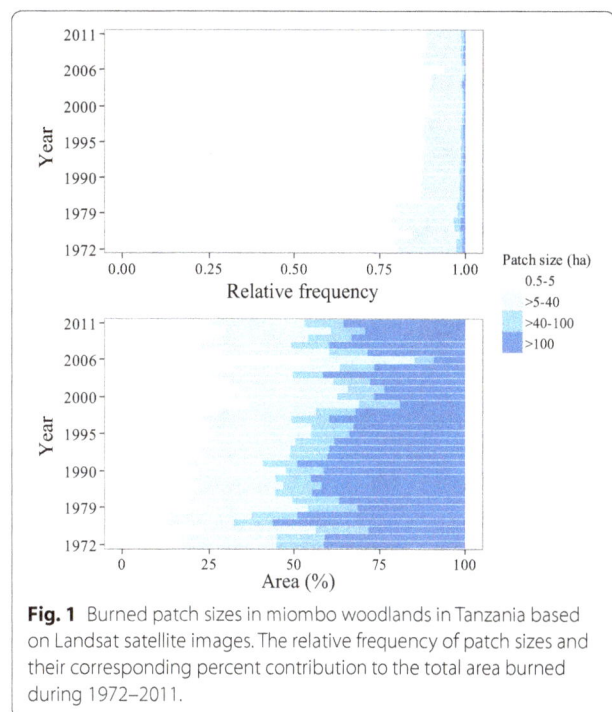

Fig. 1 Burned patch sizes in miombo woodlands in Tanzania based on Landsat satellite images. The relative frequency of patch sizes and their corresponding percent contribution to the total area burned during 1972–2011.

period, relative to other size classes. Overall, small-burned patches, which are a common occurrence over spatial and temporal scales, account for more of the total area burned than large burned patches.

Frequency-size distribution of burned patches

Frequency-size statistics of burned areas suggest a fire regime dominated by small-sized fires with scaling, $\beta = 2.16 \pm 0.04$, with $r^2 = 0.99$ for the whole woodland during 1972–2011. Wet miombo has a slightly smaller scaling, $\beta = 2.13 \pm 0.03$, with $r^2 = 0.99$ relative to dry miombo where scaling, $\beta = 2.15 \pm 0.04$, with $r^2 = 0.99$. Given the high number of annual burned patches, it was deemed relevant to analyze the annual frequency-size distributions. Annual analyses resulted in scaling ranging from $\beta = 1.89 \pm 0.04$ to $\beta = 2.53 \pm 0.15$, with $r^2 > 0.98$ for the whole miombo woodland. Similarly, wet miombo has a slightly smaller scaling than dry miombo from annual analyses, ranging from $\beta = 1.71 \pm 0.17$ to $\beta = 2.50 \pm 0.19$, with $r^2 > 0.95$ and from $\beta = 1.82 \pm 0.05$ to $\beta = 2.57 \pm 0.43$, with $r^2 > 0.94$, respectively.

Burned extents

Figure 2 presents patterns of burned areas detected from Landsat images, summarized at a 5 × 5 km grid. At this scale, fire incidences appear to be consistently within the same spatial extents. Temporal differences in the extent burned per window show an irregular spatial trend. Annually, up to 13.7% and 12.6% of the total area with available

Table 1 Omission and commission errors of burned pixels

Class	Samples		Errors	
	Correctly classified	Incorrectly classified	Omission (%)	Commission (%)
Burned	1,022	365	26.3	0.7
Not burned	7,826	7	0.1	4.5

Kappa coefficient = 0.82.

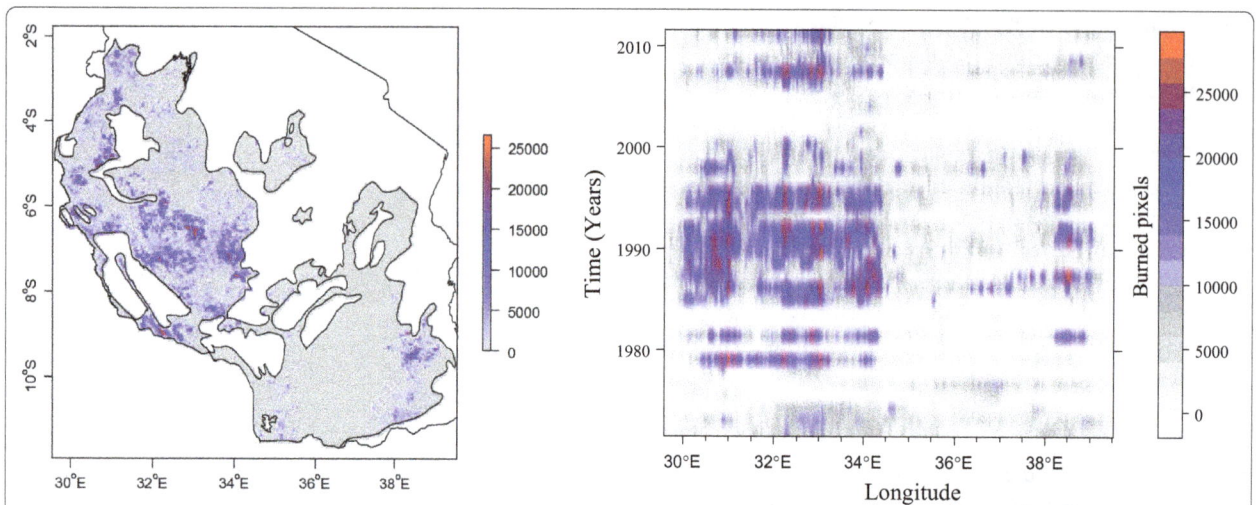

Fig. 2 Spatio-temporal patterns of annual burned areas in miombo woodlands in Tanzania. Burned areas from Landsat imagery are summarized at 5 × 5 km resolution, to show the spatial extent for a selected year (1995) on the *left* and spatial–temporal patterns (1972–2011) on the *right*. For the hovmoller diagram on the *right hand side*, *white spaces* represent missing data and areas outside miombo woodland extent while *zero* represents areas not burned.

imagery was detected as burned in wet and dry miombo, respectively. When combined with partially burned areas, up to 65.8% and 42.1% of wet and dry miombo, respectively, was detected as burned annually. For the whole miombo woodland in Tanzania up to 11.3% is burned annually, while when combined with partially burned areas, up to 50.6% of the woodland is affected by fire annually. Table 2 provides a decadal summary of the contribution of wet and dry miombo areas to the total area burned

for the whole woodland. In this table, comparisons are more reliable between dry and wet miombo for the same duration than between durations due to differences in the number of years with available data for each location.

Spatial and temporal patterns of active fires
Early and late dry season burning
A west-to-east transition of fire events from early to late burning is observed in Fig. 3. The sudden drop of

Table 2 Burned area characteristics in (a) dry and (b) wet miombo areas in Tanzania during 1972–2011

Duration	P (ha)	A (%)	M [Dry] (%)	RI	Scaling (β)	A_P
(a) Dry miombo						
1972–1979	930–53,300	3.6–12.6	5.2 [2.9]–11.3 [9.8]	1.6	1.82–2.11	NA
1980–1989	5.4–27,270	3.7–6.5	4.5 [2.2]–8.6 [3.6]	1.8	1.97–2.57	14.3–26.9
1990–1999	1,649–64,650	0.6–8.0	0.8 [0.4]–10.0 [5.1]	2.4	2.00–2.15	10.0–34.1
2000–2011	676.4–21,090	0.7–6.1	1.0 [0.5]–6.6 [4.3]	2.8/3.0[a]	1.93–2.29	4.3–23.6

Duration	P (ha)	A (%)	M [Wet] (%)	RI	Scaling (β)	A_P
(b) Wet miombo						
1972–1979	1,199–29,050	6.7–11.0	7.6 [2.2]–5.2 [2.3]	1.4	1.83–2.12	NA
1980–1989	8.2–15,760	5.9–11.2	4.5 [2.3]–8.6 [5.0]	1.6	2.04–2.50	16.3–31.3
1990–1999	1,166–36,530	1.6–13.7	0.8 [0.4]–10.0 [4.9]	1.4	2.02–2.14	13.3–52.1
2000–2011	9.1–15,160	0.2–7.7	0.8 [0.1]–6.6 [2.3]	2.0/2.1[a]	1.71–2.26	1.2–25.3

Values in square brackets represents the percent contribution of dry/wet miombo to *M*.

P the range of largest annual burned patches detected, may have aggregated with time during the fire season, *A* the range of the total area burned in dry/wet miombo as a percentage of the dry/wet miombo area with data, A_P the total area partially burned in dry/wet miombo as percentage of the total dry/wet miombo area with data, *M* the total area burned in miombo at the same time when *A* was recorded, as percentage of miombo area with data, *RI* Fire return interval observed for every 2,500 ha from Landsat satellite images.

[a] Based on MODIS detected fires for every 314 ha for the period 2001–2013.

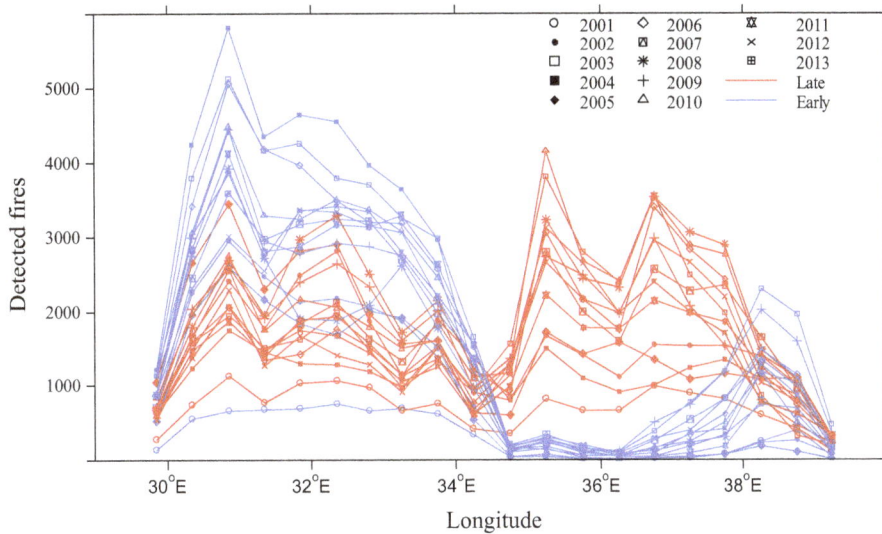

Fig. 3 Spatial distribution of MODIS detected fires in miombo woodlands in Tanzania. July was defined to mark the end of early dry season burning for the entire woodland; local variations are likely to occur.

incidences at ~35.75°E is partly explained by the extent of miombo woodland areas (see Fig. 4) and it marks a distinction between an early dry season burning dominated west to a late dry season burning dominated east.

Based on the number of detected fires each month the fire season peaks during the first part of the dry season in July (Fig. 5). To investigate the effect of early dry season burning on late dry season fires, fire radiative power (FRP) values of late dry season fires were compared for those fires which were either close (within 1 km; i.e. approximately within the same fire pixel) or far (>1 km) from early dry season fires during the same fire season. There is no significant reduction, at 95% confidence level, of FRP values in the late dry season fires, which were close to early dry season burned areas than those far from them.

Fire activity
The combined characteristics of detected active fires provide a composite estimate of fire activity given in Fig. 6. Fire activity is consistently high in the western part of the woodland with the exception of areas along its northeastern border. An increasing systematic westward reduction in fire activity is observed along this border during 2001–2013 (Fig. 6). This reduction is associated with the expansion of croplands when interpreted in the context of the GLC-Share land cover types [51]. On the other hand, there is a shift from high to low fire activity between years on the central, south and eastern parts of the woodland.

Fire return interval
The mean fire return interval for a circular area of ~314 ha, centered at the location of MODIS detected fires was 2.7 years (range 1–13 years) between 2001 and 2013. When the analysis was performed for every 2,500 ha during 1972–2011, based on burned areas detected from Landsat images, the interval was reduced to 2.1 years.

Discussion
Historical fire regimes are best reconstructed from long-term consistent ground records, charcoal deposition in soils or fire scars on trees with annual growth rings [52–55]. In the absence of these, fire history in miombo woodlands of Tanzania was documented from Landsat satellite images and MODIS detected active fires for the past ~40 years. Both fire prevalence and burned extents have recently decreased (Table 2). This decrease is likely an outcome of a number of contributing factors, including a reduction in miombo woodland coverage through e.g. conversion of the woodland into permanent cultivated fields and fire management practices in some parts of the woodland. Burned areas and detected active fire events are consistently within the same spatial coverage (Figs. 2, 6), at the scale of the current analysis. The lack of an independent burned area perimeter for validation restricted our analysis to burned pixels with the highest confidence, which underestimate the total area burned. When thoroughly validated, an analysis including partially burned areas might increase fire activity and shorten fire return intervals in some parts of the woodland. This however, will not affect the general patterns presented in this study.

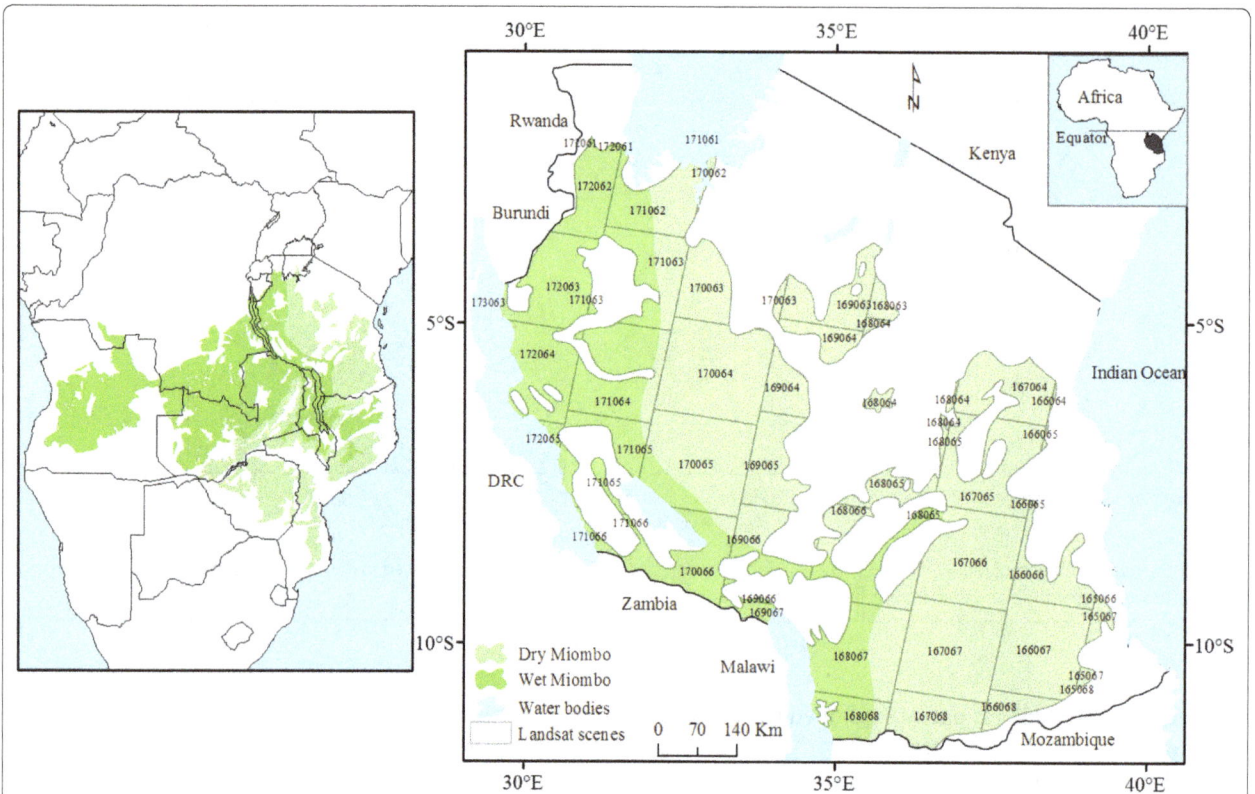

Fig. 4 Distribution and classification of miombo woodlands in Tanzania. The map is based on White's vegetation map of Africa [74]. *Numbers* show the identification (path and row numbers) and extents of Landsat TM/ETM+ scenes.

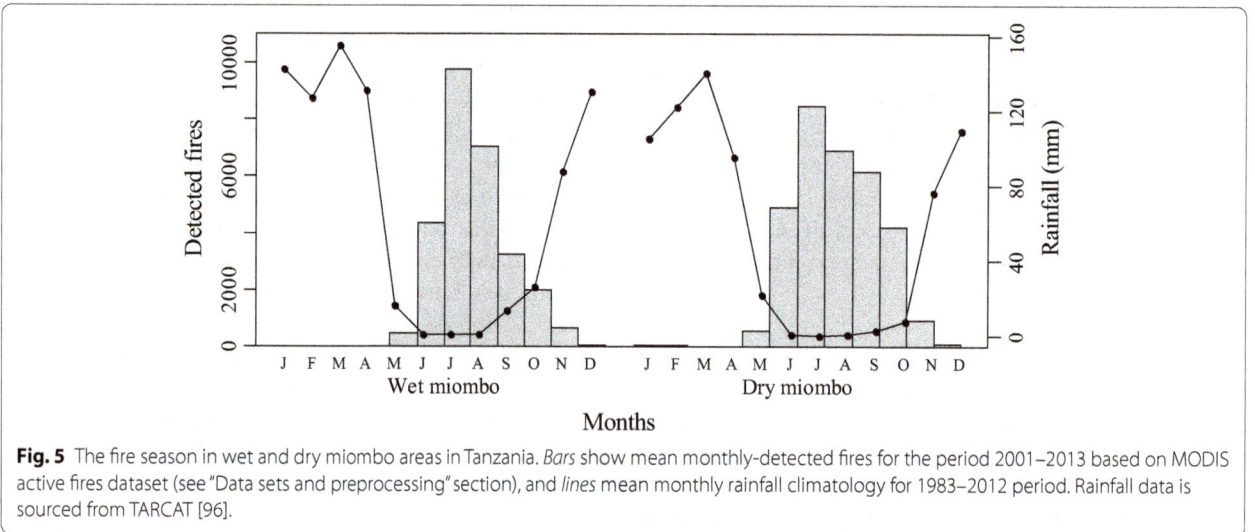

Fig. 5 The fire season in wet and dry miombo areas in Tanzania. *Bars* show mean monthly-detected fires for the period 2001–2013 based on MODIS active fires dataset (see "Data sets and preprocessing" section), and *lines* mean monthly rainfall climatology for 1983–2012 period. Rainfall data is sourced from TARCAT [96].

Fire prevalence

About 46% of the woodland area had a mean fire return interval of <2 years for every 314 ha during 2001–2013 period. Field observations in a dry miombo site have shown a mean fire return interval of 1.6 years in Zambia

[23], while a return interval of 3 years on a regional scale was observed based on satellite data [24]. In a global study, the fire return interval for African savannas and grasslands has been reduced, from 4.8 years early in 1900 to 3.6 years towards 2000 [15]. Fire return interval

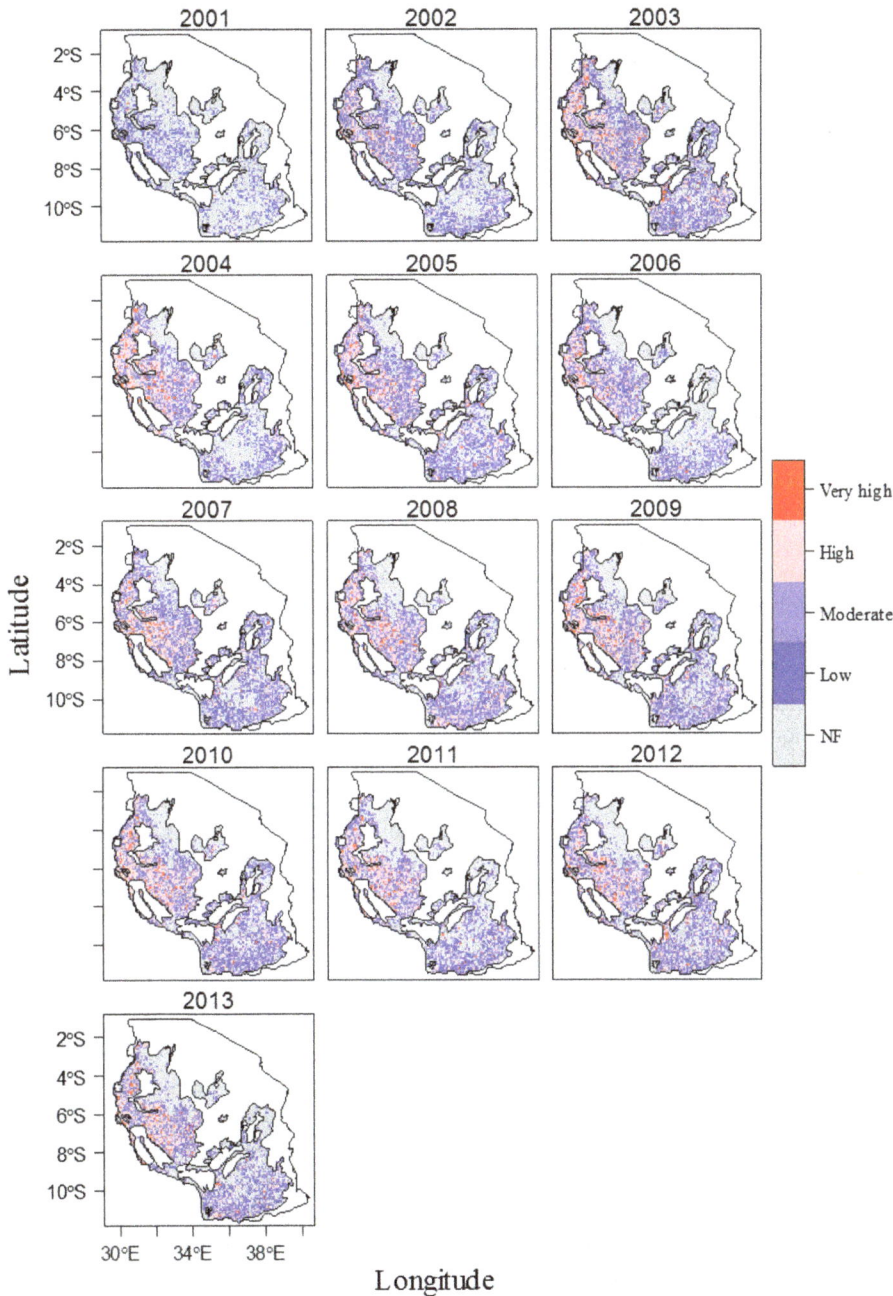

Fig. 6 Fire activity in miombo woodlands in Tanzania. Fire activity based on density, proximity and fire radiative power of MODIS detected fire events, and annual length of the fire season, at a 5 × 5 km resolution. *NF* represents areas with no fire activity.

and burning seasonality have selective effects on different components of the woodland, as they influence the intensity of fires and extents burned. Results from a study that combined field observations and modelling, from miombo sites in Zimbabwe and Mozambique, show that at least 2 years are required between successive low intensity burns to allow tree establishment and development [25]. Although spatial variation is expected within the scale at which fire activity and return intervals are estimated in the present study, our results indicate that almost half of the woodland is ignited at a return interval that threaten the longer-term sustainability of the tree cover. However, the resulting burned patterns (see "Burned extents" section) indicates an offset, to some extent, of the effect of recurrent fires. About 74% of the woodland area had a mean return interval of <2 years for every 2,500 ha between 1972 and 2011. Therefore, frequent fires have been part of some portions of the woodland for the past ~40 years, when satellite data is available. It is important to note that there is a seasonal and inter-annual variation of burned patches within each of the 2,500 ha and thus the return interval varies at smaller scales. Shorter fire return intervals are observed in wet miombo (Table 2), which are the same areas where annual fire activity is consistently high (Fig. 6). Within dry miombo, shorter return intervals persist in western- as compared to eastern- dry miombo areas. The mean fire return interval was 2.5 and 3.8 years for western- and eastern- dry miombo areas, respectively, for the period 2001–2013. On wider spatial and temporal extents, western- and eastern- dry miombo areas have 1.8 and 2.9 years of fire return intervals, respectively, for the period 1972–2011. Western parts of the woodland, including both dry and wet miombo areas, have higher fire prevalence than eastern parts of the woodland, which consist mainly of dry miombo areas. The higher fire prevalence is mainly a result of the interacting effects of rainfall patterns that influence fuel availability, and ignition sources. In southern African savannas, shorter return intervals occurred in higher rainfall areas but interacted with soil properties and herbivory, over a time period encompassing fire suppression, natural fires and controlled fires [55]. Similarly, in western African savannas, higher fire prevalence occurred in relation to increasing rainfall but interacted with both vegetation type and choices by herders and farmers to burn at different times during the fire season [11]. Rainfall influences productivity of grasses that make up the fuel load, but the fire prevalence is ultimately dependent on human influences on ignitions, fire season and extents burned [6, 7, 11, 56].

The west to east dominance of early and late dry season burning, respectively (Fig. 3), might be explained by differences in the length of the dry season. Parts of the western side of the study area receives light rains during September, from north and extending southward. These light rains continues through the main rainy season, thus reducing the length of the fire season. Central and eastern parts have a unimodal rainfall pattern and thus remain relatively dry until the beginning of another season in November/December, facilitating conditions favorable for late dry season fires.

The observed reduction in fire activity from north towards west (Fig. 6) is associated with expansion of croplands. The expansion of croplands in this area is likely a response to growing mining activities and respectively settlements in the Geita and Kahama districts, north of the study area. Expansion of croplands has had similar effect in northern hemisphere African savannas, where decreasing annual burned area occurred with

increasing croplands [57]. As with fire activity, croplands had smaller burned extents when compared to vegetated cover types in the GLC-Share Database [51] for the extent of the study area. Based on GLC-Share cover types, our results show that up to 1.6% of the croplands are burned annually compared to 4.3% of grasslands, 2.6% of tree covered areas, 3.9% of shrubs covered areas and 10% of herbaceous vegetation, aquatic or regularly flooded. The datasets used to compile GLC-Share database for Tanzania is from 2001. Therefore, the values presented above are within a decadal range; from 1995 to 2005.

Burned extents
Burned patch sizes
Small-burned patches, less than five hectares in size, are the most prevalent across spatial and temporal scales. Smaller burned patches are a common occurrence across tropical savannas and are mainly associated with traditional fire management practices [18, 21, 58, 59]. Burned patches with similar sizes were associated with farm preparation in a neighboring Mozambican savanna [60]. Similarly, small fires within African savannas have been associated with agricultural activities and fragmentation of the landscape as a result of high population densities [14, 61]. These fires, burning small patches at a time progressively during the dry season, are generally a desirable management tool and are less damaging to savanna woodlands [18, 62, 63], unless they escape to burn unintended areas. Most of the cases of escaped fires are associated with clearing of new farms as opposed to burning agricultural residues in established agricultural areas. As discussed in the "Fire prevalence" section, our results show that burned extents were smaller in croplands than in other cover types. We could not quantify the effect of other sources of fire on burned patch sizes. However, Butz [21] has observed an increase is large accidental fires and a decrease in small fragmented fires, within a pastoral community in the savannas of northeastern Tanzania. In this area, a decline in nomadic pastoralism has occurred with a trend towards sedentarization and diversified livelihoods [64, 65]. Butz identified changes in rainfall patterns, population growth and fire suppression policies as the drivers of the change in the fire regime. Similar drivers of change in fire regimes persist in western African savannas [8]. In general, competition over land areas increases with a growing population, leading to changes in socioeconomic practices [64] and increasing land fragmentation. Consequently, fire regimes including the frequency, season and sizes of burned areas vary with localized adaptation to these changes in the context of the landscape pattern, and may be influenced by public policies and rainfall patterns [8, 14, 61, 66]. Fires with a higher threat are those ignited within woodland areas where tenure accessibility

and private uses are restricted. This threat is associated with increasing homogeneity and buildup of fuels with decreasing human activities [61]. In such cases, the prevailing weather regulates the spread of a fire when ignited [7], as opposed to the human control that fragments the landscape with small fires. In a recent analysis of MODIS burned area product (at 500 m resolution) for the whole of Tanzania between 2000 and 2011, up to 77% of the annual burned area in the country was detected on gazetted land [67]. Although the causes of these fires were not evaluated, it is less likely that they all stem from control burning for fire management purposes. Similarly, protected areas in the southern hemisphere African savannas had relatively larger burned areas than outside protected areas [13]. In the Llanos savannas of Columbia, relatively larger burned areas associated with hunting were observed in a national park compared to indigenous reserves and ranches [68]. In western African savannas, higher densities of fire events occurred in protected areas of Burkina Faso and lower populated areas of Mali [12, 69]. At a smaller scale, the highest tree mortality associated with fire in central Zambia occurred within an encroached part of a national park [70]. This highlights priority fire causes and affected areas that need further detailed analyses and probably a reassessment of management practices. Combining traditional and contemporary fire management practices may achieve reduction in burned extents and consequently biomass burning emissions [71].

We observed a larger scaling of frequency-size distribution, indicating a higher ratio of small relative to large fires, in this study as compared to other ecosystems, e.g. in the United States and Spain [53, 72]. In a global study of fire size distribution, Hantson et al. [14] have also observed a dominance of small fires in our study area, with a similar range of the scaling parameter, β (see "Frequency-size distribution of burned patches" section, Table 2). Small fires are recurrent in both dry and wet miombo areas but dry miombo areas experiences both smaller and larger fires than wet miombo areas (Table 2). The difference in size classes of burned patches in wet and dry miombo contributes to the slightly smaller scaling of frequency-size distribution in wet miombo, which implies large burned patches contribute slightly more to the total area burned in wet than in dry miombo. These large burned patches are partly a result of aggregation of smaller fires during the fire season. Generally, large fires are rare but small fires accumulate to cover extended areas in the course of a fire season each year. Archibald et al. [13] found similar contribution of small fires to the annual burned area in southern hemisphere Africa savannas. Small recurrent fires reduce the risk of occasional large fires, which have recently occurred in areas where fire suppression strategies are enforced.

Partially burned areas

Partially burned areas were defined to include intermixed pixels groups that are burned, partially burned and those with a diminishing char signature. They cover relatively wider extents than completely burned areas each year, ranging between 3.2 and 40.6% and 0.8–11.3%, respectively, for the whole woodland. Table 2 provides a decadal summary for wet and dry miombo areas. Between 9 and 14% of Tanzania's area was detected as burned annually during 2000–2011 from a lower (500 m) resolution burned area product [67]. Of this burned area, 69% occurred in the woodland. Miombo woodland areas covers approximately 90% of the forested areas in Tanzania, implying that much of the burned areas in the country are not detected at the lower resolution. Lower detection rates are possibly higher in the mixed burned–unburned pixels. Similar to completely burned areas, western parts of the woodland have relatively larger extents of partially burned areas, predominantly in wet miombo, than eastern parts of the woodland. Rigorous validation was not performed for partially burned areas, thus they were not further analyzed. However, they provide crucial information for understanding vegetation dynamics, which requires the season and severity of fires at specific areas.

Conclusions

We have documented the recent fire regime, for the past ~40 years, of the miombo woodland areas of Tanzania at spatial and temporal resolutions that have not been recorded before, to the best of our knowledge. The observed fire patterns for the past 40 years show that the majority of fire events occur in the western parts of miombo woodlands, consisting of wet miombo and western dry miombo areas. Fire events on the western parts of the woodland occur mainly during the first part of the dry season. Thus, an early dry season fire peak characterizes the west while a late dry season fire peak characterizes the east. Almost half of the woodland area has fire return intervals of <2 years. Return intervals are shorter in wet than in dry miombo areas. Short return intervals limit fuel loading and therefore prevents large intense fires. Human activities play a major role in shaping fire regimes. Mainly small sized fires characterize the regime across spatial and temporal scales. Occasional large fires are more frequently detected in dry than in wet miombo areas. Management strategies need to address spatially specific needs of wet and dry miombo areas, in the light of their fire regimes and socio-economic context.

Methods
Study area
The study area is miombo woodlands in Tanzania (Fig. 4). Miombo woodlands are disturbance driven moist

savannas that are shaped by natural and anthropogenic disturbances, to a larger extent, than by nutrient and water availability [1, 73]. They occur on nutrient poor soils and generally experience a warm-to-hot climate with a dry cold season [24]. The average annual rainfall ranges between 600 and 1,500 mm and falls during 5–6 months [23], followed by an extended dry period. Wet miombo areas, which receives more than 1,000 mm of average annual rainfall, are distinguished from dry miombo areas receiving less than 1,000 mm of average annual rainfall [74]. The woodland is characterized by wooded canopy species and an understory consisting of shrubs and light demanding grass species [24]. The annual production of these flammable, 0.5–2 m tall, grasses every rain season followed by accumulation of litter from the deciduous trees, makes miombo woodland highly susceptible to annual fires.

The fire season extends from the onset of the dry season to its end (Fig. 5), although isolated burning events may occur throughout the year at different localities. During the fire season, individual fires burn small patches at a time with the exception of very occasional big fires in areas where fuel load is accumulated and continuous. Towards the end of the fire season, a mosaic of burned and unburned patches occur.

Data sets and preprocessing

Landsat Level 1 Terrain (L1T) corrected product satellite images and MODIS collection 5 Level 2 MOD14/MYD14 active fire product form the major data source for this study. We derive fire patterns from the two datasets independently and compare results. Similar patterns will indicate that the datasets are representative of the fire patterns in the study area. This is important because although Landsat provides a finer spatial resolution its temporal resolution (16 days), and further limitation by cloud cover, may limit detection of savanna fires. On the other hand, MODIS detected active fires provide a more complete coverage at high temporal resolution but its temporal coverage is relatively short (since 2000) compared to that of Landsat (since 1972). In addition, MODIS can detect small active fires that are not captured by coarse resolution burned area products [45]. Thus, combining the two datasets benefits from their complementary availability, spatial and temporal characteristics [41].

All available Landsat images were downloaded from the USGS Global Visualization Viewer [75], to cover the study extent (Fig. 4) and for the period 1972–2011. Availability was constrained by image quality, predominantly percentage cloud cover within the study extent. Thus, a complete spatio-temporal dataset was difficult to achieve. For each year, processing and analysis was performed for areas where at least one image was available during the fire season. A total of 1,835 scenes, among them 234 MSS, 1,284 TM and 317 ETM+ SLC-On, were processed. Landsat TM imagery was preferred over MSS imagery for the period when both were available. Each image was converted to at-surface reflectance using the Dark Object Subtraction (DOS) method [76–78].

MODIS active fire data for the whole country were downloaded from the Fire Information for Resource Management System (FIRMS) [79], for the period between November 2000 and December 2013. MOD14/MYD14 provides, among others, coordinates of detected fires (the center of fire pixels at 1 km resolution), their acquisition date and time and respective FRP. Fire locations within miombo were retrieved and categorized as early dry season burning (January–July) or late dry season burning (August–December). Isolated fire events during the wet season were included in respective dry season burning based on the month of their detection. July was chosen to mark the end of early dry season burning for the entire woodland area, consistent with prescribed early burning between May and July in some parts of Tanzania [80–82]. This distinction was made to capture patterns of fire during the dry season, since the timing of burning influences the intensity and spread of a fire and thus its effects, such that fire management through prescribed burning is recommended during early dry season [23].

Spatial and temporal patterns of burned areas and active fires

Burned areas were detected by means of fuzzy classification of spectral indices derived from Landsat satellite images while spatial patterns of active fires were analyzed based on MODIS dataset. We limited our analysis to prevalence, burned extents and spatial patterns of active fires and burned patches. The general processing flow is summarized in Fig. 7. Analyses were performed in GRASS GIS [83] and R version 3.1.0 [84].

Training and testing of the fuzzy classification

A total of 523,092 pixels were sampled by visual interpretation from representative scenes for training and testing purposes. Burned areas were identified based on color composites with SWIR, NIR and VIS bands in RGB display, an approach that has been employed to extract image based training, and testing samples [46, 47, 85]. Active fires captured by Landsat satellite images and the analyst's field experience were utilized in line with the color composites. These formed the basis for selection of spectral indices.

Spectral Indices used for fuzzy classification

Spectral indices commonly used for burned area mapping were identified based on literature review. Eleven

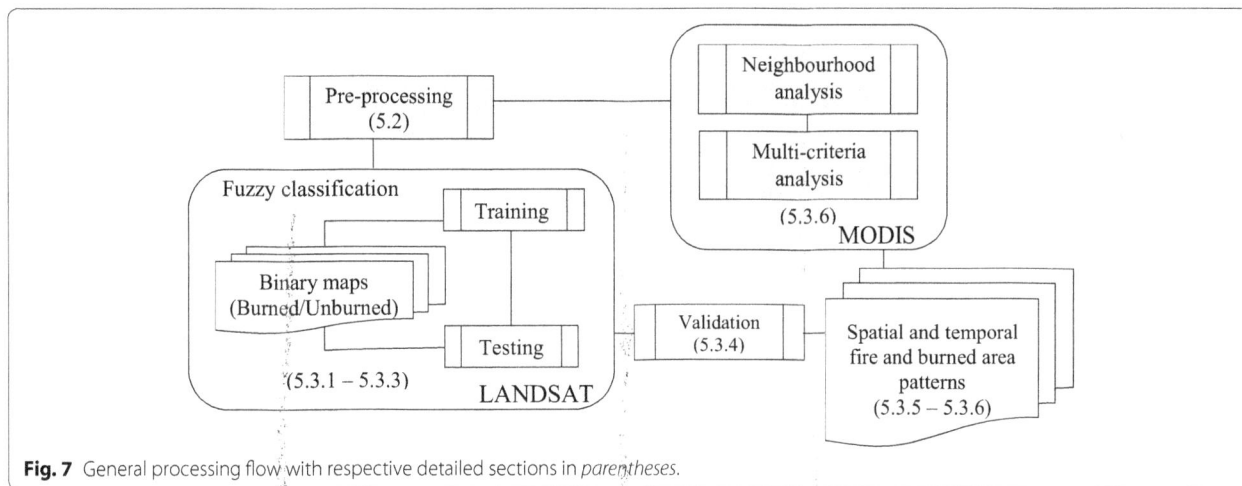

Fig. 7 General processing flow with respective detailed sections in *parentheses*.

indices with a potential for discriminating burned areas in sparsely vegetated areas were selected and tested. These included BAI [49], BAIM [86], $BAIM_L$ [46], GEMI [87], MIRBI [50], NBR_S [46], NBR_L [88], NBR_2 [47], NDVI, SARVI [89] and SAVI [90]. The range of values from individual scenes and the differences among scenes, for which burned pixels were well separated from other cover types, was determined for each index to form the base for fuzzy sets definition (see "Fuzzy membership rules" section). Penalized logistic regression was then employed to analyze the discrimination performance of burned from unburned pixels for each spectral index. An analysis combining sampled pixels from all scenes was also done to investigate how well results at scene level could be generalized.

Fuzzy membership rules

Fuzzy discrimination employs membership rules that are defined in terms of fuzzy sets [91], whose elements differentiate definite members from definite non-members and those with some level of uncertainty as to whether they are members or not. Fuzzy classification was experimented for each index individually and for different combinations of indices. Indices and combinations thereof were selected (Table 3) for fuzzy set definition based on how well they distinguished burned from unburned areas. Selected indices conformed to regression results (see "Spectral Indices used for fuzzy classification" section).

Validation of detected burned areas

Due to the lack of an independent burned area perimeter for validation, completely burned areas were distinguished from partially burned and unburned areas, based on their membership scores, for the purpose of

Table 3 Spectral indices used in fuzzy classification

Indices	Use
$BAIM_L$ and MIRBI	Detect burned areas at different post-fire conditions
$BAIM_L$ and threshold	Mask bare soil, water[a], topographic and cloud shadows
NBR_L and threshold	Distinguish active fires from other features
BAI	Detect burned areas on MSS imagery

$BAI = 1/(\rho_{c2} - \rho_2)^2 + (\rho_{c4} - \rho_4)^2; \rho_{c2} = 0.1, \rho_{c4} = 0.06.$

$BAIM_L = 1/(\rho_{c4} - \rho_4)^2 + (\rho_{c7} - \rho_7)^2; \rho_{c4} = 0.05, \rho_{c7} = 0.2.$

$MIRBI = 10 \times \rho_7 - 9.8 \times \rho_5 + 2.$

$NBR = (\rho_4 - \rho_7)/(\rho_4 + \rho_7).$

ρ_2 = Band 2 of MSS on Landsat 4-5 and Band 5 of MSS on Landsat 1-3.

ρ_4 = Band 4 of MSS on Landsat 4-5, Band 7 of MSS on Landsat 1-3 and Band 4 of TM/ETM+ .

ρ_7 = Band 7 of TM/ETM+ .

[a] Permanent water bodies were manually masked out from fuzzy classification results.

restricting further analysis to definite burned areas. Partially burned areas consisted of intermixed pixel groups of burned, partially burned and those with a diminishing char signature (Fig. 8). These areas were not included in subsequent analyses but they indicate the spatial and temporal extents of areas affected by fire each year. Validation of completely burned areas, which are referred to as burned areas, was performed based on visual interpretation of randomly selected samples, from another set of representative scenes different from those used for training and testing fuzzy classification. To validate the performance of the fuzzy classification when including also the partially burned areas, we employed visual analysis and unsupervised clustering. This approach, combining visual interpretation and unsupervised clustering, is suitable for discriminating burned areas in African savannas [42].

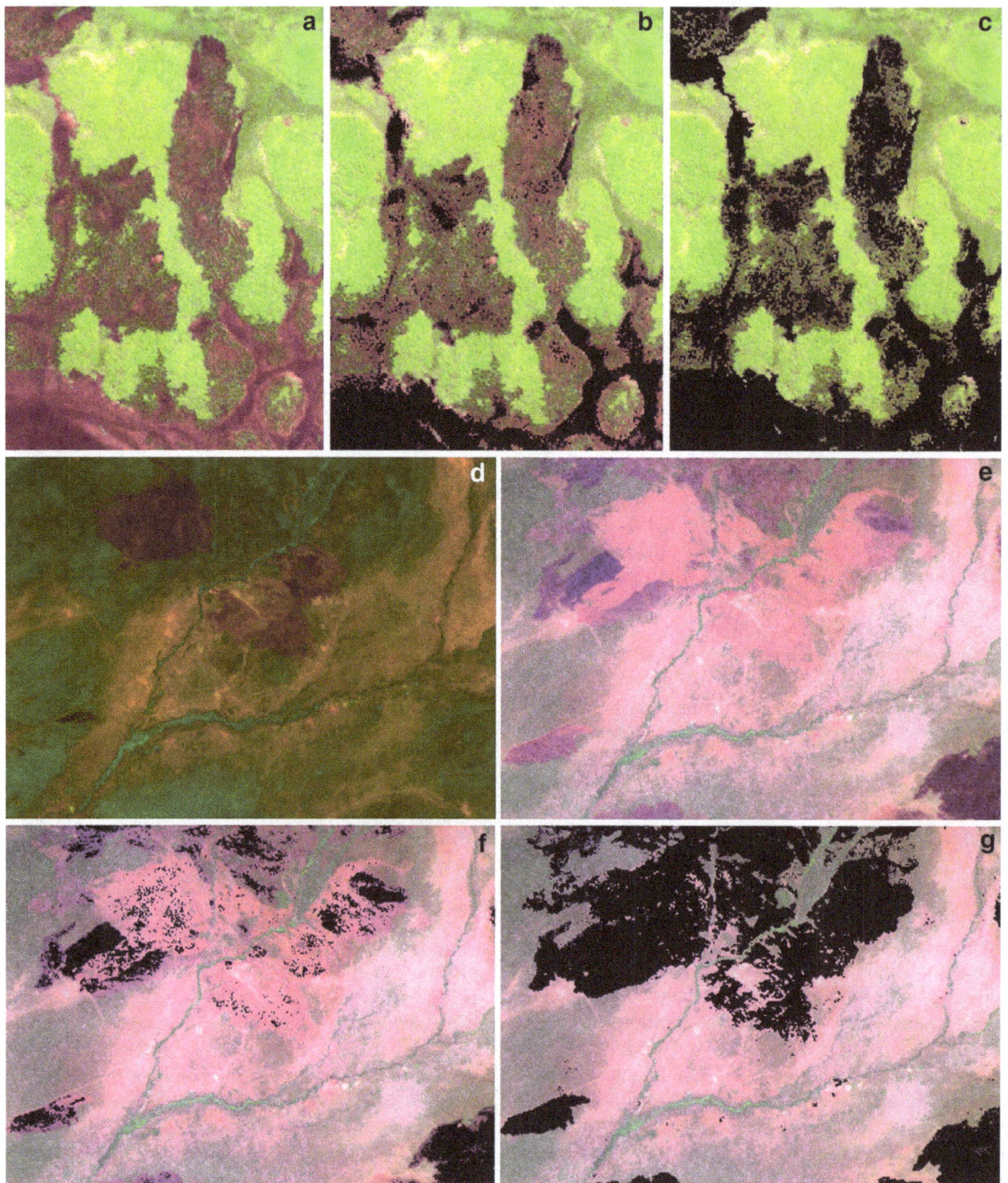

Fig. 8 Illustration of areas defined as partially burned. The *top panel* shows an area with mixed burned and unburned pixels (**a**). The *middle panel* shows burned patches at the beginning of the fire season (**d**) and the same area later (**e**) during the fire season. Burned areas (**b**) and (**f**), consist of contiguous groups of burned pixels with a definite fire scar. A mix of burned and unburned pixels and those with a diminishing fire scar were defined as partially burned, shown in (**c**) and (**g**) when combined with burned areas. Detection of burned areas with a diminishing fire scar is desirable when an image from an earlier date during the fire season is not available.

We adapted the approach described in [47] where three 1,000 × 1,000 pixels image subsets were visually interpreted to delineate burned/partially burned area perimeters. An independent image analyst examined these visually interpreted burned areas with support of false color composites (bands 432 and 741 as RGB) in combination with clustering of the bands 741 data subset, utilizing ERDAS Imagine 2014. The results were then used to validate the combined burned and partially burned area.

Burned patch sizes and spatio-temporal variation in burned extents

The sizes of burned patches were calculated based on contiguous burned pixels at scene level, while burned extents from annual mosaics after accounting for multiple detections between acquisitions. The fire return interval based on detected burned areas was determined by overlaying a 5 × 5 km grid on annual burned area maps, thus a return interval for every 2,500 ha. Grid cells containing burned patches >0.5 ha were considered affected by fire in respective years and provided a crude estimate of fire return interval for each cell. We use this estimate for consistency across the spatial and temporal extents with different Landsat data availability and for comparison with MODIS data (see "Occurrence and spatial patterns of active fires" section). The 0.5 ha threshold was selected based on reported burned patch sizes from anthropogenic fire sources in neighboring Mozambican savannas [60]. Frequency-size distributions of burned patches >0.1 and >0.4 ha for TM/ETM+ and MSS imagery, respectively, were examined for the period 1972–2011. Frequency densities of patch size classes were analyzed in log–log space, where the slope coefficient, β, provided the scaling of burned patch sizes i.e. the ratio of the number of large to small fires [53].

Occurrence and spatial patterns of active fires

The spatial association of detected fires was examined by Ripley K function for inhomogeneous spatial patterns [92]. Annual fire activity was derived at a 5 × 5 km resolution as a composite measure of active fire characteristics, including density and proximity of fires, annual duration of the fire season and range of FRP values. These combined characteristics of active fires provides a classification of the fire activity that is related to the fire regime [93]. The FRP values, for instance, are associated with the type of vegetation burned [94] and the density of fires is a good predictor of burned areas [13]. The grid resolution was based on both an optimal choice for spatial aggregation when comparing datasets with different resolution as applied in [44, 95] and for practical handling purposes. Fire return interval based on detected active fires was examined by a neighborhood analysis within a 1-km distance from each detected fire.

Thus, a return interval for an area of ~314 ha, which is the area of a circle of 1 km radius, centered at the location of detected active fires. This distance was selected to reflect the ground size of detected fire pixels. Results for each year provided fire return interval given locations of detected fires for that year and their average provided mean return interval for all years.

Abbreviations

BAI: Burned Area Index; BAIM: MODIS Burned Area Index; FIRMS: Fire Information for Resource Management System; FRP: Fire Radiative Power; GEMI: Global Environment Monitoring Index; MIRBI: Mid-Infrared Burn Index; MODIS: Moderate Resolution Imaging Spectroradiometer; NBR: Normalized Burn Ratio; NDVI: Normalized Difference Vegetation Index; SARVI: Soil and Atmospherically Resistant Vegetation Index; SAVI: Soil Adjusted Vegetation Index.

Authors' contributions

BT planned and implemented the study and prepared the manuscript. ØBD performed an independent visual analysis. TG planned some of the methods and prepared the manuscript. ØT prepared the manuscript. All authors contributed in revising the manuscript. All authors read and approved the final manuscript.

Author details

[1] Department of Ecology and Natural Resource Management, Norwegian University of Life Sciences, P.O. Box 5003, 1432 Ås, Norway. [2] Department of Geoinformatics, School of Geospatial Sciences and Technology, Ardhi University, P.O. Box 35176, Dar es Salaam, Tanzania. [3] Department of Mathematical Sciences and Technology, Norwegian University of Life Sciences, P.O. Box 5003, 1432 Ås, Norway.

Acknowledgements

We are very grateful to the Climate Change Impacts, Adaptation and Mitigation (CCIAM) Program in Tanzania for funding this study. We thank Professor Fred Midtgaard and Professor Seif Madoffe for initiating this study and for valuable comments on the manuscript. We also thank the anonymous reviewers for the very pertinent comments that greatly improved a previous version of the manuscript.

Compliance with ethical guidelines

Competing interests

The authors declare that they have no competing interests.

References

1. Sankaran M, Hanan NP, Scholes RJ, Ratnam J, Augustine DJ, Cade BS et al (2005) Determinants of woody cover in African savannas. Nature 438(7069):846–849
2. Staver AC, Archibald S, Levin S (2011) Tree cover in sub-Saharan Africa: rainfall and fire constrain forest and savanna as alternative stable states. Ecology 92(5):1063–1072
3. Lehmann CE, Anderson TM, Sankaran M, Higgins SI, Archibald S, Hoffmann WA et al (2014) Savanna vegetation–fire–climate relationships differ among continents. Science 343(6170):548–552
4. Bond W, Woodward F, Midgley G (2005) The global distribution of ecosystems in a world without fire. New Phytol 165(2):525–538
5. Archibald S, Staver AC, Levin SA (2012) Evolution of human-driven fire regimes in Africa. Proc Natl Acad Sci 109(3):847–852
6. Van Der Werf GR, Randerson JT, Giglio L, Gobron N, Dolman A (2008) Climate controls on the variability of fires in the tropics and subtropics. Glob Biogeochem Cycles 22:GB3028

7. Archibald S, Nickless A, Govender N, Scholes R, Lehsten V (2010) Climate and the inter-annual variability of fire in southern Africa: a meta-analysis using long-term field data and satellite-derived burnt area data. Glob Ecol Biogeogr 19(6):794–809

8. Laris P (2013) Integrating land change science and savanna fire models in West Africa. Land 2(4):609–636

9. Archibald S, Lehmann CE, Gómez-Dans JL, Bradstock RA (2013) Defining pyromes and global syndromes of fire regimes. Proc Natl Acad Sci 110(16):6442–6447

10. Bowman DM, Balch J, Artaxo P, Bond WJ, Cochrane MA, D'Antonio CM et al (2011) The human dimension of fire regimes on Earth. J Biogeogr 38(12):2223–2236

11. Mbow C, Nielsen TT, Rasmussen K (2000) Savanna fires in east-central Senegal: distribution patterns, resource management and perceptions. Human Ecol 28(4):561–583

12. Laris P, Caillault S, Dadashi S, Jo A (2015) The human ecology and geography of burning in an unstable savanna environment. J Ethnobiol 35(1):111–139

13. Archibald S, Scholes R, Roy D, Roberts G, Boschetti L (2010) Southern African fire regimes as revealed by remote sensing. Int J Wildland Fire 19(7):861–878

14. Hantson S, Pueyo S, Chuvieco E (2015) Global fire size distribution is driven by human impact and climate. Glob Ecol Biogeogr 24(1):77–86

15. Mouillot F, Field CB (2005) Fire history and the global carbon budget: a 1 × 1 fire history reconstruction for the 20th century. Glob Change Biol 11(3):398–420

16. Andreae MO, Atlas E, Cachier H, Cofer WR III, Harris GW, Helas G et al (1996) Trace gas and aerosol emissions from savanna fires. Biomass Burn Glob Change 1:278–295

17. van der Werf GR, Randerson JT, Giglio L, Collatz G, Mu M, Kasibhatla PS et al (2010) Global fire emissions and the contribution of deforestation, savanna, forest, agricultural, and peat fires (1997–2009). Atmos Chem Phys 10(23):11707–11735. doi:10.5194/acp-10-11707-2010

18. Bird RB, Codding BF, Kauhanen PG, Bird DW (2012) Aboriginal hunting buffers climate-driven fire-size variability in Australia's spinifex grasslands. Proc Natl Acad Sci 109(26):10287–10292

19. Moussa K, Bassett T, Nkem J (eds) (2011) Changing fire regimes in the Cote d'Ivoire savanna: implications for greenhouse emissions and carbon sequestration. In: Sustainable Forest Management in Africa: some solutions to natural forest management problems in Africa. Proceedings of the sustainable forest management in Africa Symposium. Stellenbosch, 3–7 November 2008, 2011, Stellenbosch University, Stellenbosch, South Africa

20. Laris P, Wardell DA (2006) Good, bad or 'necessary evil'? Reinterpreting the colonial burning experiments in the savanna landscapes of West Africa. Geogr J 172(4):271–290

21. Butz RJ (2009) Traditional fire management: historical fire regimes and land use change in pastoral East Africa. Int J Wildland Fire 18(4):442–450

22. GOFC-GOLD (2014) A sourcebook of methods and procedures for monitoring and reporting anthropogenic greenhouse gas emissions and removals associated with deforestation, gains and losses of carbon stocks in forests remaining forests, and forestation. GOFC-GOLD Report version COP20-1, GOFC-GOLD Land Cover Project Office, Wageningen University, The Netherlands

23. Chidumayo EN (1997) Miombo ecology and management: an introduction. Intermediate Technology Publications Ltd (ITP), London

24. Frost P (1996) The ecology of miombo woodlands. In: Campbell BM (ed) The miombo in transition: woodlands and welfare in Africa. CIFOR, Bogor, pp 11–57

25. Ryan CM, Williams M (2011) How does fire intensity and frequency affect miombo woodland tree populations and biomass? Ecol Appl 21(1):48–60

26. Trapnell C (1959) Ecological results of woodland burning experiments in northern Rhodesia. J Ecol 47(1):129–168

27. UNFCCC (2011) Report of the conference of the parties on its sixteenth session, held in Cancun from 29 November to 10 December 2010, Adendum, Part Two: action taken by the conference of the parties at its sexteenth session, FCCC/CP/2010/7/Add.1. United Nations Framework Convention on Climate Change, Bonn, Germany

28. Ahrends A, Burgess ND, Milledge SA, Bulling MT, Fisher B, Smart JC et al (2010) Predictable waves of sequential forest degradation and biodiversity loss spreading from an African city. Proc Natl Acad Sci 107(33):14556–14561

29. Barlow J, Parry L, Gardner TA, Ferreira J, Aragão LE, Carmenta R et al (2012) The critical importance of considering fire in REDD+ programs. Biol Conserv 154:1–8

30. Ryan CM, Hill T, Woollen E, Ghee C, Mitchard E, Cassells G et al (2012) Quantifying small-scale deforestation and forest degradation in African woodlands using radar imagery. Glob Change Biol 18(1):243–257

31. De Michele C, Accatino F, Vezzoli R, Scholes R (2011) Savanna domain in the herbivores-fire parameter space exploiting a tree–grass–soil water dynamic model. J Theor Biol 289:74–82

32. Accatino F, De Michele C, Vezzoli R, Donzelli D, Scholes RJ (2010) Tree-grass co-existence in savanna: interactions of rain and fire. J Theor Biol 267(2):235–242

33. Bowman DM, MacDermott HJ, Nichols SC, Murphy BP (2014) A grass–fire cycle eliminates an obligate-seeding tree in a tropical savanna. Ecol Evol 4(21):4185–4194

34. Bond WJ, Keeley JE (2005) Fire as a global 'herbivore': the ecology and evolution of flammable ecosystems. Trends Ecol Evol 20(7):387–394

35. Kikula IS (1986) The influence of fire on the composition of Miombo woodland of SW Tanzania. Oikos 46(3):317–324

36. Giglio L, Randerson J, Van der Werf G, Kasibhatla P, Collatz G, Morton D et al (2010) Assessing variability and long-term trends in burned area by merging multiple satellite fire products. Biogeosciences 7(3):1171–1186

37. Russell-Smith J, Ryan PG, Durieu R (1997) A LANDSAT MSS-derived fire history of Kakadu National Park, monsoonal northern Australial, 1980–94: seasonal extent, frequency and patchiness. J Appl Ecol 34(3):748–766

38. Trigg S, Flasse S (2000) Characterizing the spectral-temporal response of burned savannah using in situ spectroradiometry and infrared thermometry. Int J Remote Sens 21(16):3161–3168

39. Chuvieco E, Opazo S, Sione W, Valle HD, Anaya J, Bella CD et al (2008) Global burned-land estimation in Latin America using MODIS composite data. Ecol Appl 18(1):64–79

40. Armenteras D, Romero M, Galindo G (2005) Vegetation fire in the savannas of the Llanos Orientales of Colombia. World Resour Rev 17(4):531–543

41. Boschetti L, Roy DP, Justice CO, Humber ML (2015) MODIS–Landsat fusion for large area 30 m burned area mapping. Remote Sens Environ 161:27–42

42. Laris P (2005) Spatiotemporal problems with detecting and mapping mosaic fire regimes with coarse-resolution satellite data in savanna environments. Remote Sens Environ 99(4):412–424

43. Giglio L, Randerson JT, van der Werf GR (2013) Analysis of daily, monthly, and annual burned area using the fourth-generation global fire emissions database (GFED4). J Geophys Res Biogeosci 118(1):317–328. doi:10.1002/jgrg.20042

44. Roy DP, Boschetti L (2009) Southern Africa validation of the MODIS, L3JRC, and GlobCarbon burned-area products. IEEE Trans Geosci Remote Sens 47(4):1032–1044

45. Randerson J, Chen Y, Werf G, Rogers B, Morton D (2012) Global burned area and biomass burning emissions from small fires. J Geophys Res Biogeosci (2005–2012) 117:G04012

46. Bastarrika A, Chuvieco E, Martín MP (2011) Mapping burned areas from Landsat TM/ETM+ data with a two-phase algorithm: balancing omission and commission errors. Remote Sens Environ 115(4):1003–1012

47. Stroppiana D, Bordogna G, Carrara P, Boschetti M, Boschetti L, Brivio P (2012) A method for extracting burned areas from Landsat TM/ETM+ images by soft aggregation of multiple Spectral Indices and a region growing algorithm. ISPRS J Photogramm Remote Sens 69:88–102

48. Barbosa PM, Grégoire J-M, Pereira JMC (1999) An algorithm for extracting burned areas from time series of AVHRR GAC data applied at a continental scale. Remote Sens Environ 69(3):253–263

49. Chuvieco E, Martin MP, Palacios A (2002) Assessment of different spectral indices in the red-near-infrared spectral domain for burned land discrimination. Int J Remote Sens 23(23):5103–5110

50. Trigg S, Flasse S (2001) An evaluation of different bi-spectral spaces for discriminating burned shrub-savannah. Int J Remote Sens 22(13):2641–2647. doi:10.1080/01431160110053185

51. Latham J, Cumani R, Rosati I, Bloise M (2014) FAO global land cover (GLC-SHARE) Beta-Release 1.0 Database, Division LaW

52. Kasin I, Blanck Y, Storaunet KO, Rolstad J, Ohlson M (2013) The charcoal record in peat and mineral soil across a boreal landscape and possible linkages to climate change and recent fire history. Holocene 23(7):1052–1065

53. Malamud BD, Millington JD, Perry GL (2005) Characterizing wildfire regimes in the United States. Proc Natl Acad Sci USA 102(13):4694–4699

54. Scott AC (2000) The Pre-Quaternary history of fire. Palaeogeogr Palaeoclimatol Palaeoecol 164(1):281–329

55. Van Wilgen B, Biggs H, O'regan S, Mare N (2000) Fire history of the savanna ecosystems in the Kruger National Park, South Africa, between 1941 and 1996. S Afr J Sci 96(4):167–178

56. Le Page Y, Oom D, Silva J, Jönsson P, Pereira J (2010) Seasonality of vegetation fires as modified by human action: observing the deviation from eco-climatic fire regimes. Glob Ecol Biogeogr 19(4):575–588

57. Andela N, van der Werf GR (2014) Recent trends in African fires driven by cropland expansion and El Nino to La Nina transition. Nature Climate Change 4(9):791–795

58. Laris P (2002) Burning the seasonal mosaic: preventative burning strategies in the wooded savanna of southern Mali. Human Ecol 30(2):155–186

59. Mistry J, Berardi A, Andrade V, Krahô T, Krahô P, Leonardos O (2005) Indigenous fire management in the cerrado of Brazil: the case of the Krahô of Tocantíns. Human Ecol 33(3):365–386

60. Shaffer LJ (2010) Indigenous fire use to manage savanna landscapes in Southern Mozambique. Fire Ecol 6(2):43–59. doi:10.4996/fireecology.0602043

61. Archibald S, Roy DP, Wilgen V, Brian W, Scholes RJ (2009) What limits fire? An examination of drivers of burnt area in Southern Africa. Glob Change Biol 15(3):613–630

62. Laris P (2011) Humanizing savanna biogeography: linking human practices with ecological patterns in a frequently burned savanna of southern Mali. Ann Assoc Am Geogr 101(5):1067–1088

63. Brockett B, Biggs H, Van Wilgen B (2001) A patch mosaic burning system for conservation areas in southern African savannas. Int J Wildland Fire 10(2):169–183

64. McCabe JT, Leslie PW, DeLuca L (2010) Adopting cultivation to remain pastoralists: the diversification of Maasai livelihoods in northern Tanzania. Human Ecol 38(3):321–334. doi:10.1007/s10745-010-9312-8

65. Nkedianye D, de Leeuw J, Ogutu JO, Said MY, Saidimu TL, Kifugo SC et al (2011) Mobility and livestock mortality in communally used pastoral areas: the impact of the 2005–2006 drought on livestock mortality in Maasailand. Pastoralism 1(1):1–17. doi:10.1186/2041-7136-1-17

66. Hudak AT, Fairbanks DH, Brockett BH (2004) Trends in fire patterns in a southern African savanna under alternative land use practices. Agric Ecosyst Environ 101(2):307–325

67. FAO (2013) A fire baseline for Tanzania. Sustainable forest management in a changing climate. FAO-Finland Forestry Programme. Dar es Salaam, Tanzania

68. Romero-Ruiz M, Etter A, Sarmiento A, Tansey K (2010) Spatial and temporal variability of fires in relation to ecosystems, land tenure and rainfall in savannas of northern South America. Glob Change Biol 16(7):2013–2023

69. Caillault S, Ballouche A, Delahaye D (2014) Where are the 'bad fires' in West African savannas? Rethinking burning management through a space–time analysis in Burkina Faso. Geogr J. doi:10.1111/geoj.12074

70. Chidumayo E (2002) Changes in miombo woodland structure under different land tenure and use systems in central Zambia. J Biogeogr 29(12):1619–1626

71. Russell-Smith J, Cook GD, Cooke PM, Edwards AC, Lendrum M, Meyer C et al (2013) Managing fire regimes in north Australian savannas: applying Aboriginal approaches to contemporary global problems. Front Ecol Environ 11(s1):e55–e63

72. Moreno M, Malamud B, Chuvieco E (2011) Wildfire frequency–area statistics in Spain. Procedia Environ Sci 7:182–187

73. Bond W, Midgley G, Woodward F (2003) What controls South African vegetation-climate or fire? S Afr J Bot 69(1):79–91

74. White F (1983) The vegetation of Africa: a descriptive memoir to accompany the UNESCO/AETFAT/UNSO vegetation map of Africa by F White. Natural Resources Research Report XX, UNESCO, Paris, France

75. USGS Global Visualization Viewer. http://glovis.usgs.gov/. Accessed 13 April 2015

76. Chavez PS (1996) Image-based atmospheric corrections-revisited and improved. Photogramm Eng Remote Sens 62(9):1025–1036

77. Song C, Woodcock CE, Seto KC, Lenney MP, Macomber SA (2001) Classification and change detection using Landsat TM data: when and how to correct atmospheric effects? Remote Sens Environ 75(2):230–244

78. Tizado EJ (2013) i.landsat.toar: calculates top-of-atmosphere radiance or reflectance and temperature for Landsat MSS/TM/ETM+/OLI. In: GRASS Development Team (ed) Geographic Resources Analysis Support System (GRASS 7) user's manual: open source geospatial foundation project. http://grass.osgeo.org

79. Fire Information for Resource Management System. https://earthdata.nasa.gov/data/near-real-time-data/firms. Accessed 13 April 2015

80. Nssoko E (2004) Community-based fire management in the Miombo woodlands: a case study from Bukombe District, Shinyanga, Tanzania. Aridlands No 55, May/June 2004

81. Luoga E, Witkowski E, Balkwill K (2005) Land cover and use changes in relation to the institutional framework and tenure of land and resources in eastern Tanzania miombo woodlands. Environ Dev Sustain 7(1):71–93

82. Hassan SN, Rija AA (2011) Fire history and management as determinant of patch selection by foraging herbivores in western Serengeti, Tanzania. Int J Biodivers Sci Ecosyst Serv Manage 7(2):122–133

83. GRASS Development Team (2012) Geographic Resources Analysis Support System (GRASS) Software. Open Source Geospatial Foundation Project. http://grass.osgeo.org

84. R Core Team (2014) R: a language and environment for statistical computing. R Foundation for Statistical Computing, Vienna, Austria. http://www.R-project.org/

85. Koutsias N, Karteris M (2000) Burned area mapping using logistic regression modeling of a single post-fire Landsat-5 Thematic Mapper image. Int J Remote Sens 21(4):673–687

86. Martín M, Gómez I, Chuvieco E (eds) (2005) Performance of a burned-area index (BAIM) for mapping Mediterranean burned scars from MODIS data. In: Proceedings of the 5th international workshop on remote sensing and GIS applications to forest fire management: fire effects assessment. Universidad de Zaragoza, GOFC GOLD, EARSeL, Paris

87. Pinty B, Verstraete M (1992) GEMI: a non-linear index to monitor global vegetation from satellites. Vegetatio 101(1):15–20

88. Key CH, Benson NC (1999) Measuring and remote sensing of burn severity: the CBI and NBR. Poster Abstract. In: Neuenschwander LF, Ryan KC (eds) Proceedings joint fire science conference and workshop, vol II, Boise, ID, 15–17 June 1999. University of Idaho and International Association of Wildland Fire, p 284

89. Huete A, Liu H, Batchily K, Van Leeuwen W (1997) A comparison of vegetation indices over a global set of TM images for EOS-MODIS. Remote Sens Environ 59(3):440–451

90. Huete AR (1988) A soil-adjusted vegetation index (SAVI). Remote Sens Environ 25(3):295–309

91. Jasiewicz J (2011) A new GRASS GIS fuzzy inference system for massive data analysis. Comput Geosci 37(9):1525–1531

92. Baddeley AJ, Møller J, Waagepetersen R (2000) Non-and semi-parametric estimation of interaction in inhomogeneous point patterns. Stat Neerl 54(3):329–350

93. Chuvieco E, Giglio L, Justice C (2008) Global characterization of fire activity: toward defining fire regimes from Earth observation data. Glob Change Biol 14(7):1488–1502

94. Giglio L, Csiszar I, Justice CO (2006) Global distribution and seasonality of active fires as observed with the terra and aqua moderate resolution imaging spectroradiometer (MODIS) sensors. J Geophys Res Biogeosci (2005–2012) 111:G02016

95. Eva H, Lambin EF (1998) Remote sensing of biomass burning in tropical regions: sampling issues and multisensor approach. Remote Sens Environ 64(3):292–315

96. Maidment RI, Grimes D, Allan RP, Tarnavsky E, Stringer M, Hewison T et al (2014) The 30 year TAMSAT African rainfall climatology and time series (TARCAT) data set. J Geophys Res Atmos 119(18):10619–10644. doi:10.1002/2014JD021927

4

Local discrepancies in continental scale biomass maps: a case study over forested and non-forested landscapes in Maryland, USA

Wenli Huang[1]*⦿, Anu Swatantran[1], Kristofer Johnson[2], Laura Duncanson[1], Hao Tang[1], Jarlath O'Neil Dunne[3], George Hurtt[1] and Ralph Dubayah[1]

Abstract

Background: Continental-scale aboveground biomass maps are increasingly available, but their estimates vary widely, particularly at high resolution. A comprehensive understanding of map discrepancies is required to improve their effectiveness in carbon accounting and local decision-making. To this end, we compare four continental-scale maps with a recent high-resolution lidar-derived biomass map over Maryland, USA. We conduct detailed comparisons at pixel-, county-, and state-level.

Results: Spatial patterns of biomass are broadly consistent in all maps, but there are large differences at fine scales (RMSD 48.5–92.7 Mg ha^{-1}). Discrepancies reduce with aggregation and the agreement among products improves at the county level. However, continental scale maps exhibit residual negative biases in mean (33.0–54.6 Mg ha^{-1}) and total biomass (3.5–5.8 Tg) when compared to the high-resolution lidar biomass map. Three of the four continental scale maps reach near-perfect agreement at ~4 km and onward but do not converge with the high-resolution biomass map even at county scale. At the State level, these maps underestimate biomass by 30–80 Tg in forested and 40–50 Tg in non-forested areas.

Conclusions: Local discrepancies in continental scale biomass maps are caused by factors including data inputs, modeling approaches, forest/non-forest definitions and time lags. There is a net underestimation over high biomass forests and non-forested areas that could impact carbon accounting at all levels. Local, high-resolution lidar-derived biomass maps provide a valuable bottom-up reference to improve the analysis and interpretation of large-scale maps produced in carbon monitoring systems.

Keywords: Temperate deciduous forest, Lidar, Aboveground biomass, Carbon

Background

Accurate maps of forest aboveground biomass are critical for reducing uncertainties in the carbon cycle and informing carbon management decisions [1–3]. While no method provides direct measurements of biomass over large scales, a combination of remotely sensed data and a well established field inventory is considered suitable for monitoring programs such as REDD+ [4, 5]. Data inputs for biomass estimation have varied widely with tradeoffs between

availability, cost and coverage. Accuracy of estimated biomass has also varied with the sensitivity of data to forest structure, spatial resolution, choice of statistical models, and the accuracy of field training data. Regardless, biomass estimates from different maps seem to agree at very coarse scales [4]. For example, Mitchard et al. [4] found that pantropical biomass maps converged at regional scales even though they varied locally. They concluded that uncertainties were largely related to spatial patterns of forest cover change. Langner et al. [5] evaluated pan-tropical biomass maps and successfully combined them into a framework for deriving REDD+ Tier 1 carbon storage estimates. While these findings are encouraging for national and

*Correspondence: wlhuang@umd.edu
[1] Department of Geographical Sciences, University of Maryland, College Park, USA
Full list of author information is available at the end of the article

continental scale reporting, there is a need to examine local discrepancies more closely as errors or uncertainty at fine-scales can complicate the use of coarse scale maps in local planning and decision making.

Almost all large-area biomass maps are derived from two-dimensional remote sensing data that have wide coverage but are generally less sensitive to canopy structure, particularly in moderate to high biomass forests (e.g. multispectral and single polarized SAR). Furthermore, they do not currently include fine scale variations in tree cover because of their coarse spatial resolution. Lidar instruments measure three-dimensional canopy structure which improves the accuracy of biomass maps [3] but lidar datasets have limited coverage and are expensive to acquire. An alternative is to use high-resolution lidar derived biomass maps, where available, to evaluate existing coarse scale maps, and make them more compatible for decision-making.

In 2010, NASA initiated the Carbon Monitoring System (CMS) to quantify carbon sources and sinks for an improved understanding of the global carbon cycle [6]. The program combines top-down continental scale approaches with bottom-up local scale approaches. The top-down approach relies on satellite observations to quantify carbon storage and terrestrial fluxes for national reporting. The bottom-up approach focusses on mapping carbon stocks and uncertainties at fine scales. Within the US, continental scale maps use Forest Inventory Analysis (FIA) plot data for model development, and biomass estimates are in turn compared with FIA county or regional averages as a type of validation [7–9]. However, these validations are not based on independent data, and often lack constraints at high spatial resolution. Moreover, field inventories generally do not include trees outside forests [9]. Continental scale maps therefore do not predict biomass outside forested areas and may significantly underestimate carbon balances [10].

A thorough understanding of local-scale discrepancies requires an independently derived high-resolution estimate. Recently, such a map was produced for the state of Maryland as part of CMS [11, 12]. Biomass estimates were derived from lidar data in conjunction with non-FIA field data using machine-learning approaches. The 30 m biomass maps incorporated tree canopy cover at the 1 m resolution, thus including forested and non-forested trees in the process. This local scale effort provides a reference for evaluating existing coarse scale maps.

We present results from a detailed comparison of the biomass map produced over Maryland (hereafter referred to as CMS_RF) with four national scale biomass maps: (A) NBCD2000 [13], (B) Blackard [14], (C) Wilson [15], and (D) Saatchi [16] at the pixel-, county- and state-level. We quantify the degree and spatial patterns of differences

to gain an improved understanding of map discrepancies and their impacts on carbon accounting.

Methods

Study area and field data

Maryland has a land area of ~25,600 km^2 (Fig. 1) and can be divided into 3 major physiographic provinces (or ecoregions) based on species-composition and environmental gradients. These are the Eastern Coastal Plain (hereafter, "Eastern Shore"), the combined Western Coastal Plain and Piedmont (hereafter, "Piedmont") and the combined Blue Ridge, Valley and Central Appalachians (hereafter, "Appalachian"). The wide variability in topography, forest types, and environmental gradients makes it a suitable test-bed for national map comparisons.

We first generated a biomass map using existing lidar data and independent field estimates. Field data were collected in 848 variable and fixed radius plots selected through a stratified sampling of NLCD land cover (evergreen, deciduous, wetlands, mixed and non-forest) and lidar canopy heights (Fig. 1). Tree measurements of diameter at breast height (dbh) and species were recorded in each plot. Allometric estimates of aboveground biomass (Mg ha^{-1}) were calculated for each tree using equations from Jenkins et al. [17] and appropriate blow up factors were applied to estimate biomass density for the variable radius plots. For more details on field data collection, refer to [11, 18]. In addition to these new plots, FIA data were obtained from across the state and used for model validation only [9].

Local scale CMS_RF biomass map

Leaf-off, discrete return lidar data were obtained from the Maryland Department of Natural Resources (DNR) and individual counties. Tree canopy cover and canopy height were mapped at 1 m resolution using a combination of Lidar and high-resolution leaf-on multispectral imagery for every county and seamlessly across the entire state [19, 20]. Lidar canopy height models were masked using high-resolution tree cover to obtain canopy heights over forested and non-forested areas. Lidar metrics such as height percentiles, densities, and canopy cover were calculated within 30 m grid cells corresponding to the NLCD land cover dataset. Field based estimates of biomass were then related to the lidar metrics using Random Forests regression models [21, 22]. Three separate empirical models were developed, one for each physiographic region, and were applied to predict biomass for counties within the region. Predictions over individual counties were merged into a statewide biomass map at 30 m resolution (CMS_RF map). Details of the biomass estimation are available in [18] and [12].

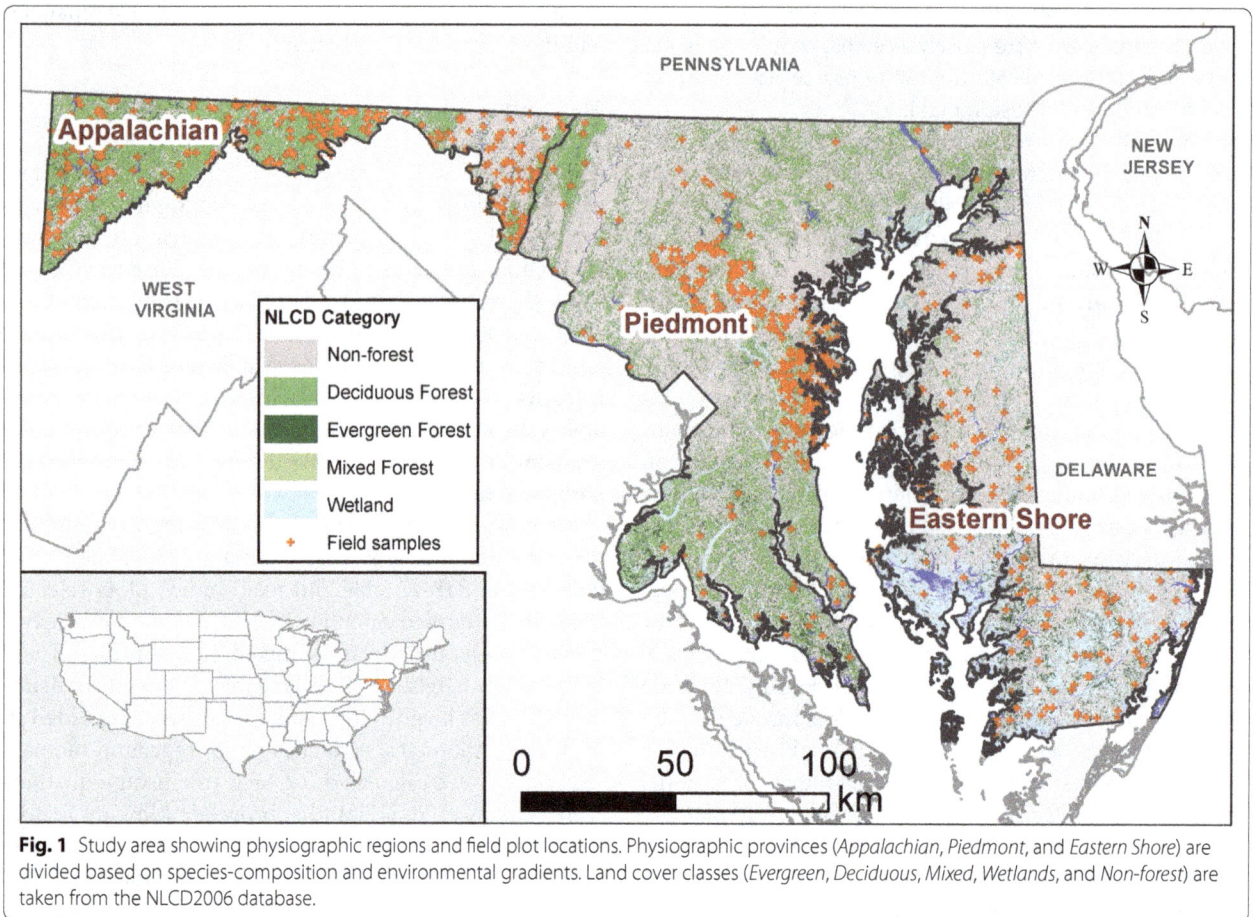

Fig. 1 Study area showing physiographic regions and field plot locations. Physiographic provinces (*Appalachian*, *Piedmont*, and *Eastern Shore*) are divided based on species-composition and environmental gradients. Land cover classes (*Evergreen*, *Deciduous*, *Mixed*, *Wetlands*, and *Non-forest*) are taken from the NLCD2006 database.

Continental scale biomass maps

Four national biomass products (Fig. 2; Table 1) were compared to the CMS_RF map. Each of these maps was derived using medium to coarse resolution satellite imagery. The NBCD2000 was the first 30 m national product developed using InSAR data from the 2000 Shutter Radar Topography Mission (STRM) and Landsat ETM+ data [13, 23]. NBCD2000 provided two versions of biomass: (A) NBCD_FIA map in which tree-level biomass estimates were obtained from tree tables in the FIA database (FIADB); and (B) NBCD_NCE or National Consistent allometric Equations in which biomass estimates were derived from equations developed by Jenkins et al. [17]. We used the NBCD_NCE version for consistency with our field biomass estimates, which were also derived from national allometric equations.

The Blackard map was developed at the 250 m spatial resolution [14] using tree-based regression (i.e., Cubist). It was developed by relating FIA plot data to multi-variable geospatial predictors, including Moderate Resolution Imaging Spectrometer (MODIS) data in 2001, percent tree cover and land cover proportions (from the NLCD

1992 product), topographical variables, and annual climate parameters, etc.

The Wilson map, also developed at the 250 m spatial resolution, was derived from MODIS imagery data from 2002 to 2008 and FIA field plots using a Phenological Gradient Nearest Neighbor (PGNN) imputation approach and canonical correspondence analysis (CCA) models [15, 24]. The Wilson map is a newer and improved version of the Blackard map.

The Saatchi map is a CMS national-scale map derived using a combination of NASA remote sensing data, forest inventory and ancillary data (the same method as [25]). Waveforms from the Geoscience Laser Altimeter System (GLAS) lidar were used to derive Lorey's height, which was then related to FIA biomass. The GLAS shots with predicted biomass estimates were used as ground truth (i.e., biomass plot samples) and related to multiple remote sensing inputs, including MODIS, PALSAR, and Landsat imagery using Maximum Entropy (MaxEnt) models for predicting biomass at the continental scale. An updated version of the Saatchi map (Saatchi et al., personal communication) reported improvements such

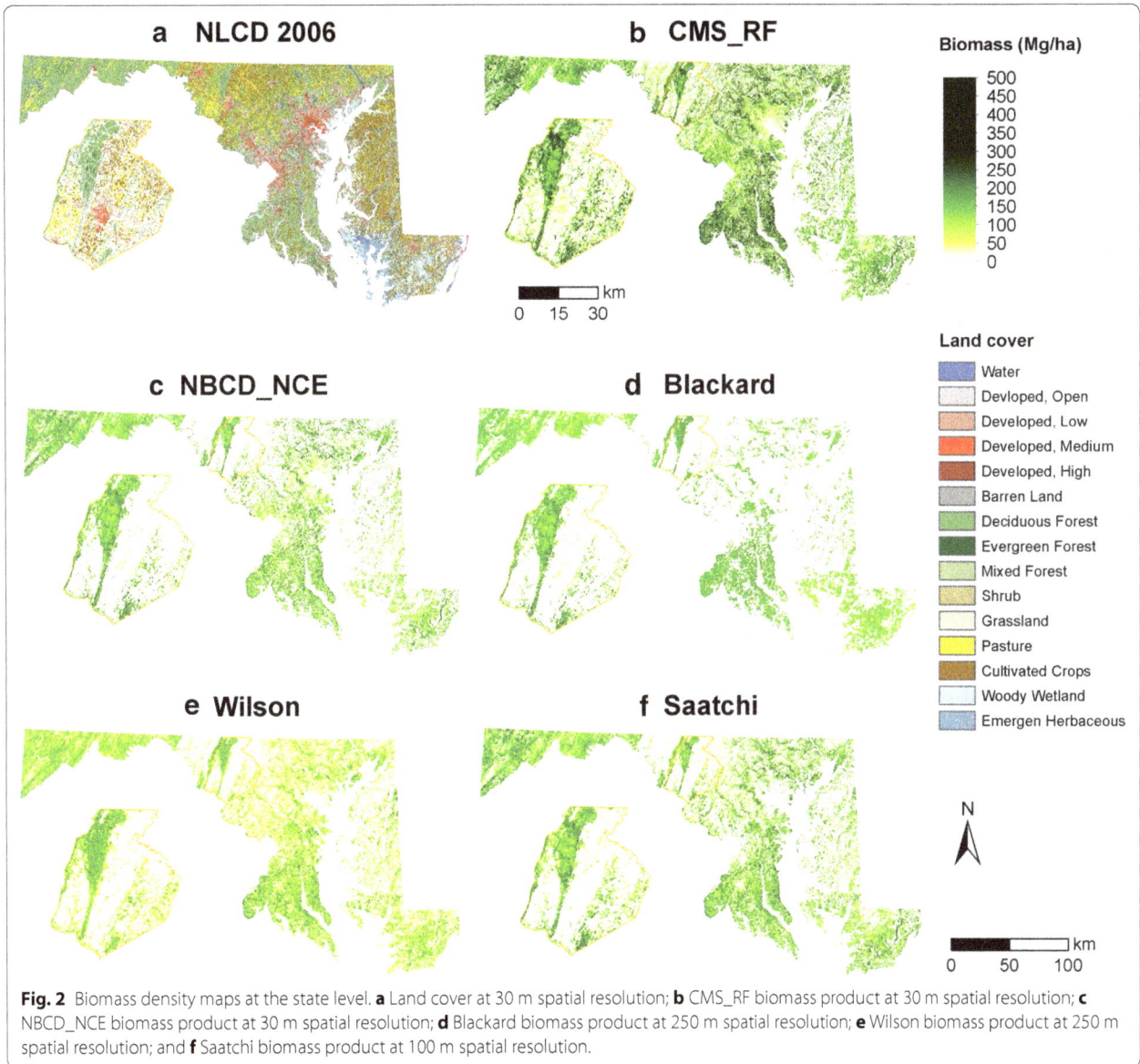

Fig. 2 Biomass density maps at the state level. **a** Land cover at 30 m spatial resolution; **b** CMS_RF biomass product at 30 m spatial resolution; **c** NBCD_NCE biomass product at 30 m spatial resolution; **d** Blackard biomass product at 250 m spatial resolution; **e** Wilson biomass product at 250 m spatial resolution; and **f** Saatchi biomass product at 100 m spatial resolution.

as: (A) reprocessed GLAS data, (B) 15 allometric equations that include three forest types (deciduous, coniferous, and mixed) for 5 regions of the US, and (C) NLCD non-vegetated gaps filled by PALSAR and Landsat data. We present results from the updated version but also include a comparison of the old and new versions in the supplement (Additional file 1: Figure S1 and Additional file 2: Figure S2).

Map comparisons

All maps were warped to a common frame of reference (UTM 18N NAD 83) ensuring minimum distortion to the native projections. Maps were matched to the same extents and pixel sizes. The 30 m biomass density maps

(e.g. CMS_RF and NBCD_NCE) were aggregated to 250 m and coarser resolutions (e.g. 500 m, 1 km, and 4 km). The Wilson, Blackard (originally 250 m), and Saatchi maps (originally ~90 m) were each aggregated to 500 m, 1 km, and 4 km.

A canopy cover mask was used to differentiate between forested and non-forested areas in our comparisons. The mask was created from the NLCD 2006 dataset for consistency with the land cover used in the CMS_RF stratification [18] and [12]. The mask included deciduous forest (41), evergreen forest (42), mixed forest (43), woody wetlands (90) and emergent herbaceous wetlands (95) from the NLCD dataset. NLCD defines forest as more than 20 percent of areas dominated by trees. Therefore, a 20 %

Table 1 Summary of biomass products used in this study

Product	Sensor and year	Field data and year	Resolution	Forest mask	Approach	Uncertainty map	References
CMS_RF	DRL 2004–2012	2011–2014	30 m	NAIP high-res tree canopy cover	Random forest, regression tree	Percentile error (QRF)	[12, 18]
NBCD_NCE	Landsat + SRTM 2000	2000	30 m	NLCD 2001	Random forest, regression tree	Quality voids	[13]
Blackard	MODIS 2001[a]	2005–2009	250 m	NLCD 1992	Cubist, regression tree	Relative error	[14]
Wilson	MODIS 2002–2008	2005–2009	250 m	NLCD 2001 percent tree canopy 25 %	PGNN, kNN		[15]
Saatchi	MODIS + PAL-SAR + Landsat	2005	~250 m v1[b] ~90 m v2[b]	NLCD 2006	MaxEntropy, parametric	Percent error	[16]

[a] Year is national maps in eastern US.

[b] Original maps are in lat/lon, where v1 with 0.00222222 ≅ 250 m and v2 with 0.00083333 deg ≅ 90 m.

DRL Discrete Return Lidar, 1–2 m small footprint lidar aggregate, *NAIP* National Agriculture Imagery Program, *QRF* Quantile Regression Forests, *SRTM* Shuttle Radar Topography Mission, *PALSAR* Phased Array type L-band Synthetic Aperture Radar, *PGNN* Phenological Gradient Nearest Neighbor, *kNN* k-nearest neighbor.

threshold was set while aggregating the mask from 30 to 250 m and other coarse resolutions. Comparisons were made over: forested areas only; non-forested areas only; and over forested and non-forested areas combined.

Statistical indicators such as coefficient of determination (R^2), root mean squared difference (RMSD), RMSD% or CV (coefficient of variation of the RMSD), and mean bias error (MBE) were used to compare the CMS_RF product with the four national maps. The Fuzzy Numerical Index (FNI) is a valuable quantitative descriptor of the spatial similarities and differences between maps and was included in our comparisons, following [26].

$$R^2 = 1 - \frac{\sum_{i=1}^{n}(C_i - M_i)^2}{\sum_{i=1}^{n}(M_i - \overline{M})^2} \quad (1)$$

$$RMSD = \sqrt{\sum_{i=1}^{n}\frac{(C_i - M_i)^2}{n}} \quad (2)$$

$$RMSD\% = \frac{RMSD}{\overline{C}} \times 100 \quad (3)$$

$$MBE = \frac{\sum_{i=1}^{n}(C_i - M_i)}{n} \quad (4)$$

$$FNI = \frac{\sum_{i=1}^{n} 1 - \frac{|C_i - M_i|}{\max(C_i, M_i)}}{n} \quad (5)$$

M_i is the value of national map; C_i is the CMS_RF predicted value; i is the sample index; $\overline{C}$ and $\overline{M}$ are the means of CMS_RF and national map respectively; and n is the sample size.

Results

Spatial patterns of biomass were consistent with land cover and physiographic gradients in visual comparisons. Within forested areas, all maps showed distinct dendritic patterns corresponding to riparian zones that had higher biomass than surrounding areas. Similar spatial patterns of biomass were also noted along ridges, valleys and forested patches with high structural variability.

Although spatial patterns were similar, biomass densities and levels of detail varied considerably (Fig. 2). The CMS_RF biomass map provided greater detail over urban/suburban landscapes (Fig. 3, e.g. trees along roadsides, hedges and backyards) when compared visually with high-resolution [1 m] land cover map and high-resolution imagery (Google Earth). The other maps predicted little or no biomass in non-forested areas. Differences over heterogeneous areas were particularly large (Fig. 3). Results ranged between 36,600 and 119,679 Mg, showing wide local-scale differences.

FNI provides a spatial representation of similarities and differences when calculated at a pixel-level. However, it does not capture the positive and negative deviations with respect to the CMS_RF map. We therefore calculated a mean FNI value for each map comparison with values ranging from 0 (perfect dissimilarity) to 1 (perfect similarity). A combination of map differences and FNI index values provided additional spatial and quantitative understanding of map discrepancies (Fig. 4; Table 2). Differences between maps were prominent in the Piedmont region, over counties in southern Maryland and along the Appalachians in the West. The Saatchi map was most similar to the CMS_RF map (FNI = 0.53) while the Blackard Map (FNI = 0.26) was the most dissimilar. The Wilson map had almost an equal proportion of similar

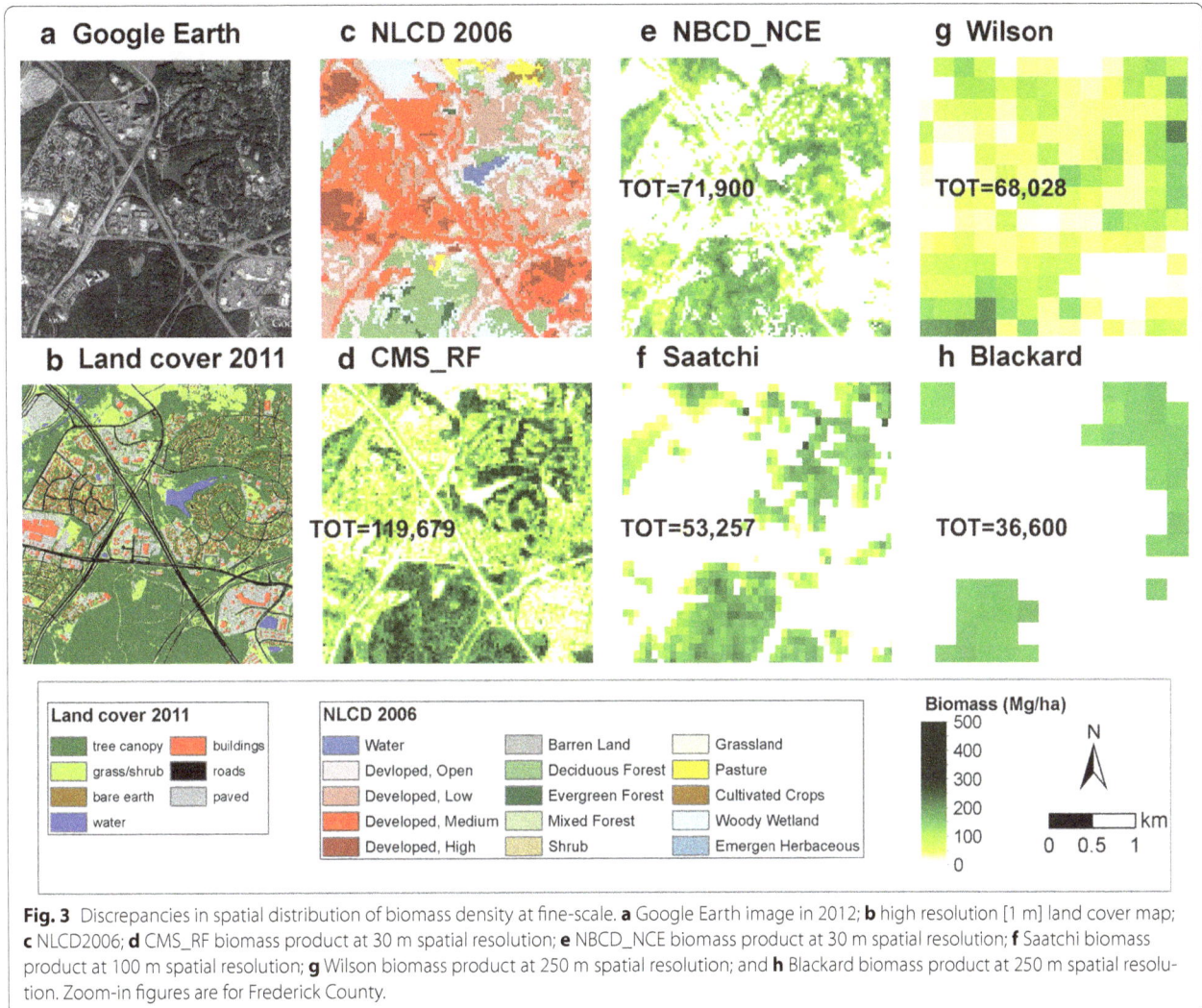

Fig. 3 Discrepancies in spatial distribution of biomass density at fine-scale. **a** Google Earth image in 2012; **b** high resolution [1 m] land cover map; **c** NLCD2006; **d** CMS_RF biomass product at 30 m spatial resolution; **e** NBCD_NCE biomass product at 30 m spatial resolution; **f** Saatchi biomass product at 100 m spatial resolution; **g** Wilson biomass product at 250 m spatial resolution; and **h** Blackard biomass product at 250 m spatial resolution. Zoom-in figures are for Frederick County.

and dissimilar pixels (FNI = 0.49) while the NBCD map was slightly lower with an FNI of 0.48.

Comparisons at the pixel level

(i)   High-resolution comparisons [30 m]

Pixel level comparisons between the NBCD_NCE and CMS_RF biomass products showed wide scatter with a large number of zero biomass predictions from the NBCD map (Fig. 5). Most areas that did not have biomass values on the NBCD map had predictions in the CMS_RF map. The NBCD biomass values were biased lower than the 1:1 line with an overall RMSD of 75.0 Mg ha^{-1}. Biomass distributions from CMS_RF and NBCD_NCE maps showed large differences over total and non-forested regions (Fig. 6). The NBCD_NCE distribution was bimodal with modes shifted toward the left (or lower biomass ranges). The CMS_RF map had higher and more widely distributed values over non-forested regions. The NBCD_NCE dataset did not predict biomass outside forests but the non-forest histograms had some high biomass values. This was because an older NLCD (2000) forest/non-forest mask was used to generate the NBCD_NCE map. The time lag between the maps and the difference in forest/non-forest masks complicated the comparisons but did not affect the overall trend in the forested and non-forested scatter plots (Fig. 5).

(ii)   Comparisons with field data

We compared predictions from the CMS_RF and NBCD_NCE maps with biomass estimates from FIA data (average of four sub-plots)

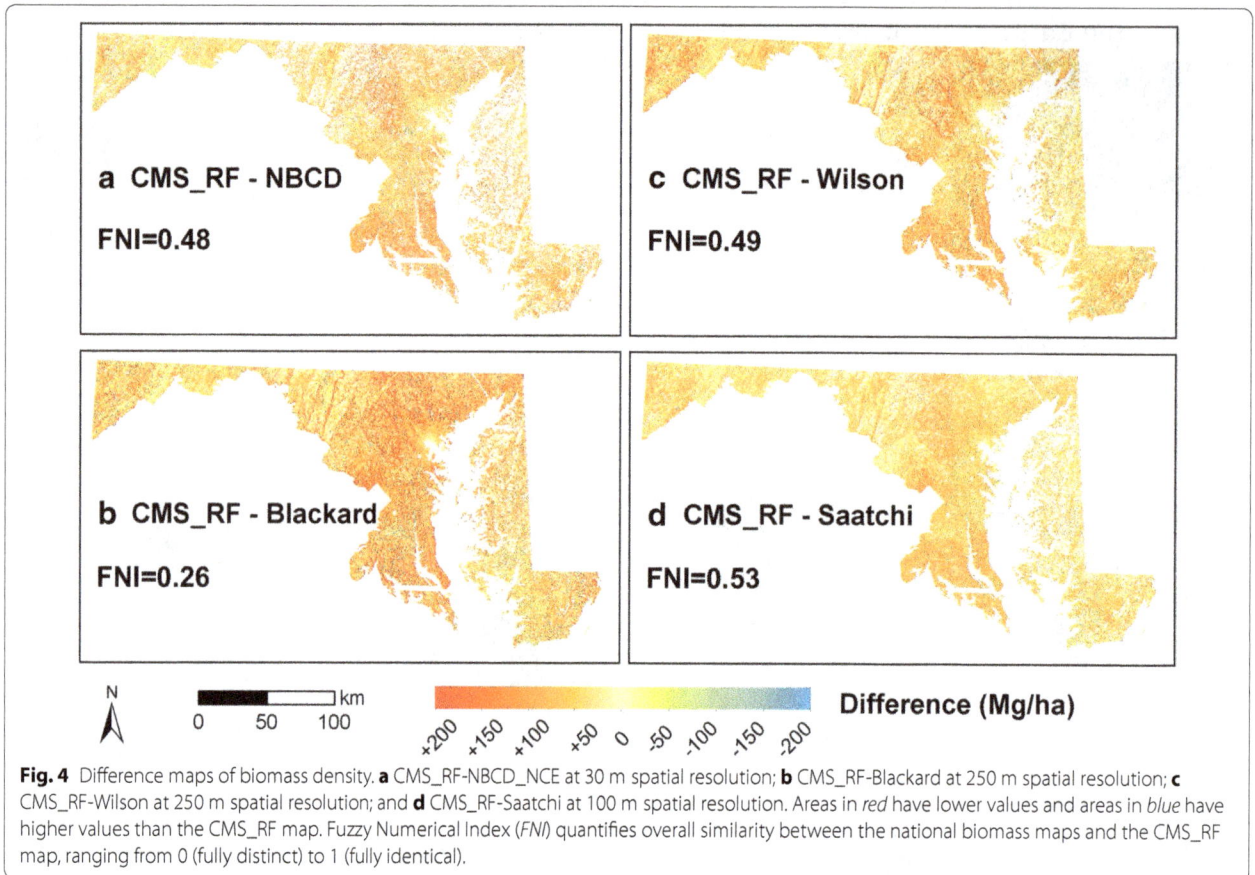

Fig. 4 Difference maps of biomass density. **a** CMS_RF-NBCD_NCE at 30 m spatial resolution; **b** CMS_RF-Blackard at 250 m spatial resolution; **c** CMS_RF-Wilson at 250 m spatial resolution; and **d** CMS_RF-Saatchi at 100 m spatial resolution. Areas in *red* have lower values and areas in *blue* have higher values than the CMS_RF map. Fuzzy Numerical Index (*FNI*) quantifies overall similarity between the national biomass maps and the CMS_RF map, ranging from 0 (fully distinct) to 1 (fully identical).

Table 2 Mean Fuzzy Numeric Index

Name	All	Forest	Non-forest
NBCD_NCE	0.48	0.62	0.22
Blackard	0.26	0.38	0.04
Wilson	0.49	0.52	0.41
Saatchi	0.53	0.69	0.25

Values calculated from maps at 250 m resolution.

(Fig. 7a) and our variable radius field plots (Fig. 7b, c). The Random Forests model used to generate the CMS_RF map explained ~50 % variability in biomass from variable radius field plots ($R^2 = 0.49$, RMSE = 89.3 Mg ha^{-1}, n = 848). A cross-validation of the CMS_RF map with plot level FIA data showed higher agreement, partly due to higher sample number ($R^2 = 0.69$, RMSE = 58.2 Mg ha^{-1}, n = 1,055). On the other hand, a cross validation of the NBCD_NCE map with variable radius estimates resulted in substantially weaker relationships ($R^2 = 0.14$, RMSE = 125.1 Mg ha^{-1}, n = 433).

(iii) Comparisons at the pixel level [250 m]

Large disagreements were observed in the scatter plots and associated errors at the 250 m resolution (Fig. 8). Overall RMSD values ranged between 48.5 and 92.7 Mg ha^{-1}. The RMSD values ranged between 55.0 and 90.0 Mg ha^{-1} over forested regions, and between 33.9 and 103.9 Mg ha^{-1} over non-forested regions. The Saatchi and NBCD maps agreed more closely with the CMS_RF map with fewer zero biomass values after spatial aggregation. The updated version of Saatchi map agreed closely with the NBCD and CMS_RF map, while the original version showed a large difference (Additional file 1: Figure S1 & Additional file 2: Figure S2). The Blackard map was the least correlated with the CMS_RF map while the Wilson map had a large scatter around the 1:1 line.Histograms of biomass in intervals of 10 Mg ha^{-1} were generated and analyzed over the entire range (0–400 Mg ha^{-1}) (Fig. 9). There was little agreement among the maps across the entire range of predicted values. The only similarities were between the NBCD_NCE and the Saatchi map

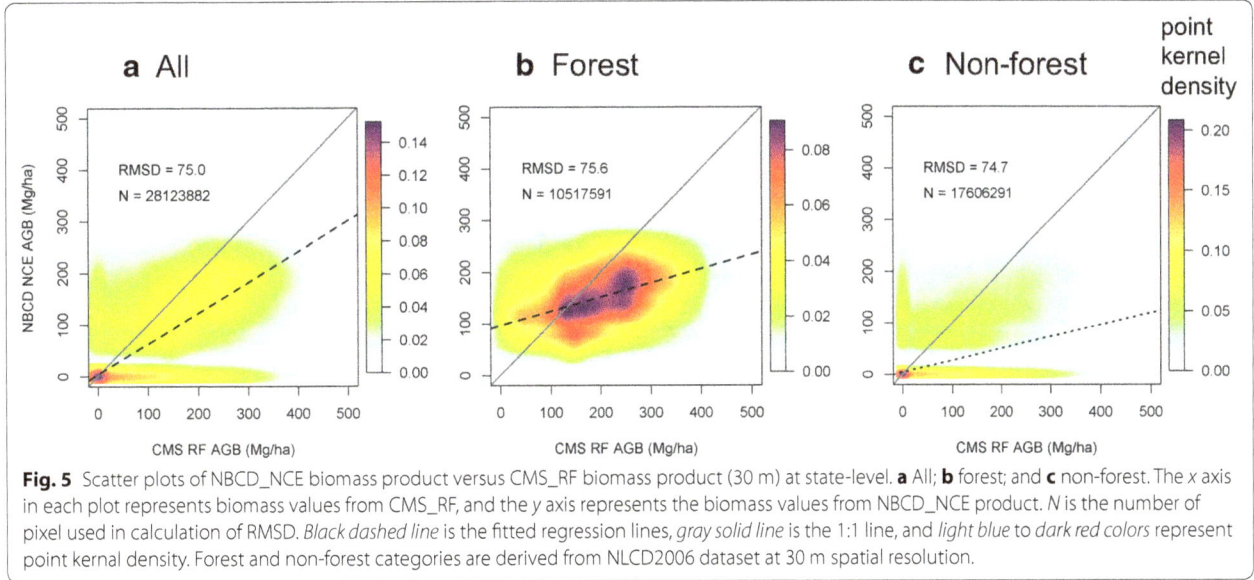

Fig. 5 Scatter plots of NBCD_NCE biomass product versus CMS_RF biomass product (30 m) at state-level. **a** All; **b** forest; and **c** non-forest. The x axis in each plot represents biomass values from CMS_RF, and the y axis represents the biomass values from NBCD_NCE product. N is the number of pixel used in calculation of RMSD. *Black dashed line* is the fitted regression lines, *gray solid line* is the 1:1 line, and *light blue* to *dark red colors* represent point kernal density. Forest and non-forest categories are derived from NLCD2006 dataset at 30 m spatial resolution.

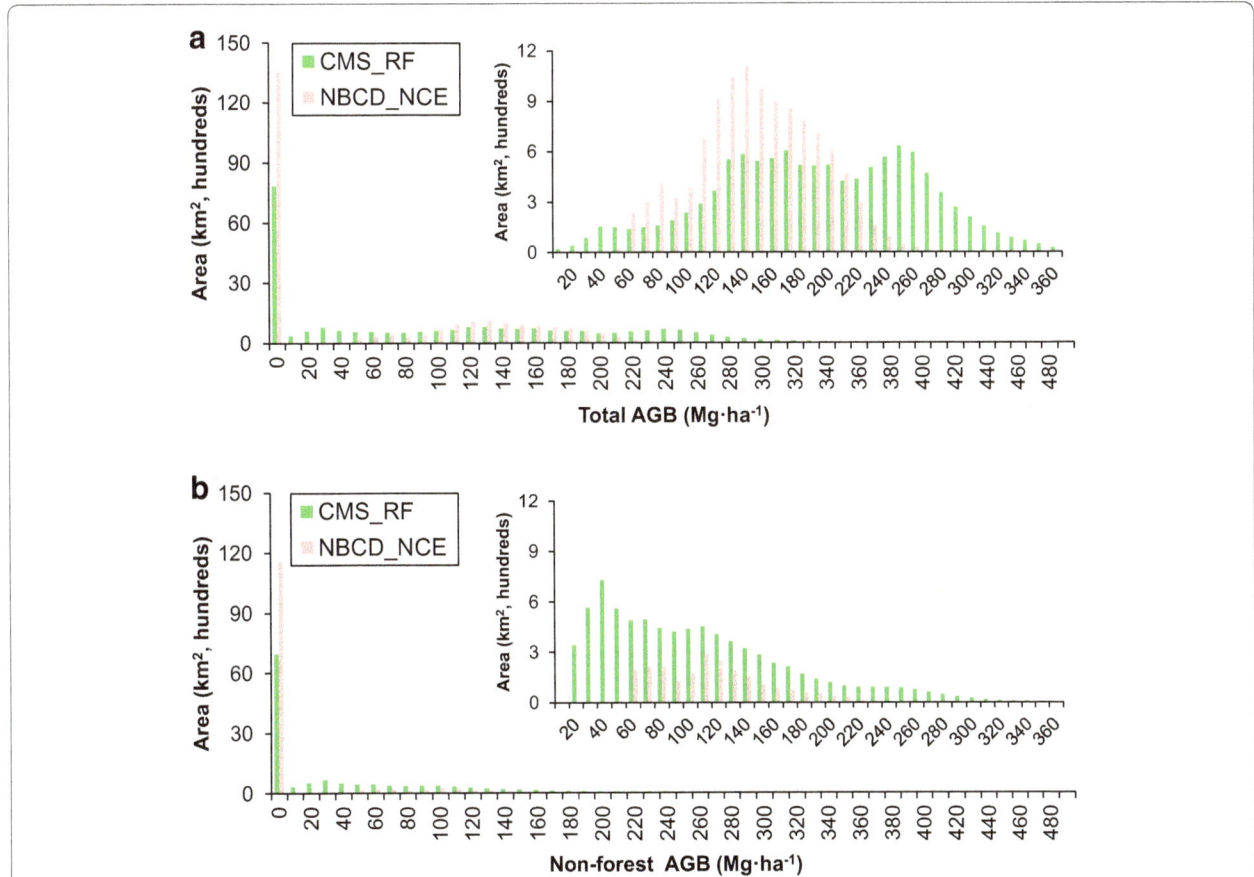

Fig. 6 *Histograms* showing the biomass distribution of CMS_RF and NBCD_NCE products over the state of Maryland at 30 m resolution in 10 Mg ha^{-1} bins. **a** All and **b** non-forest. Note that zero values are ignored in the *inset* plots. Non-forest category is derived from NLCD2006 dataset.

a CMS_RF vs. FIA   **b** CMS_RF vs. Field   **c** NBCD_NCE vs. Field

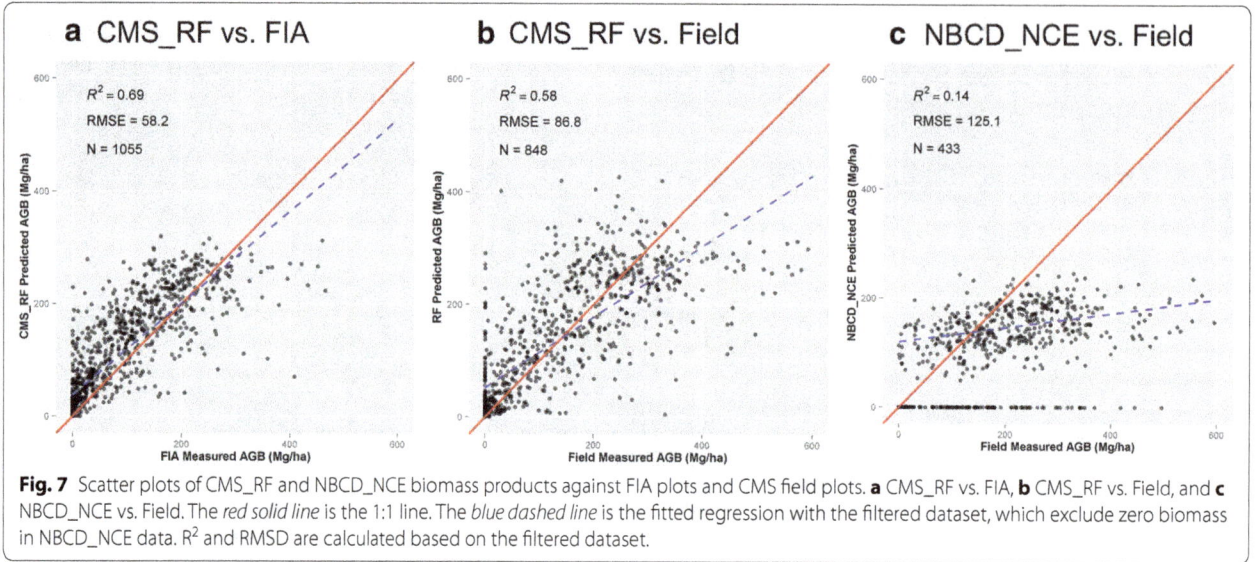

Fig. 7 Scatter plots of CMS_RF and NBCD_NCE biomass products against FIA plots and CMS field plots. **a** CMS_RF vs. FIA, **b** CMS_RF vs. Field, and **c** NBCD_NCE vs. Field. The *red solid line* is the 1:1 line. The *blue dashed line* is the fitted regression with the filtered dataset, which exclude zero biomass in NBCD_NCE data. R^2 and RMSD are calculated based on the filtered dataset.

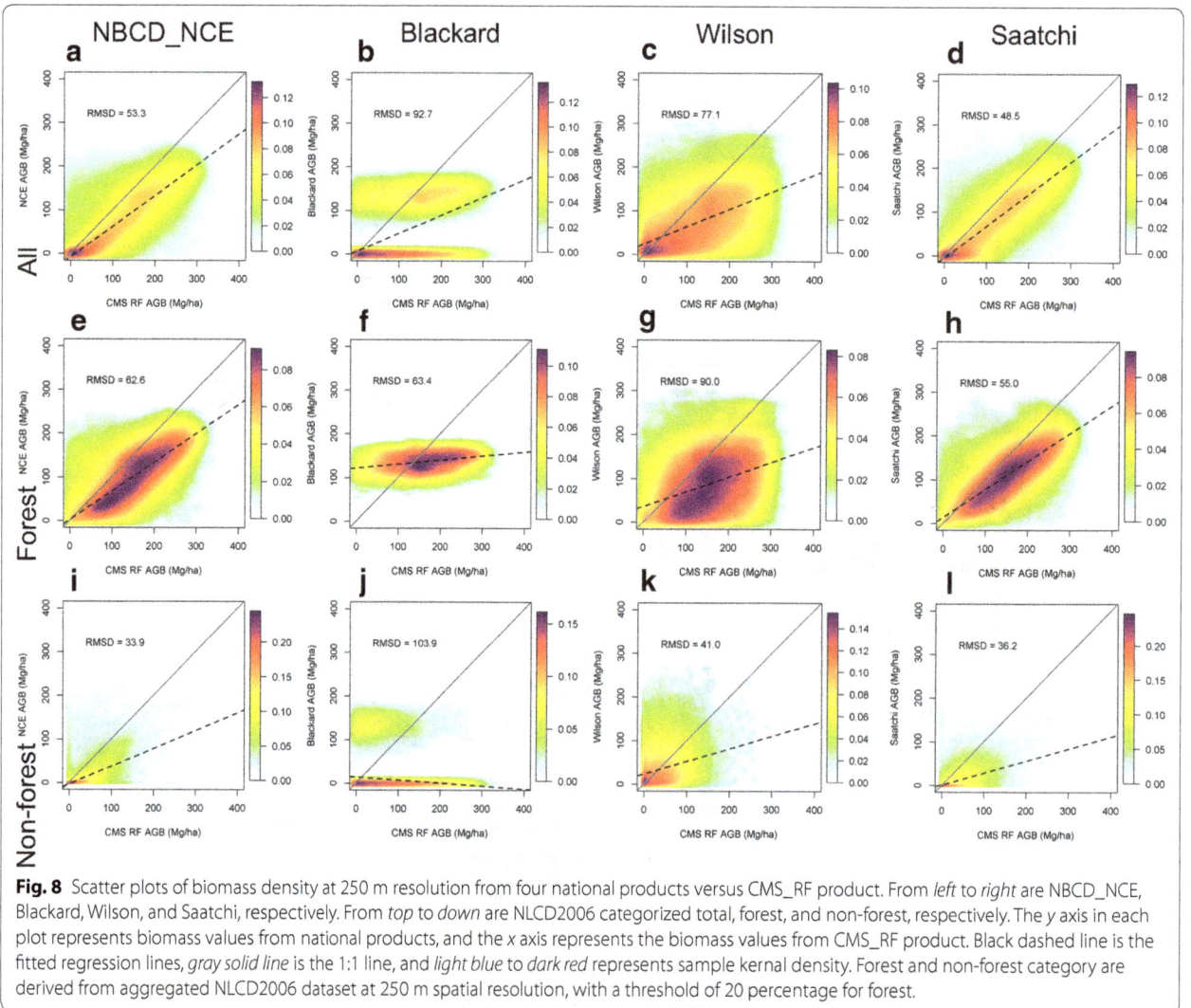

Fig. 8 Scatter plots of biomass density at 250 m resolution from four national products versus CMS_RF product. From *left* to *right* are NBCD_NCE, Blackard, Wilson, and Saatchi, respectively. From *top* to *down* are NLCD2006 categorized total, forest, and non-forest, respectively. The y axis in each plot represents biomass values from national products, and the x axis represents the biomass values from CMS_RF product. Black dashed line is the fitted regression lines, *gray solid line* is the 1:1 line, and *light blue* to *dark red* represents sample kernal density. Forest and non-forest category are derived from aggregated NLCD2006 dataset at 250 m spatial resolution, with a threshold of 20 percentage for forest.

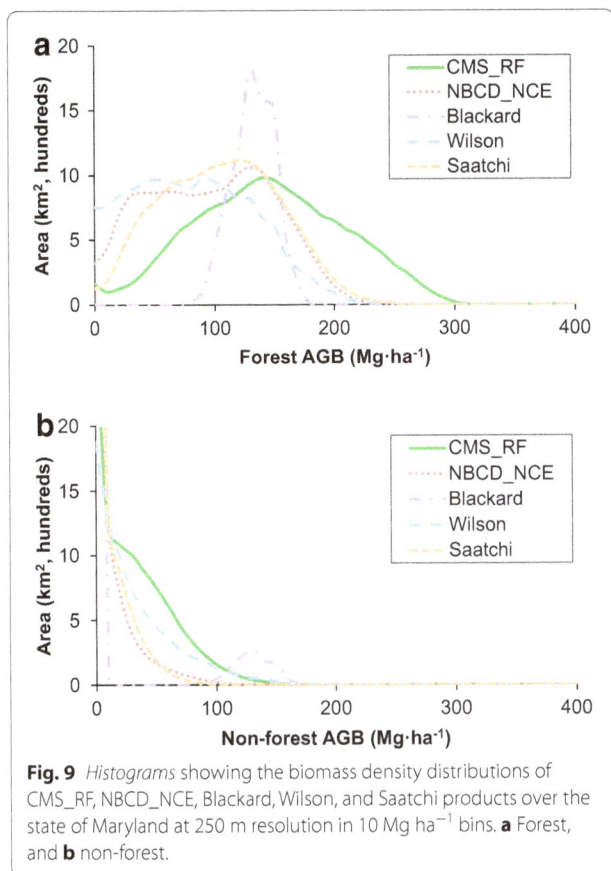

Fig. 9 *Histograms* showing the biomass density distributions of CMS_RF, NBCD_NCE, Blackard, Wilson, and Saatchi products over the state of Maryland at 250 m resolution in 10 Mg ha^{-1} bins. **a** Forest, and **b** non-forest.

above 125 Mg ha^{-1}. The distribution of biomass in different ranges was also vastly different. Biomass values in the Blackard map were predominantly between 100 and 150 Mg ha^{-1} while all other datasets had values less than 250 Mg ha^{-1}. Only the CMS_RF maps had predictions in ranges greater than 250 Mg ha^{-1}.

Comparisons at the county level

At the county level, the four maps showed improved correlation with the CMS_RF map in both mean (Fig. 10) and total biomass (Fig. 11). Among the three physiographic regions, the counties in Appalachian region were closer to 1:1 line in all four products. Counties in Piedmont region had more evenly distributed biomass values in all products except the Blackard map. Counties in Eastern Shore region were more clustered, ranging between 40.1 and 79.2 Mg ha^{-1} for mean, and 4.6 and 7.8 Tg for total biomass respectively. Despite the improved correlation, the MBE was high in all four products, ranging between −33.0 and −54.6 Mg ha^{-1} for mean, and −3.5 and −5.8 Tg for total biomass respectively.

County totals from the continental scale maps and the CMS_RF map were also compared with FIA totals (Fig. 12). For this comparison, we used the gap-filled

Jenkins estimate from FIA data as it includes non-forested biomass [9]. Continental scale maps were strongly correlated with FIA at county level and had high coefficients of determination (0.63–0.80), but consistently underestimated biomass with a negative bias, ranging between −3.4 and −1.1 Tg for total biomass (Fig. 12a–d). The CMS_RF map showed good agreement too but had a positive bias and overestimated biomass, particularly in counties that had many low biomass areas such as in the Piedmont (Fig. 12e).

Comparisons at the state level

There were significant differences between the biomass totals at the state level (Fig. 13). The national maps estimated state totals between 126.0 and 170.6 Tg and seemed to converge but were much lower when compared to the CMS_RF map. A detailed breakdown of mean and total biomass from all the maps is provided in Tables 3 and 4. The CMS_RF had higher mean (Tables 3, 4) and total biomass values (Fig. 13) over both forested and non-forested regions. The CMS_RF map also had higher total biomass than what is traditionally reported by FIA (164 Tg, 2008–2012 collection period) (Additional file 3: Table S1, [27]). However, we note that FIA does not measure trees in areas defined as "non-forest" and the allometric approach used by FIA to calculate tree biomass is known to give lower estimates in this region [9]. Adjusting for these nuances in the FIA data achieved better agreement with CMS_RF, although the FIA estimate was still lower by 43 Tg (Additional file 3: Table S1, [28]).

Lastly, we examined the coefficient of determination and corresponding errors as a function of resolution to detect trends and convergence between the maps (Fig. 14). The R^2 values for both total and mean biomass increased with decreasing resolution, gradually moving closer to 0.90. Correspondingly, RMSD values decreased gradually stabilizing at ~35 Mg ha^{-1}. The NBCD_NCE, Saatchi, and Wilson maps converged with near perfect agreement at around 4 km and onward. The Blackard map showed similar trends but less convergence with other products. Despite the improved agreement, the maps did not converge with the CMS_RF map at any scale considered in this study.

Discussion

Spatial patterns of similarities and differences were consistent with land cover and physiographic gradients. Geographically, the greatest spatial discrepancies were in the Piedmont region. This is not unexpected, given the urban development and suburban sprawl in the region. Coarse scale maps did not capture the heterogeneity of urban-suburban landscapes as finely as the CMS_RF map, hence the difference (Fig. 3). Distinct spatial patterns of

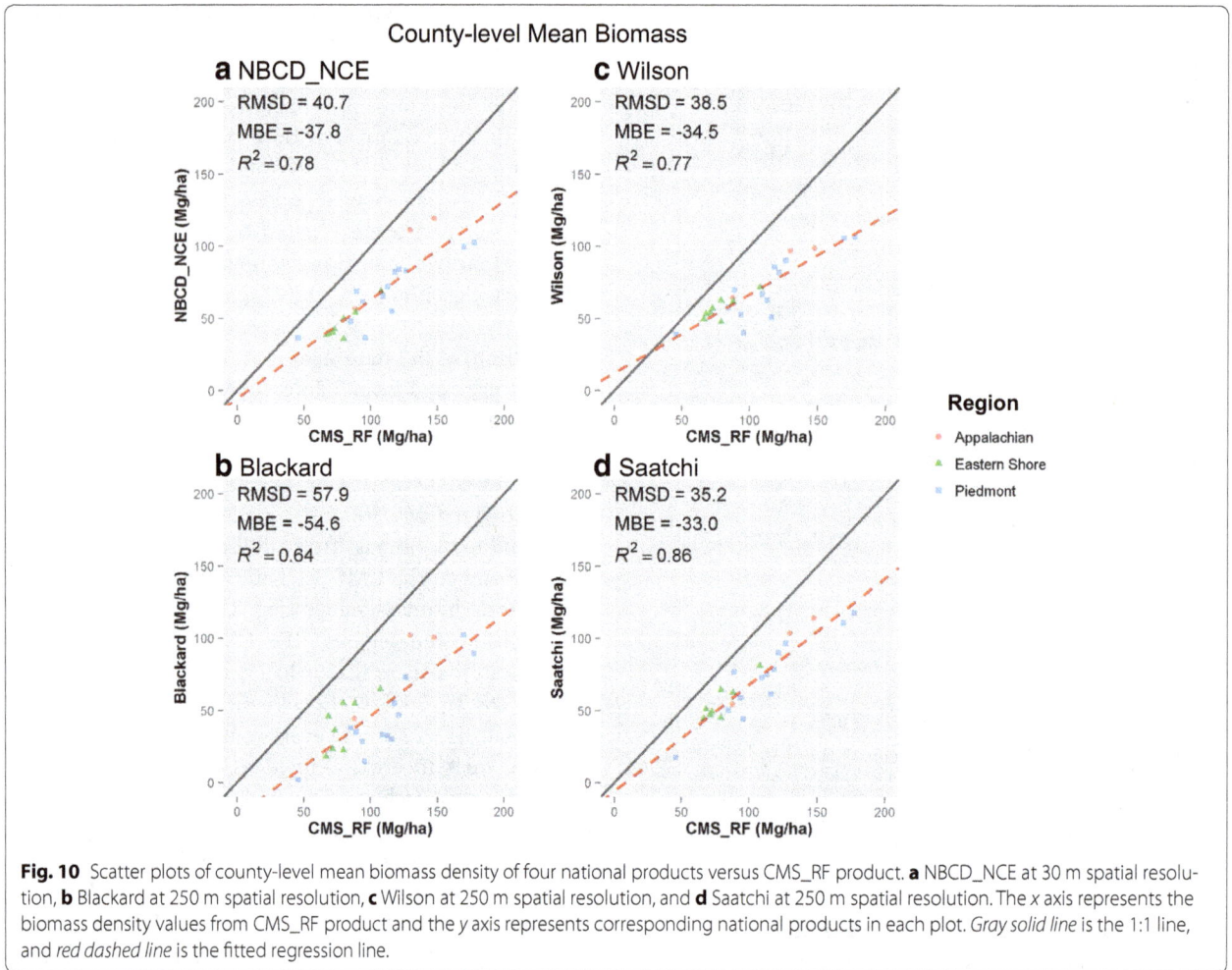

Fig. 10 Scatter plots of county-level mean biomass density of four national products versus CMS_RF product. **a** NBCD_NCE at 30 m spatial resolution, **b** Blackard at 250 m spatial resolution, **c** Wilson at 250 m spatial resolution, and **d** Saatchi at 250 m spatial resolution. The x axis represents the biomass density values from CMS_RF product and the y axis represents corresponding national products in each plot. *Gray solid line* is the 1:1 line, and *red dashed line* is the fitted regression line.

differences were also observed in Western Maryland and Southern Piedmont. These areas corresponded to dense forests where estimates from all the continental scale maps were lower. The Eastern shore had fewer discrepancies, probably because of sparse tree cover and lower biomass. However, unusually high values were noted in the national maps over several low-lying areas. This could be because of the mixed reflectance of water and vegetation over wetlands that is not easily separated in coarse resolution imagery [29].

We expected the 30 m NBCD_NCE map to be most similar to the CMS_RF map because it closely matched the spatial patterns in the CMS_RF map and had finer details than the other maps. However, the enhanced Saatchi map agreed more closely (Figs. 8, 4), despite having a coarse resolution (~90 m) and fewer predictions beyond 250 Mg ha^{-1}. This was probably because the Saatchi map had more predictions in the 50–100 Mg ha^{-1} range than the NBCD_NCE map. The NBCD_NCE map had many pixels with very low biomass

values (Fig. 9) which reduced its overall agreement with the CMS_RF map. Another surprising digression was the 250 m Wilson map that had a higher overall similarity index (FNI) than the NBCD_NCE map and the best agreement (Table 2) with the CMS_RF map over non-forested regions. A closer examination revealed that the Wilson map had better predictions in non-forested areas than any other map because it did not include a forest/non-forest mask and was developed using different models for areas greater than and less than 50 % NLCD forest cover. Thus, the agreement of continental scale maps with high-resolution estimates is not necessarily a function of spatial resolution but depends more on modeling approaches, time-lags and forest/non-forest definitions.

Choice of statistical/modelling approach was less critical in the CMS_RF estimation [18] but affected biomass predictions in other maps. The Blackard and Wilson maps used similar inputs yet had entirely different spatial distributions and histograms (Fig. 9) because of the difference in the regression models (Table 1). Similarly,

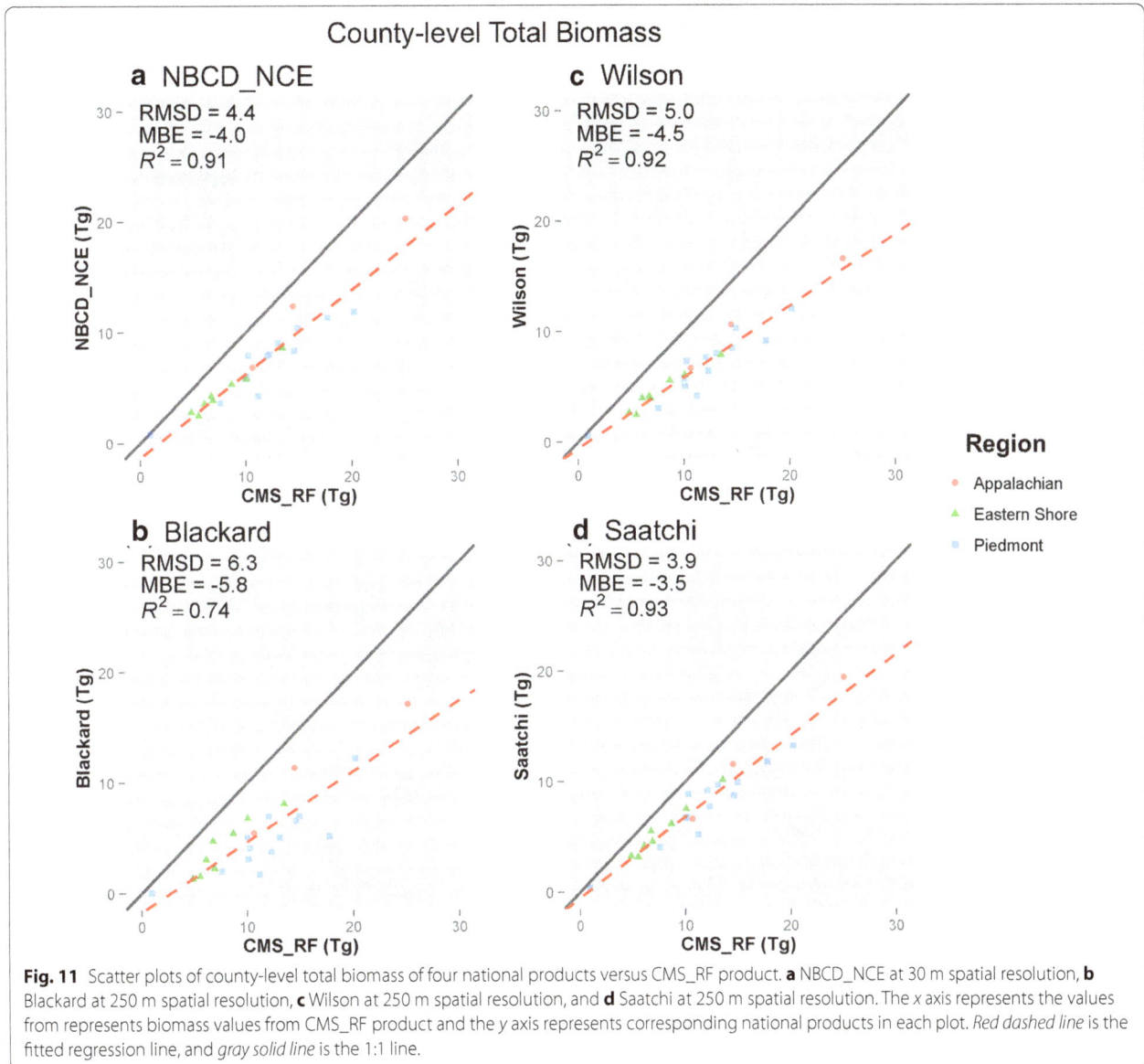

Fig. 11 Scatter plots of county-level total biomass of four national products versus CMS_RF product. **a** NBCD_NCE at 30 m spatial resolution, **b** Blackard at 250 m spatial resolution, **c** Wilson at 250 m spatial resolution, and **d** Saatchi at 250 m spatial resolution. The *x* axis represents the values from represents biomass values from CMS_RF product and the *y* axis represents corresponding national products in each plot. *Red dashed line* is the fitted regression line, and *gray solid line* is the 1:1 line.

we noted a strong influence of the MaxEnt model in the form of stratified predictions (Additional file 1: Figure S1) from the original Saatchi map. Such discrepancies are not easily detected in a broad comparison but are evident in a pixel-by-pixel comparison, as demonstrated in this study.

Continental scale maps (except the Wilson map) did not predict values outside forested areas because of limited FIA field plots for model development. This reduced their total biomass estimates and increased pixel-level discrepancies with the CMS_RF map. While we acknowledge that a fair comparison cannot be made over non-forested regions, we quantified the effect of excluding non-forest biomass on county and state level totals. Our results indicate that the underestimation is non-trivial, particularly in heterogeneous landscapes such as our

study area. We provide further corroboration to findings of [10] and support the need for including biomass outside forests in carbon reporting.

Some apparent non-forested biomass values crept into the national map totals (Fig. 13) because of time lags between maps and inconsistencies in forest/non-forest masks from the NLCD product. We noticed high values (greater than 100 Mg ha^{-1}) in non-forest histograms from some national maps (Fig. 9). These could be artifacts of forested areas that were converted since the production of the maps or differences in NLCD classifications over time. Some non-forest biomass was a result of edge effects in the coarse scale maps. Discrepancies could also be attributed to canopy cover thresholds used for comparisons (e.g. 20 % in this study). Larger

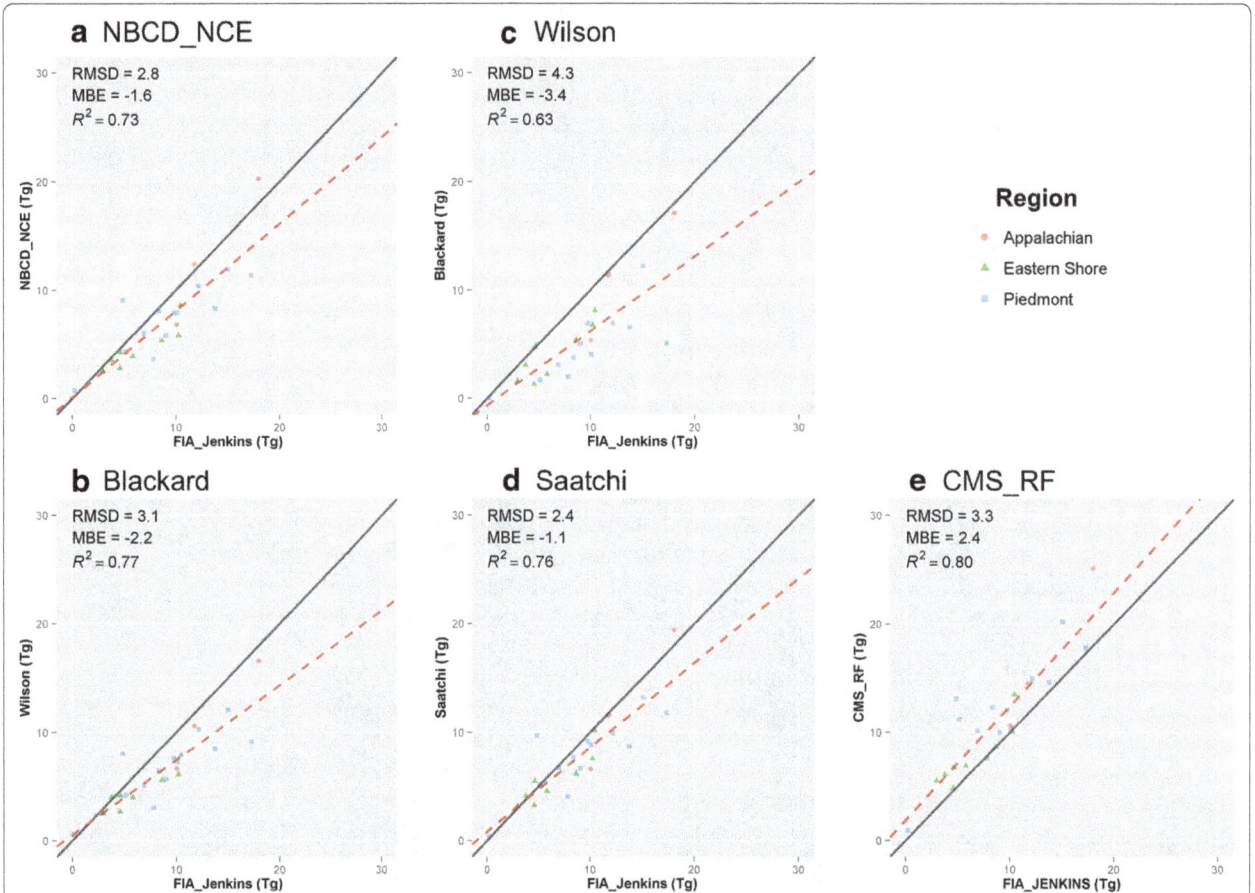

Fig. 12 Scatter plots of county-level total biomass of four national products and CMS_RF against estimates from FIA_Jenkins. **a** NBCD_NCE at 30 m spatial resolution, **b** Blackard at 250 m spatial resolution, **c** Wilson at 250 m spatial resolution, **d** Saatchi at 250 m spatial resolution, and **e** CMS_RF at 30 m spatial resolution. The x axis represents the biomass totals from FIA_Jenkins, and the y axis represents corresponding national products in each plot. *Red dashed line* is the fitted regression line, and *gray solid line* is the 1:1 line. *FIA_Jenkins* represents biomass estimates using Jenkins allometrics and gap-filled for non-forest biomass.

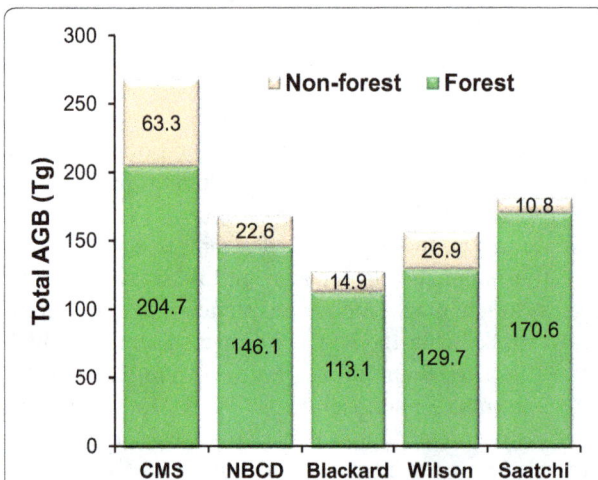

Fig. 13 Comparison of total biomass at the state level from the CMS_RF map and the four national products over forested and non-forested areas. The forest/non-forest mask is aggregated from NLCD2006 with a threshold of 20 percentage.

Table 3 Mean and total biomass for CMS_RF and NBCD_NCE products at 30 m resolution by forest and non-forest class

Type	CMS_RF		NBCD_NCE		Area compared[a]
	Mean (Mg ha^{-1})	Total (Tg)	Mean (Mg ha^{-1})	Total (Tg)	(km^2)
Forest	175.8	204.7	125.5	146.1	11,642
Non-forest	46.3	63.3	16.6	22.6	13,670
All	105.9	268.0	66.7	168.8	25,312
	Mg ha^{-1}	Tg	Mg ha^{-1}	Tg	km^2

[a] Summarized from CMS_RF and NBCD_NCE products at 30 m resolution.

thresholds can lead to lower non-forest biomass and vice versa. Some of these inconsistencies can be reduced by including sub-pixel estimates of tree cover [30] instead of a forest/non-forest mask in future continental scale mapping projects similar to the Wilson map [15]. This may

Table 4 Mean and total biomass for three national products at 250 m resolution by forest and non-forest class

Type	Blackard		Wilson		Saatchi		Area compared
	Mean (Mg ha^{-1})	Total (Tg)	Mean (Mg ha^{-1})	Total (Tg)	Mean (Mg ha^{-1})	Total (Tg)	(km^2)
Forest	97.2	113.1	111.4	129.7	146.6	170.6	11,642
Non-forest	10.9	14.9	19.7	26.9	7.9	10.8	13,670
All	50.6	128.0	61.9	156.6	71.7	181.5	25,312
	Mg ha^{-1}	Tg	Mg ha^{-1}	Tg	Mg ha^{-1}	Tg	km^2

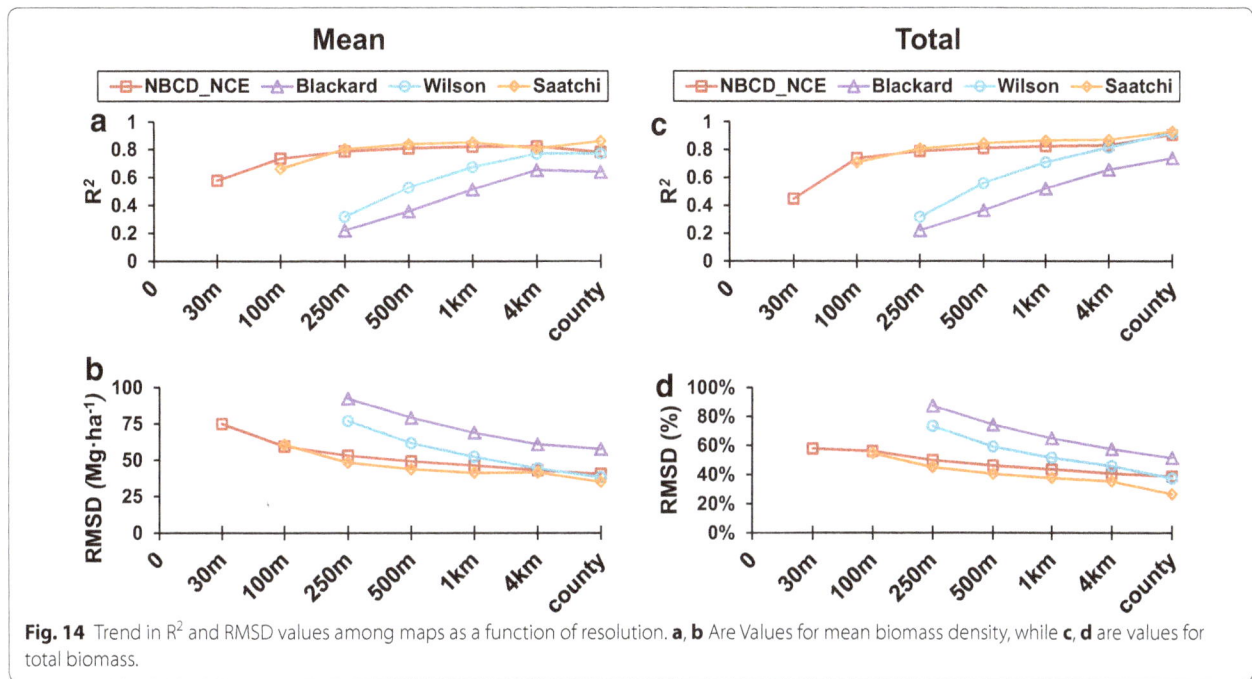

Fig. 14 Trend in R^2 and RMSD values among maps as a function of resolution. **a**, **b** Are Values for mean biomass density, while **c**, **d** are values for total biomass.

greatly improve the agreement among maps, particularly in the 0–250 Mg ha^{-1} range (Fig. 8).

Another important difference between the continental scale maps and CMS_RF map was in high biomass forests. Continental scale maps had few predictions greater than 250 Mg ha^{-1}. This was because they were developed using passive multispectral/radar data that were not sensitive enough to canopy structure in medium to high biomass ranges [31]. We expected some improvement in the enhanced Saatchi map as it included space-borne lidar data but did not observe any. This was probably because lidar data were used for model calibration rather than prediction. Biomass predictions were therefore influenced more by the 2D remote sensing data than the lidar inputs. One way of improving estimates beyond the 250 Mg ha^{-1} range is by including lidar measurements with higher resolution such as those from GEDI (expected launch in 2018) [32] or ICESAT-2 (expected

launch in 2017) as predictor variables individually or through fusion with other datasets.

Some discrepancy in total biomass values between the different maps can be attributed to the differences in allometric models applied to the field dataset used to develop the maps. For example, re-calculating tree biomass from field data in Maryland with Jenkins equations [17] instead of the Component Ratio Method (CRM) [33] that is currently used by FIA, increases the total biomass by 11 %. Therefore, it is possible that the difference between the CMS_RF map and the maps derived from field data that applied the CRM (i.e., Blackard, Wilson, and Saatchi), could be somewhat lower than calculated in Table 4.

In general, the CMS_RF map had higher values than all the other maps because the discrete-return airborne lidar were effective in predicting biomass beyond 250 Mg ha^{-1} and the high-resolution tree cover mask ensured estimates for virtually all trees in the State. We noted some overestimation, particularly in the low biomass ranges, when we compared the CMS_RF map to FIA county totals (Fig. 12). This could be attributed to the Random Forests model that predicted higher biomass in areas with very low canopy height and cover or limitations with FIA estimates. More research is needed to understand these differences but the cross-validation of the CMS_RF map with FIA data at plot level was strong (Fig. 7) indicating the overall robustness of the CMS_RF map and its suitability as a reference for map comparisons.

Interestingly, all maps showed better agreement at the county scale despite large discrepancies at finer scales. One reason for this could be that all maps captured some of the variability in biomass as a function of canopy structure, land cover type and physiographic gradient, irrespective of inputs and modeling approaches. This was evident from similarities in spatial patterns and FNI values. All maps (except the Blackard map) had greater than 45 % similarity with the CMS_RF map at the pixel level, contributing to the agreement. Secondly, all continental scale maps were developed using statistical regressions with FIA data which is meant to provide an unbiased state-wide estimate. A regression model, by nature, estimates the mean of the predictor data. This is also applicable to the Random Forest model used to generate the CMS_RF map. Since all maps were more or less accurate in predicting mean biomass, there was a fundamental agreement despite the variability at fine scales. Outliers reduced on spatial aggregation and the agreement between maps increased, as observed from the decreasing RMSD and increasing goodness of fit (Fig. 14).

Continental scale maps showed increasing agreement at coarse scales and converged between 4 km and 10 km (county-scale). This is similar to trends observed in Mitchard et al. [4] and Avitabile et al. [26]. However, the agreement is misleading, as these maps do not converge with the CMS_RF map at any scale considered in this study (Fig. 14). The mismatch was because of a relatively constant negative bias in all the continental scale estimates when compared to the CMS_RF and was as high as 30 % (Figs. 10, 11) at county scale. The negative bias was primarily because of the underestimation in high biomass ranges and lack of predictions in non-forested areas. Since this difference does not diminish with coarsening resolution, we argue that local-scale discrepancies may affect carbon reporting at all levels and should not be ignored.

Conclusion

A detailed validation with high-resolution estimates can be valuable in identifying discrepancies and making continental-scale maps truly applicable to carbon accounting applications. We demonstrated one example over temperate forested and non-forested landscapes in Maryland. More studies across different biomes are required to confirm these findings. Armed with a comprehensive understanding from such validations, we can improve and integrate multi-source datasets to inform carbon monitoring efforts.

Additional files

Additional file 1: Figure S1. Scatter plots of biomass at 250 m resolution from old Saatchi (v1) and new Saatchi (v2) maps versus CMS_RF product.

Additional file 2: Figure S2. Histograms showing the distribution of forest biomass from old Saatchi (v1), new Saatchi (v2), and CMS_RF maps.

Additional file 3: Table S1. Total Maryland biomass, in Tg, for CMS_RF at 30 m resolution and FIA calculations (2008–2012 cycle) by forest and non-forest classification. FIA forest and non-forest definitions do not follow NLCD landcover classes, rather by on the ground plot conditions. Non-forest biomass in the FIA dataset was calculated by methods described in [28].

Authors' contributions
The study was designed by AS, WH and RD. WH performed the analysis and produced the figures using data layers and tables provided by KJ, LD, and HT. WH and AS wrote the manuscript and were co-first authors. All co-authors provided valuable feedback, interpretation of results and manuscript revisions. All authors read and approved the final manuscript.

Author details
[1] Department of Geographical Sciences, University of Maryland, College Park, USA. [2] USDA Forest Service, Northern Research Station, Newtown Square, PA, USA. [3] Rubenstein School of the Environment and Natural Resources, University of Vermont, Burlington, USA.

Acknowledgements
This work was funded by the NASA's CMS project (NNX10AT74G—PI, Ralph Dubayah). We are thankful to Sassan Saatchi for providing us the old and new versions of the CMS continental scale maps. We are thankful to Katelyn Dolan for her valuable inputs and comments that made the manuscript stronger. The datasets used in this study are available on line from: (a) CMS_RF [18] is available at http://carbonmonitoring.umd.edu/data.html, (b) NBCD [13] is available at http://www.whrc.org/mapping/nbcd/, (c) Blackard et al. [14] is available at http://webmap.ornl.gov/biomass/biomass.html, (d) Wilson et al. [15] at http://www.fs.usda.gov/rds/archive/Product/RDS-2013-0004, (e) Saatchi et al. [16] is available at http://carbon.jpl.nasa.gov/data/dataMain.cfm.

Compliance with ethical guidelines

Competing interests
The authors declare that they have no competing interests.

References
1. Houghton R, Lawrence K, Hackler J, Brown S (2001) The spatial distribution of forest biomass in the Brazilian Amazon: a comparison of estimates. Glob Chang Biol 7(7):731–746
2. Lu D (2006) The potential and challenge of remote sensing-based biomass estimation. Int J Remote Sens 27(7):1297–1328
3. Goetz S, Dubayah R (2011) Advances in remote sensing technology and implications for measuring and monitoring forest carbon stocks and change. Carbon Manag 2(3):231–244

4.   Mitchard E, Saatchi S, Baccini A, Asner G, Goetz S, Harris N et al (2013) Uncertainty in the spatial distribution of tropical forest biomass: a comparison of pan-tropical maps. Carbon Balance Manag 8(1):10

5.   Langner A, Achard F, Grassi G (2014) Can recent pan-tropical biomass maps be used to derive alternative Tier 1 values for reporting REDD+ activities under UNFCCC? Environ Res Lett 9(12). doi:10.1088/1748-9326/9/12/124008

6.   Hurtt G, Wickland D, Jucks K, Bowman K, Brown M, Duren R et al (2014) NASA Carbon Monitoring System: prototype monitoring, reporting, and verification, 1–37

7.   Zhang X, Kondragunta S (2006) Estimating forest biomass in the USA using generalized allometric models and MODIS land products. Geophys Res Lett 33(9). doi:10.1029/2006gl025879

8.   Cartus O, Santoro M, Kellndorfer J (2012) Mapping forest aboveground biomass in the Northeastern United States with ALOS PALSAR dual-polarization L-band. Remote Sens Environ 124:466–478. doi:10.1016/j.rse.2012.05.029

9.   Johnson K, Birdsey R, Finley A, Swantaran A, Dubayah R, Wayson C et al (2014) Integrating forest inventory and analysis data into a LIDAR-based carbon monitoring system. Carbon Balance Manag 9(1):3

10. Jenkins J, Riemann R (2003) What does nonforest land contribute to the global C balance? In: Proceedings of the 3rd annual forest inventory and analysis symposium. U.S. Department of Agriculture, Forest Service, North Central Station

11. Dubayah R, Swatantran A, Johnson K, Hurtt G, Zhao M, Finley A (2014) High resolution carbon estimation using remote sensing and ecosystem modeling in NASA's carbon modeling system. ForestSAT2014 open conference system

12. Swatantran A, Huang W, Duncanson L, Johnson K, Dunne JON, Hurtt G et al. High-resolution aboveground biomass mapping for carbon monitoring in Maryland (manuscript in preparation)

13. Kellndorfer J, Walker W, LaPoint E, Bishop J, Cormier T, Fiske G et al (2012) NACP aboveground biomass and carbon baseline data, V. 2 (NBCD 2000), U.S.A., 2000. Data set. Oak Ridge, Tennessee, U.S.A.: ORNL DAAC

14. Blackard J, Finco M, Helmer E, Holden G, Hoppus M, Jacobs D et al (2008) Mapping U.S. forest biomass using nationwide forest inventory data and moderate resolution information. Remote Sens Environ 112(4):1658–1677. doi:10.1016/j.rse.2007.08.021

15. Wilson BT, Woodall CW, Griffith DM (2013) Imputing forest carbon stock estimates from inventory plots to a nationally continuous coverage. Carbon Balance Manag 8(1):1. doi:10.1186/1750-0680-8-1

16. Saatchi S, Yifan Y, Fore A, Nuemann M, Chapman B, Nguyen et al (2005) CMS biomass pilot project: US Forest Biomass Maps

17. Jenkins JC, Chojnacky DC, Heath LS, Birdsey RA (2003) National-scale biomass estimators for United States tree species. For Sci 49(1):12–35

18. Dubayah R (2012) County-scale carbon estimation in NASA's carbon monitoring system. Biomass Carbon Storage

19. O'Neil-Dunne JPM, MacFaden SW, Royar AR, Pelletier KC (2013) An object-based system for LiDAR data fusion and feature extraction. Geocarto Int 28(3):227–242. doi:10.1080/10106049.2012.689015

20. O'Neil-Dunne J, MacFaden S, Royar A, Reis M, Dubayah R, Swatantran A (2014) An object-based approach to statewide land cover mapping. Proceedings of ASPRS 2014 annual conference, 23–28 March 2014, Louisville, KY, USA

21. Breiman L (2001) Random forests. Mach Learn 45(1):5–32

22. Cutler DR, Edwards TC Jr, Beard KH, Cutler A, Hess KT, Gibson J et al (2007) Random forests for classification in ecology. Ecology 88(11):2783–2792. doi:10.2307/27651436

23. Kellndorfer JM, Walker WS, LaPoint E, Kirsch K, Bishop J, Fiske G (2010) Statistical fusion of lidar, InSAR, and optical remote sensing data for forest stand height characterization: a regional-scale method based on LVIS, SRTM, Landsat ETM+, and ancillary data sets. J Geophys Res 115:G00E8. doi:10.1029/2009jg000997

24. Wilson BT, Lister AJ, Riemann RI (2012) A nearest-neighbor imputation approach to mapping tree species over large areas using forest inventory plots and moderate resolution raster data. For Ecol Manag 271:182–198. doi:10.1016/j.foreco.2012.02.002

25. Saatchi SS, Harris NL, Brown S, Lefsky M, Mitchard ETA, Salas W et al (2011) Benchmark map of forest carbon stocks in tropical regions across three continents. Proc Natl Acad Sci 108(24):9899–9904. doi:10.1073/pnas.1019576108

26. Avitabile V, Herold M, Henry M, Schmullius C (2011) Mapping biomass with remote sensing: a comparison of methods for the case study of Uganda. Carbon Balance Manag 6(1):7

27. Miles P (2014) Forest Inventory EVALIDator web-application version 1.5.1.06. In: http://apps.fs.fed.us/Evalidator/evalidator.jsp. U.S. Department of Agriculture, Forest Service, Northern Research Station, St. Paul, MN

28. Johnson K, Birdsey RA, Cole J, Swantaran A, O'Neil-Dunne J, Dubayah R et al. Integrating LiDAR and forest inventories to fill the trees outside forests data gap. Environ Monit Assess (in review)

29. Adam E, Mutanga O, Rugege D (2010) Multispectral and hyperspectral remote sensing for identification and mapping of wetland vegetation: a review. Wetl Ecol Manag 18(3):281–296

30. Sexton JO, Song X-P, Feng M, Noojipady P, Anand A, Huang C et al (2013) Global, 30-m resolution continuous fields of tree cover: landsat-based rescaling of MODIS vegetation continuous fields with lidar-based estimates of error. Int J Digit Earth 6(5):427–448

31. Lu D, Chen Q, Wang G, Liu L, Li G, Moran E (2014) A survey of remote sensing-based aboveground biomass estimation methods in forest ecosystems. Int J Digit Earth 1–64. doi:10.1080/17538947.2014.990526

32. Dubayah R, Goetz S, Blair JB, Luthcke S, Healey S, Hansen M et al (2014) The Global Ecosystem Dynamics Investigation (GEDI) Lidar. ForestSAT2014 open conference system

33. Heath L, Hansen M, Smith J, Miles P, Smith W (2008) Investigation into calculating tree biomass and carbon in the FIADB using a biomass expansion factor approach. In: McWilliams W, Moisen G, Czaplewski R (eds) Proceedings of Forest Inventory and Analysis Symposium 2008. USDA Forest Service, Rocky Mountain Research Station, Fort Collins, Colorado, USA; Park City, Utah

5

Carbon accretion in unthinned and thinned young-growth forest stands of the Alaskan perhumid coastal temperate rainforest

David V. D'Amore[1]*[ID], Kiva L. Oken[2], Paul A. Herendeen[3], E. Ashley Steel[4] and Paul E. Hennon[1]

Abstract

Background: Accounting for carbon gains and losses in young-growth forests is a key part of carbon assessments. A common silvicultural practice in young forests is thinning to increase the growth rate of residual trees. However, the effect of thinning on total stand carbon stock in these stands is uncertain. In this study we used data from 284 long-term growth and yield plots to quantify the carbon stock in unthinned and thinned young growth conifer stands in the Alaskan coastal temperate rainforest. We estimated carbon stocks and carbon accretion rates for three thinning treatments (basal area removal of 47, 60, and 73 %) and a no-thin treatment across a range of productivity classes and ages. We also accounted for the carbon content in dead trees to quantify the influence of both thinning and natural mortality in unthinned stands.

Results: The total tree carbon stock in naturally-regenerating unthinned young-growth forests estimated as the asymptote of the accretion curve was 484 ($\pm$26) Mg C ha^{-1} for live and dead trees and 398 ($\pm$20) Mg C ha^{-1} for live trees only. The total tree carbon stock was reduced by 16, 26, and 39 % at stand age 40 y across the increasing range of basal area removal. Modeled linear carbon accretion rates of stands 40 years after treatment were not markedly different with increasing intensity of basal area removal from reference stand values of 4.45 Mg C ha^{-1} year^{-1} to treatment stand values of 5.01, 4.83, and 4.68 Mg C ha^{-1} year^{-1} respectively. However, the carbon stock reduction in thinned stands compared to the stock of carbon in the unthinned plots was maintained over the entire 100 year period of observation.

Conclusions: Thinning treatments in regenerating forest stands reduce forest carbon stocks, while carbon accretion rates recovered and were similar to unthinned stands. However, that the reduction of carbon stocks in thinned stands persisted for a century indicate that the unthinned treatment option is the optimal choice for short-term carbon sequestration. Other ecologically beneficial results of thinning may override the loss of carbon due to treatment. Our model estimates can be used to calculate regional carbon losses, alleviating uncertainty in calculating the carbon cost of the treatments.

Keywords: Carbon sequestration, Young-growth, Stand management, Ecosystem productivity, Natural resource management

Background

Forests play a key role in the global carbon cycle, containing an estimated 861 Pg C and providing a sink of 1.1 Pg C year^{-1} [1]. Forests are critical sinks for atmospheric greenhouse gases [2], and carbon fluxes occur across many carbon pools in forests, including live biomass, soils, and woody debris [3, 4]. The terrestrial carbon stock is generally stable over time scales of decades and can only slowly alter the total terrestrial carbon balance through gains or losses [4]. Disturbances that alter forest stands can provide dramatic departures from this characteristic pattern. An example is removal of carbon due to clearcut harvesting of forests, leading

---
*Correspondence: ddamore@fs.fed.us
[1] U.S. Department of Agriculture, Forest Service, Pacific Northwest Research Station, Juneau Forestry Sciences Laboratory, 11175 Auke Lake Way, Juneau, AK 99801, USA
Full list of author information is available at the end of the article

to a large loss of terrestrial carbon. The increase in biomass, or carbon accretion, as stands regenerate and grow after harvest is unknown in many forests. Thinning is a common silvicultural practice for increasing growth of individual trees and maintaining or increasing wildlife habitat. However, the influence of thinning on the carbon balance in young forests is uncertain in southeast Alaska. Carbon fluxes need to be evaluated across a range of management options to understand and estimate the short and long-term impacts of silvicultural treatments on carbon pools.

Estimates of carbon flux in young-growth stands are needed to address land management planning goals and regional, national [5] and international carbon accounting protocols [6]. Mandates to understand the potential for forests to mitigate increasing concentrations of atmospheric CO_2 require accurate accounting of forest carbon fluxes. The USDA Forest Service, for example, has prioritized understanding carbon dynamics in forests

as part of an overall strategy to protect the long-term health of forests [7]. Necessary information about carbon cycling is particularly lacking in the perhumid coastal temperate rainforests (PCTR) of the northeast pacific coastal margin [8] (Fig. 1).

Widespread commercial forest harvest has occurred across southeast Alaska for over 50 years. However, there is no estimate of the potential carbon sequestration across the ~452,000 ha [9] of young-growth forests in the region. Natural regeneration in PCTR forests is generally vigorous and leads to rapid and nearly complete occupation of space by conifer seedlings and saplings [10]. Densely-stocked stands can produce wood products similar to thinned stands [11], but the loss of light and density of overstory trees degrades the wildlife habitat [12, 13]. A common management intervention to alleviate the high stand density is thinning [14]. Felling of a portion of the stand basal area across a specific or variable [15] spacing can be applied to achieve maximum

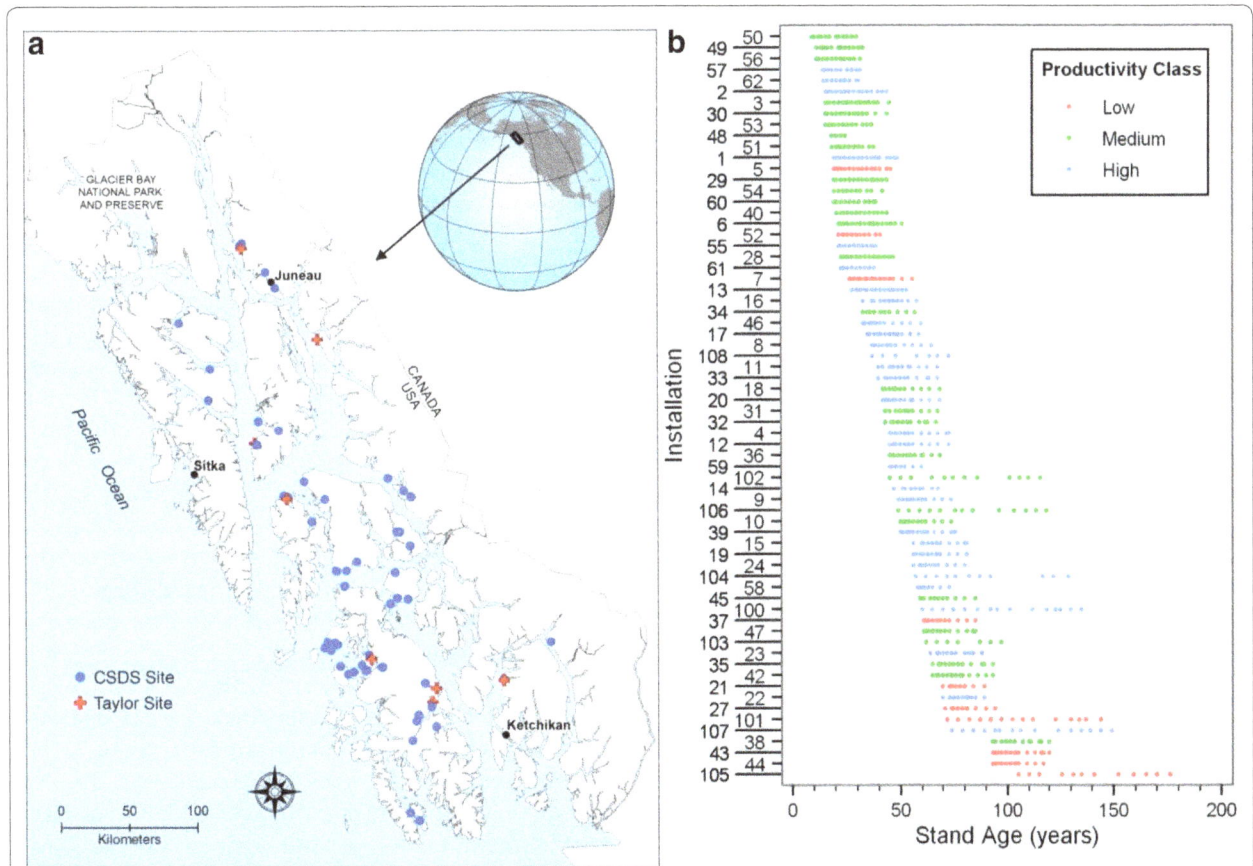

Fig. 1 Locations of the 68 Farr and 12 Taylor installations in southeast Alaska (a). Each CSDS ("Farr") installations consists of four plots: a control plot, a low-intensity thinned plot, a medium-intensity thinning plot, and a heavily-thinned plot. The Taylor installations consist of unthinned plots only and are generally in older stands. For a full description of the plots and thinning treatments see [19]. Data from the CSDS study arranged by age of stand at time of plot establishment (b). *Numbers* on Y axis refer to installation number with Farr plots <100 and Taylor plots ≥100. Productivity classes are the tertiles of the observed range for these sites as reported in [19]. Each *symbol* in an installation represents a measurement

individual tree growth. However, thinning also alters the carbon accretion trajectory of the stand [4]. When left on site, the carbon content of thinned trees, and any trees that die naturally, can be accounted for by estimating decomposition rates. The impact of stand thinning and subsequent loss of biomass via decomposition are key components in calculationg a carbon sequestration rate for use in land management planning.

Quantifying the effects of young-growth forest management on carbon storage is challenging. Allometric equations linked to direct tree measurements can be used to estimate aboveground biomass production [16–18] across stand age, and this can be converted to carbon accretion. Estimation of the long-term differences between forests with varying management treatments requires remeasurement of the same plots over decades. Long-term plots provide an excellent source of information on biomass accretion over time where plots have been maintained and re-measured.

Experimental plots maintaned by the USDA Forest Service Pacific Northwest Research Station [19] offer an opportunity to estimate carbon change over time with varying levels of thinning. This dataset includes 284 plots across 68 installation sites, remeasured over several decades and spanning stand ages up to 161 years. The temporal and geographic breadth of these experimental plots provides an excellent foundation for investigating carbon standing stocks and carbon accretion rates in young-growth forests of the PCTR. In addition, the plot system allows analysis of the effects of forest thinning on carbon storage through the combination of allometric equations and repeated tree measurements over decades. We designed this study to address the critical need for an improved understanding of carbon storage in young-growth forest of the PCTR and to quantify the effects of thinning on carbon gain or loss. We hypothesized that while thinning may increase carbon accretion in individual trees, across whole stands thinning will have a neutral to negative impact on the sequestration of carbon, depending on the intensity of thinning.

Methods overview

We utilized data from two long-term silvicultural datasets young-growth forests of southeast Alaska to estimate total tree carbon stock and accretion rate. One set of plots was started in the 1920's and were not thinned ("Taylor plots", 12 of 284). The other plot system included unthinned controls and thinning treatments applied at three intensities in a randomized block design ("Farr plots", 272 of 284). Plot measurements included both live and dead trees, so estimates for both pools were calculated to account for the loss of dead tree carbon

decomposing over time in both unthinned and treated forest stands. A new allometric model for small diameter trees was developed to fill a needed information gap in determination of carbon in small trees.

Results

Live and dead tree carbon pools in naturally-regenerating young-growth stands

Live-tree carbon increased in unthinned young-growth stands across the stand age gradient and reached an asymptote of 398 (±20) Mg C ha^{-1} based on a best fit, non-linear mixed effect model (NLME) (Fig. 2a). The estimated asymptotic maximum carbon stock in the stands increased to 484 (±26) Mg C ha^{-1} with the inclusion of dead-tree carbon (Table 1) Dead trees in unthinned plots typically represent suppression mortality as tree density decreases through time. However, these mean carbon stock estimates for the measured plots have a great deal of uncertainty. A prediction interval was derived by considering observed variability within- and among-plots, in addition to the parameter uncertainty around the asymptote described above. The 90 % prediction intervals for the asymptotic carbon stock ranged from 145 to 653 Mg C ha^{-1} for the live-tree carbon model and 161–808 Mg C ha^{-1} for the model including both live- and dead-tree carbon.

We plotted carbon accretion as the change in the carbon pool over time in plots with only live tree carbon and calculated a peak at age 34.7 years (±0.5, bootstrap SE; Fig. 2b). The carbon accretion peaked at 39.3 years (±0.5, bootstrap SE; (Fig. 2c) for the model with both live and dead tree carbon. These carbon accretion rates varied dramatically across the chronosequence of measurements in the sampled stands (Fig. 2b, c). The high variability makes it difficult to estimate quantities with any reasonable level of precision directly from accretion data. While carbon accretion was more variable than carbon stock estimates, carbon accretion can also be estimated as the derivative of carbon stocks over time. The general shape of the data cloud suggests that accretion rates peak at 39 years and then decreases, tapering off at about 100 years. The shape of the accretion curves (Fig. 2d, e) derived directly from the fitted model for the total carbon stock (Fig. 2a) indicates that accretion peaks in young stands between 35 and 40 years and then tapers off as the stands age. The estimated weighted average carbon accretion rate based on the fitted model to total carbon [45] was 3.53 (±0.17) Mg C ha^{-1} year^{-1} for the live-tree carbon model and 3.81 (±0.20) Mg C ha^{-1} year^{-1} for the model that included live- and dead-tree carbon over the 150 years age span of measured trees.

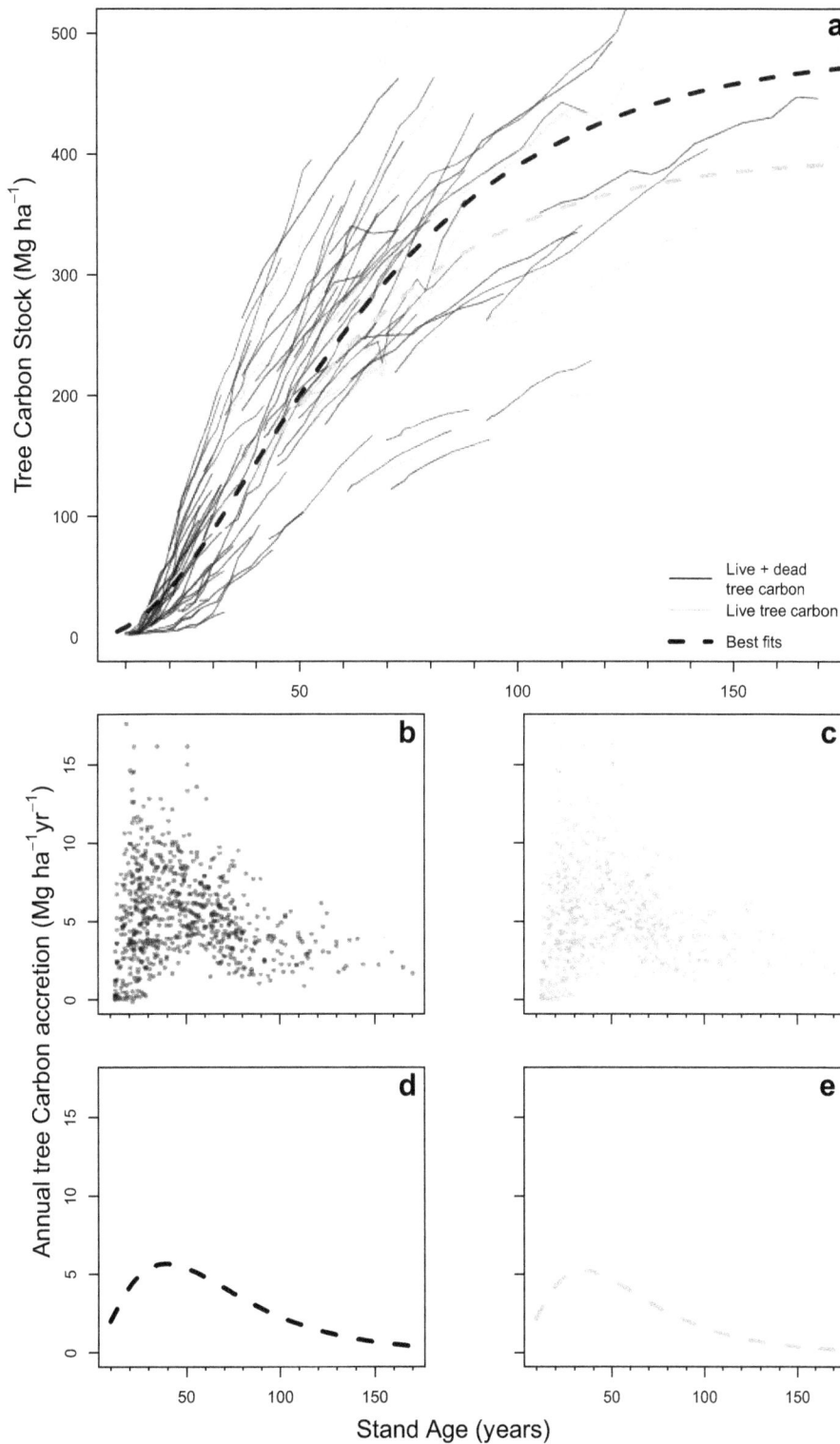

Fig. 2 Carbon stock and carbon accretion in naturally-regenerating plots (Farr study control plots and Taylor plots). **a** Measured and modeled carbon stock through time. *Solid lines* describe carbon accretion measurements within individual plots across a range of ages. *Dashed lines* are NMLE best fit models for all tree carbon (live and dead trees) and for live tree carbon. Note that the plots with low carbon stock values are all located in the same sites which occurred on the lowest productivity areas that were sampled. Observed carbon accretion rates across stand age for **b** for all carbon (live and dead trees) and for **c** for live trees only. Implied carbon accretion (derivative of the NMLE model) as a function of stand age for **d** for all carbon (live and dead trees) and for **e** for live trees only

Table 1 Parameter estimates (±SE) for best fit of carbon accretion in unthinned control stands for live tree only and for live + dead tree components of carbon using NMLE model

Model components	B_0	B_1	B_2
Control (live + dead trees)	484.58 (25.90)	0.026 (0.0008)	0.636 (0.011)
Control (live trees only)	398.24 (20.03)	0.029 (0.0009)	0.634 (0.0119)

B_0 is an estimate of the asymptote based on Eq. 3 where DBH is replaced by age and Mg C is calculated rather than height. B_1 and B_2 describe growth rate and the inflection point of the modeled relationship, respectively

Influence of thinning on carbon accretion in young-growth stands

There was a systematic decrease in the total stand carbon correlated with increasing intensity of thinning (Fig. 3). The portion of the carbon stock data in untreated young-growth stands that is nearly linear (20–100 years) was used as a basis for comparison between treated stands. The estimated average carbon pool in the unthinned control plots (Farr plots only, see methods) was greater than the estimated average carbon pool in any of the three thinning treatments at 40 years (Table 2); estimated

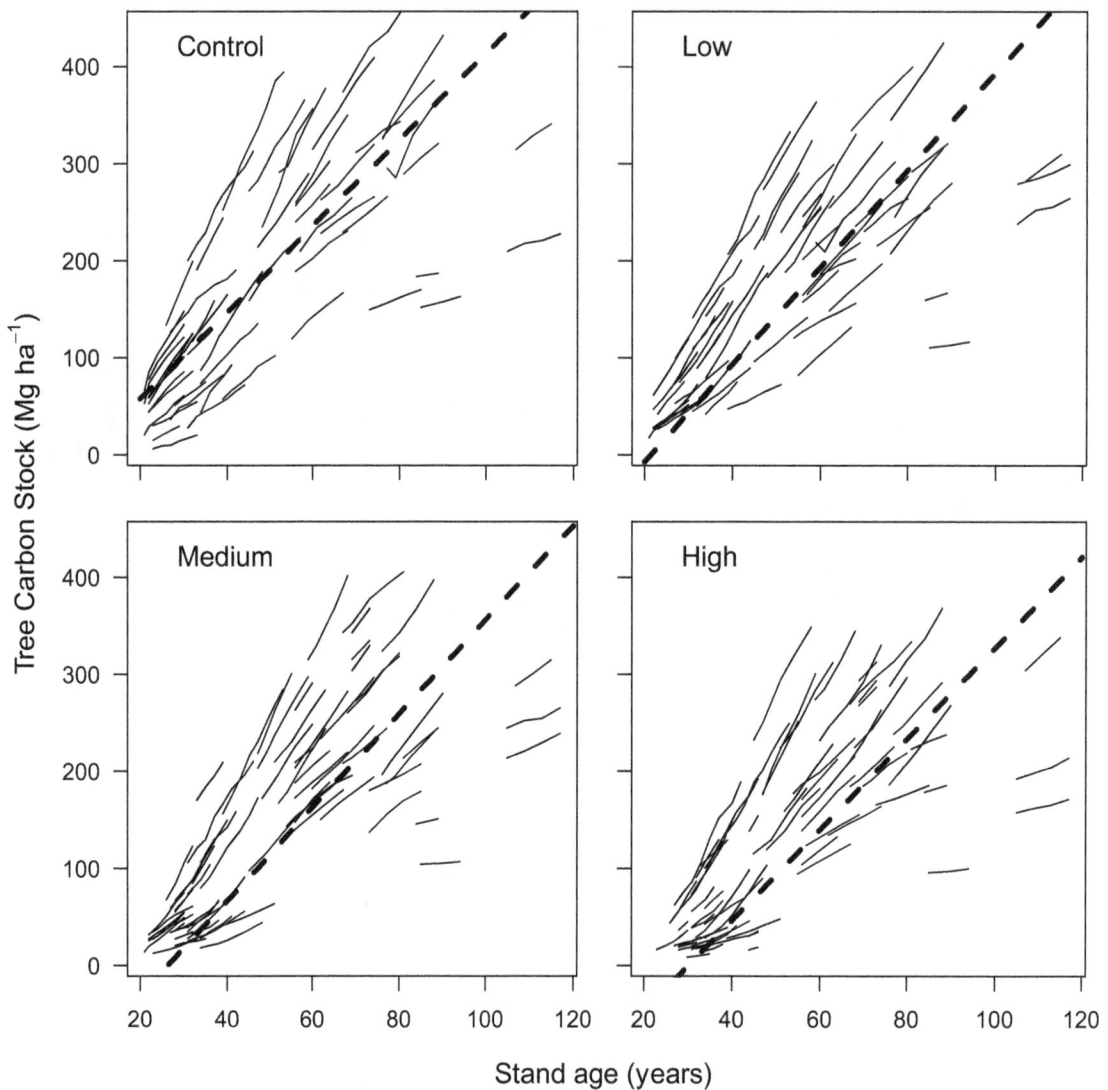

Fig. 3 Modeled live tree carbon accretion 10-years post-thinning until end of study across treatment. *Dashed lines* are the estimated slopes for each line and describe accretion rates over time; however we note that much of the difference between treatments is accounted for in the intercept which describes live carbon stock 10 years post-thinning

Table 2 Estimates of accretion rate and mean carbon density 40 y after thinning based on live tree carbon only

Treatment	Carbon accretion rate Mg C ha^{-1} y^{-1}	Among plot standard error of accretion rate	Residual standard error	Carbon density at 40 y Mg C ha^{-1}
Control	4.45 (0.31)	2.16	1.09	146.9 (7.4)
Low	5.01 (0.25)	1.87	1.28	93.4 (7.0)
Medium	4.83 (0.27)	2.20	1.04	66.3 (6.8)
Heavy	4.68 (0.31)	2.49	1.10	46.0 (6.2)

Low, medium, and heavy treatments refer to thinning intensity. Values in parentheses are parameter standard errors

average carbon pools at a given age consistently decreased with thinning intensity of treatments from low to high (Fig. 3; Table 2). The slope of the linear model fit to the data describes the stand-scale accretion rate. This accretion rate systematically decreased with thinning when decomposition of cut trees and any trees that died naturally is included and the total carbon stock was reduced by 16, 26, and 39 % across the low to high intensity thinning treatments at 40 years (Tables 2, 3). However, no major pattern between accretion rate and thinning intensity was noted with only live trees (Table 2; Fig. 4). We note that several plots, both control and treatment, displayed particularly low accretion rates and these plots were generally all located on one set of sites (Figs. 3, 4).

The residual model error in the linear models fit was similar across treatments for live trees using all plots (Tables 2, 3). This was also the case in models for live trees, cut trees, and natural mortality using plots for which cut tree data were available (221 of 284). This residual model error describes variability in carbon stocks within a plot over time after accounting for the effects of stand age and treatment. This standard error among plots for the accretion rate increased somewhat predictably across the three treatments suggesting that at more intensive levels of management it might be more difficult to predict accretion rates for an individual plot. Control plots were intermediate in their across-plot variability. We also note that residuals for both the live-tree and live-tree plus cut and natural dead tree models showed no trends over stand age, indicating that the linear model accurately described the underlying effect

of stand age on carbon stocks, but residuals did show a somewhat increasing trend over chronological time indicating a potential increase in variability of carbon stocks in recent years.

Simulation of stand carbon dynamics immediately after thinning

We simulated a hypothetical carbon accretion scenario under different thinning intensities, all of which occur when stands reach 20 years of age, based on our fitted statistical models (Fig. 5). The simulated carbon stock at the plot scale accumulates at a rapidly accelerating pace in all plots until the stands are subjected to a simulated thinning at age 20. This thinning leads to the rapid drop in the carbon stock of thinned stands, as we only accounted for the carbon in the remaining live trees in the stands for the simulation. Stands in all four treatments begin accumulating carbon again after the thinning treatment is applied according to the linear models. We expect that increased growth rates of individual trees lead to a more rapid rate of carbon accretion after thinning on a per-tree basis. Note, however, that while individual trees may accrete carbon at a more rapid rate after thinning due to increased growth rates, there are many fewer trees accreting carbon in a thinned plot. Overall, at the plot scale, there is an initial loss of carbon in the thinned stands and a similar accretion rate to the control stands (Table 2). The simulated carbon stock in all thinning treatments remains lower than that of unthinned plots up to 100 years (Fig. 5).

Table 3 Estimates of accretion rate and mean carbon density 40 y after thinning based on live and dead tree carbon

Treatment	Carbon accretion rate (Mg C ha^{-1} y^{-1})	Among plot standard error of accretion rate	Residual standard error	Carbon density at 40 years (Mg C ha^{-1})
Control	5.27 (0.32)	2.26	0.93	144.3 (8.1)
Low	5.16 (0.28)	1.97	1.09	120.7 (6.9)
Medium	5.00 (0.31)	2.27	0.94	107.0 (6.8)
Heavy	4.78 (0.32)	2.39	1.15	88.1 (5.9)

These estimates are based on the 164 of 215 study plots for which cut tree data were recorded. Low, medium, and heavy treatments refer to thinning intensity. Values in parentheses are parameter standard errors

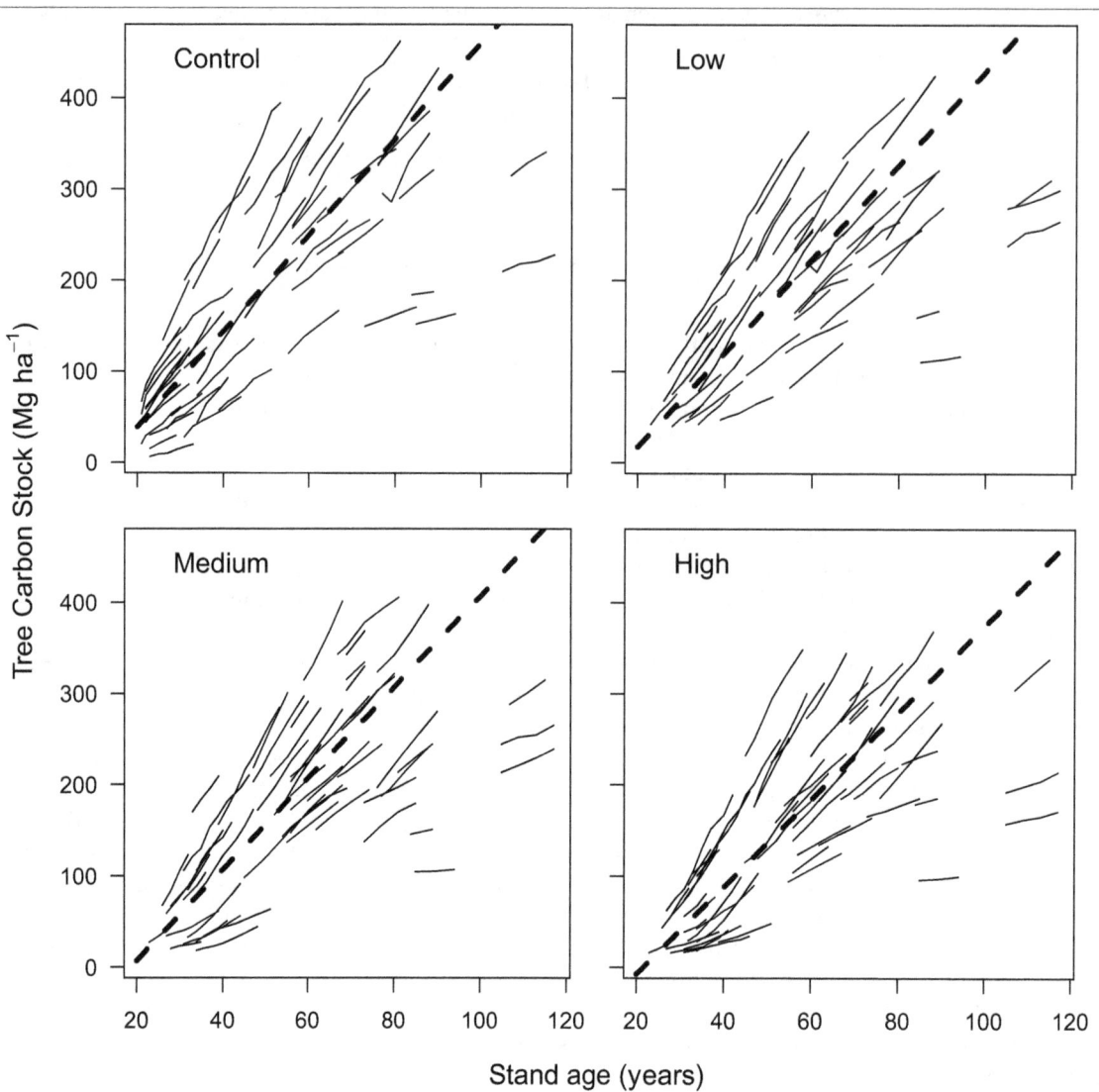

Fig. 4 Modeled live and dead tree carbon accretion 10-years post-thinning until end of study across treatment types for the 164 of 215 study plots for which cut tree data were recorded. Dead tree carbon includes thinned trees and those that died naturally. *Dashed lines* are the estimated slopes for each line and describe accretion rates over time

Discussion
Carbon balance in unthinned forest stands
The rate and location of terrestrial carbon sinks is critical to understanding the global carbon balance. Young-growth forests sequester carbon in biomass, but at widely varying rates and over different timeframes. The calculation of total carbon stock and estimated accretion rates across the age gradient of the naturally regenerating young-growth forests of southeast Alaska fills a critical information gap for this region. The loss of live carbon after thinning in naturally-regenerating stands must be considered in calculating carbon sequestration estimates for young-growth forests. Thinning treatments are

applied to achieve many ecosystem services in addition to carbon sequestration goals; therefore, our quantitative estimates of the loss of carbon after thinning enable evaluation of the carbon cost of a range of management actions for young-growth stand improvement.

Model calibration is essential for obtaining accurate carbon balance estimates across large regions [20]. Forest carbon models need to consider the entire range of stand types and ages to accurately portray the balance of carbon stock across the landscape [21]. Mature forest stands (>200 years) can accumulate carbon at an estimated 2.4 Mg C ha^{-1} year^{-1} [22]. The carbon stock in young-growth stands is particularly critical in these estimates as these stands are generally the

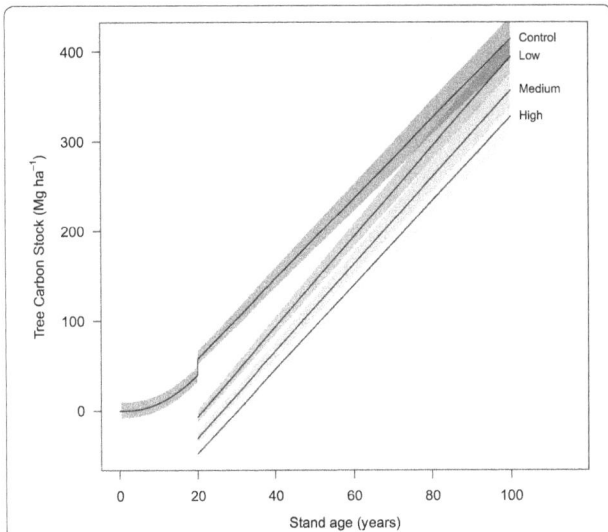

Fig. 5 Scenario of live tree carbon trajectory for first 100 years of growth. Carbon accretion begins slowly, and then accelerates. This concave up curve comes from the non-linear model fit only to the control plots. When the stand reaches 20 years, we assume the plot was treated by thinning, and compare the carbon trajectories predicted by the four different treatment models. At 20 years, the treated plots immediately lose a large quantity of live carbon due to removal of woody material that was felled during the thinning from the live carbon pool. Ribbons cover the 50 % prediction interval, but do not include random effect variance among plots. The discontinuity in the control plots occurs because two separate models were used in this hypothetical scenario; the jump is due to the structural uncertainty in the models

most active zones of carbon change on the landscape due to rapid biomass accumulation and carbon storage in trees [23]. The estimated mean accretion rate of 3.53 Mg C ha^{-1} year^{-1} over 150 year in our study area confirms the strong net gain in carbon in young-growth stands in the Alaskan PCTR. This rate is higher than the 40 years mean of 2.71 Mg C ha^{-1} year^{-1} estimated in young-growth stands in the PCTR of British Columbia [24]. Frustratingly, the uncertainty in determining the response of an individual stand is high, which limits the usefulness of model predictions for site specific estimates of carbon stock, often needed to evaluate specific management scenarios. Our models are most appropriately applied across an entire population of stands for regional and national carbon assessments. Site-specific descriptors (e.g., site productivity) that might help stratify the data and provide more accurate predictions of carbon pools will need to be applied in order to help refine our predictions of carbon accretion rates in particular locations.

Carbon balance in thinned young-growth stands

Maximizing the carbon stored in forests is a key goal of climate change mitigation programs [25]. The majority of the young-growth forest in the Alaskan PCTR result

from harvest that occurred from 1960 to 1990 [9]. Thinning young-growth stands in the PCTR is a common management strategy to improve stand structure and wood production [14] and to improve wildlife forage production [13, 26]. Renewable energy recommendations for the Alaskan PCTR highlight the potential for wood energy projects using this stock of young-growth forest [27]. However, the usual management scenario for these young-growth stands is thinning at 15–20 years [13] and nearly half of the 25–50 year old stands have been pre-commercially thinned [9, 14]. Therefore, recognizing the tradeoff between thinning for stand improvement, biomass energy, and carbon sequestration in young-growth forest stands is important for making land management decisions. A key finding in our study is that thinning persistently reduces the carbon stock in young growth stands. The rate of carbon accretion in thinned stands is higher than control plots after the initial carbon loss; but, the gap created by the initial carbon loss is maintained and the total stock of carbon in thinned stands does not equal the stock of carbon in the control plots over the entire 100 year period of observation. This is consistent with the observation that the reduction in total stand carbon stock may not change the net ecosystem exchange between pre- and post-thinning [28]. The maintenance of tree growth would explain the similarity in the trajectory of carbon accretion among the treatments after the initial period of disturbance.

Reduced carbon stocks due to thinning have been recognized in other forests [4, 29, 30], but is not often included in forest carbon accounting or management actions due to the lack of adequate stand response data. Our quantification of the reduction in carbon stock across a range of thinning treatments allows estimates of the effects of thinning on regional carbon stocks. The systematic variation in the carbon stock related to thinning intensity may offer a mitigation measure for achieving benefits for wildlife, wood quality, or understory abundance and diversity in managed stands. The enhanced growth of understory plants after thinning represents a tradeoff of energy from trees to forest floor and a reduction in overstory carbon compared to unthinned stands. Benefits of thinning young growth need to be balanced with the desire to maximize carbon storage in forests. For example, the less intensive thinning treatments maintain more carbon, but still provide a benefit for other desired conditions in a stand. As demonstrated by our comparison, the unthinned option provides the greatest carbon accretion of all of the thinning prescription options.

Limitations of analysis and information gaps

The carbon values provided in our study will be critical for estimating the carbon stock in the pool of young-growth forest in southeast Alaska, but, there is still considerable uncertainty in the range of carbon accretion values among the stands in our analysis. Therefore, site specific projects will need an improved model that is able to better reflect local conditions to carbon flux values. Factors that influence the variability in forest productivity among the sites or the response to thinning were included as random effects, but not specifically as predictive variables. Possible interactions with temperature [28], geology [31], soil saturation [32], nutrients [33] or other site-specific factors may play a role in site productivity. This uncertainty might be addressed by obtaining further information on the site factors that may influence the productivity of the plots such as soil, hydrology, or climate variables.

Potential alternate trajectories in the carbon accretion of thinned stands may arise that lead to different conclusions related to unthinned stands. We applied the same allometric equations to both unthinned and thinned stands in our analysis. It is possible that tree growth forms differ by thinning treatment and so biomass allocation would change in thinned stands. We are not aware of any existing allometric models for thinned stands of the PCTR. Therefore, we rely on the literature from other regions to support our conclusions and highlight that thinning has been found to primarily impact the biomass of the bole [34] and crown [35] of the thinned trees. Thinned stands can shift biomass accumulation from branch to leaf, but measured changes in bole biomass have been demonstrated to be small [36] unless very heavily thinned [37]. These observations provide some confidence that the total biomass calculated by our approach will not substantially change, but may be redistributed within the tree after thinning.

The residual trees left after thinning grow at an accelerated rate, but these trees are generally left in a condition where they do not maximize the growing space for many years. Thinning goals such as increased individual tree growth and allocation of energy to the forest floor for plant diversity lead to lower overstory biomass accumulation in thinned plots. While the growth rate for individual trees is greater in these plots, the amount of biomass accumulation that would be required by the individual residual trees to match the loss in biomass of similar unthinned stands would be physiologically difficult to attain. The difference is illustrated in our evaluation of the stands at 40 years in Tables 2 and 3. There could be cases where a light thinning leaves a higher density than other thinning treatments, in which case, the thinned stand may accumulate biomass similar to unthinned stands due to the additional growth of the residual trees. However, this scenario is unlikely to be applied under most operational applications.

Conclusions

Knowledge of the stock and rate of carbon accretion greatly enhances the understanding of carbon dynamics in the coastal forests of Alaska. The loss of carbon due to thinning can be used in the evaluation of management scenarios that address young-growth stand improvement. Regional carbon budgets will also be improved with estimates that include the carbon pool in young-growth stands of the PCTR.

Methods
Source of data

This study used data from the Cooperative Stand Density Study (CSDS; Fig. 1), comprised of two long-term silvilcultural field studies, previously compiled and published [19, 38, 39] and an earlier study implemented by Ray Taylor ("Taylor plots"; Fig. 1). Most data (272 of 284 plots) were from a study of thinning treatments on even-aged young-growth (<100 years) stands begun in 1974 and with remeasurements continuing until 2003 ("Farr plots"). The remaining 12 plots ("Taylor plots") were located in older even-aged stands initiated by windthrow or early timber harvest in the late 19th century. The original intent of the studies was to measure sites that represented commercially harvested forests. Both the harvested landscapes and the plots in this study are weighted towards higher productivity classes. The Taylor plots were first measured in the late 1920s, with remeasurement occurring periodically through 2003. The Farr plots were established to examine growth and yield and how regenerating forest stands were impacted by light (mean 47.7 % BA removal); medium (mean 60.9 % BA removal); and heavy (mean 73.5 % BA removal), thinning at varying stand ages across varying productivity classes (Fig. 1). A complete description of thinning prescriptions is available in [19]. Most of these stands initiated following clear-cut harvest, with a smaller number of the older stands initiated by windthrow. All plots were dominated by western hemlock (*Tsuga heterophylla (Raf.) Sarg*) and Sitka spruce (*Picea sitchensis (Bong.) Carr*), with small amounts of western redcedar (*Thuja plicata*) and red alder (*Alnus rubra*). Stand age at thinning treatment ranged from 10 to 93 years (Fig. 2b). In general, the four treatments (control, light, medium, and heavy thinning) were applied in a randomized block design across 62 installations. Plot age, productivity class, and remeasurement dates are shown in Fig. 1b.

Estimating biomass of live trees

Tree species and diameter at breast height (DBH) were recorded for each tree in the original study and at each remeasurement interval (roughly 2–5 years). A subset of trees (7308 of the 27562) was measured for height during each remeasurement using a clinometer and tape or laser. Tree heights were estimated from diameter and height relationships for the remaining trees (Additional file 1: Appendix A). DBH and height were used to estimate carbon using allometric equations of the form:

$$B = b_0 + b_1 d^2 h \qquad (1)$$

where d is the diameter at breast height (DBH, in meters), h is the height above breast height (m), and B is the dry biomass (kg) for all the aboveground and belowground components of the tree [17]. The constant b_0 is the biomass of a tree at breast height and b_1 is related to the tree's density. The constants b_0 and b_1 are species-specific. We separately accounted for red alder (*Alnus rubra* Bong.), Shore pine (*Pinus contorta* var. *contorta* Douglas ex. Loudon), western redcedar (*Thuja plicata* Donn ex D. Don), Sitka spruce (*Picea sitchensis*) and western hemlock (*Tsuga heterophylla*). Calculations for any other species were done with the western hemlock equations from [17]. Note that Sitka spruce and western hemlock account for more than 98 % of all tree measurements.

The equations developed by Standish et al., [19] had a minimum tree diameter of 3.1 to 5.3 cm, and due to the large intercept terms, did not accurately estimate the biomass of small trees. The presence of many small diameter trees in our database required the development of a new equations We developed allometirc biomass equations for small trees by sampling 60 small diameter Sitka spruce and western hemlock and calculating the total biomass based on whole tree harvest and weighing (Additional file 2: Appendix B). These empirical biomass relationships for small diameter trees were based on Sitka spruce and western hemlock trees (<7.5 cm dbh) sampled in three locations arrayed across the geographic region of the database (Additional file 2: Appendix B). The dbh threshold for using our empirical biomass estimates for small trees versus the constants from Standish et al. [19], suitable for larger trees, was defined by the intersection of our local parameterization curve and the Standish parameterization under the assumed height-diameter relationship (Additional file 2: Appendix B). Because the height-diameter relationship and allometric parameterizations were species-specific, the diameter threshold that determined which biomass equation to apply was also unique to each species.

Estimating biomass of dead trees
Dead trees, both those cut during thinning and left on site and those that died from natural mortality, are often ignored in estimates of forest carbon pools and fluxes. In our analysis all cut trees were considered to be left on site to decompose. Cut trees were recorded in 164 of 215 treatment plots. Most plots missing cut tree data were reported in [19] as lacking pre-thinning data. The

exceptions are the 16 treatment plots of installation 62 ("Staney Creek"), for which no explanation of the missing cut tree data is given. In all cases, analysis that considered the effect of management on dead trees was based on the 164 plots for which cut tree data were available.

We estimated carbon content of dead trees using the following deterministic relationship previously parameterized for the region in a study of the decomposition rate of thinning slash [40]:

$$B = B_0[0.3870 \exp(-0.1429\, t) \\ + 0.6198 \exp(-0.00223\, t)] \qquad (2)$$

where B_0 is the estimated biomass at the time of death in kilograms, and t is time since tree death in years. This equation was used for both trees that were cut at the beginning of the study during initial thinning and left on site as well as for trees that died of natural causes, typically from suppression, at some point during the study's duration. For the latter case, we assumed the tree died and began decomposition at the midpoint between the date on which the last live measurement was taken and the date on which it was marked as dead.

Estimating carbon at the plot level from individual tree biomass
We assumed that carbon made up 48 % of the dry biomass [41] of an individual tree for both live and dead trees and that the root to aboveground biomass ratio was 0.2 [17]. Carbon estimates over all trees within a plot were aggregated into a single estimate of megagrams of carbon per hectare.

Ingrowth
Due to irregular inclusion of ingrowth measurements, our analysis of carbon estimates did not account for biomass additions due to ingrowth of new trees. We evaluated the potential impact of excluding ingrowth in our carbon estimates for plots with available ingrowth measurements. In 95 % of the measurements, the contribution of ingrowth was <5 % of total plot carbon. However, the error from excluding ingrowth likely increases with stand age as these forests begin to reach the understory re-initiation phase [42].

How does carbon accretion change with stand age in naturally-regenerating forests?
To understand basic underlying carbon dynamics of young growth stands in the PCTR, we first evaluated naturally-regenerating plots. By combining data from the Farr control plots and the Taylor plots, none of which were thinned, we had a very long chronosequence of naturally-regenerating plots (Fig. 1b), measured between 1926 and 2000 that were 10 to 170 years of age. We fit an

asymptotic nonlinear equation to relate carbon content to stand age [43]:

$$TC = A\left[1 - \exp\left(b_1\,age\right)\right]^{\frac{1}{1-b_2}}, \tag{3}$$

where TC is total carbon in a stand ($Mg\ ha^{-1}$) and stand age (*age*) is measured in years. We used non-linear mixed effects models to account for correlation among repeated measures within plots, thereby allowing the stand index to implicitly enter the model as a random feature of each plot. The random effect was placed on the asymptotic amount of carbon in the plot, consistent with the idea that the random effect reflects differences in site productivity index. Models were fit using the nlme package in R [44]. The model was first fit using estimates of carbon from live trees only. We then fit the model again to estimate carbon based on both live and dead trees.

We estimated the weighted average rate of carbon accretion as:

$$\frac{Ab_1}{2b_2 + 2}$$

Using this equation [45], we weight the instantaneous rate of accretion, so that the steeper portion the curve, is most influential when accounting for overall carbon. Estimates of parameter uncertainty were derived using parametric bootstrapping.

How does carbon accretion change with thinning?

We examined the impact of the three thinning treatments on carbon accretion using the 272 Farr plots. We did not include data from the Taylor plots in this analysis as there were no equivalent examples of older thinned plots. Carbon dynamics in the first 10 years after thinning were nonlinear due to the rapidly decelerating pace of decomposition of cut trees. These early data describe a different ecological process than data from >10 years post-thinning and were therefore excluded from our model. We excluded the first 10 years of measurements from control plots in the same blocks to balance the design. Within this age range of approximately 20–100 year-old stands, the carbon stock increased linearly among all four treatments. Therefore, we fit a linear mixed effects model to this data set. A random effect was placed on both the intercept and the slope, which was supported by likelihood ratio test, P < 0.001. These slopes describe the estimated average carbon accretion rate for stands within each treatment.

Authors' contributions
DVD and PEH designed and implemented the study. KLO and PAH organized the database, prepared files for analysis, and created figures. KLO analyzed the data. EAS contributed to the design of the data analysis and interpreta-tion of the model results. DVD wrote the initial draft of the manuscript with methods provides by KLO and PAH. All contributed to writing and editing drafts and preparing the final manuscript. All authors read and approved the final manuscript.

Author details
[1] U.S. Department of Agriculture, Forest Service, Pacific Northwest Research Station, Juneau Forestry Sciences Laboratory, 11175 Auke Lake Way, Juneau, AK 99801, USA. [2] Quantitative Ecology and Resource Management, University of Washington, Box 355020, Seattle, WA 98195, USA. [3] Graduate Degree Program in Ecology, Colorado State University, Fort Collins, CO 80523, USA. [4] U.S. Department of Agriculture, Forest Service, Pacific Northwest Research Station, 400 N 34th Street, Suite 201, Seattle, WA 98103, USA.

Acknowledgements
We would like to acknowledge the work of Ray Taylor, Bill Farr, and Mike McClellan for providing stewardship of the CSDS plots and data over 80 years. We would like to thank Dave Bassett and other workers for their dedication to re-measurement of the plots, and Mark Nay for comments on an earlier version of this manuscript and two anonymous reviwers for their review of the manuscript. We also thank Frances Biles for assistance with Fig. 1.

Competing interests
The authors declare that they have no competing interests in this manuscript.

References
1. Pan Y, Birdsey RH, Fang J, Houghton R, Kauppi PE, Kurz WA, Phillips OL, Shvidenko A, Lewis SL, Canadell JG, Ciais P, Jackson RB, Pacala SW, McGuire AD, Piao S, Rautianinen A, Sitch S, Hayes D. A large and persistent carbon sink in the world's forests. Science. 2011;333:988–93.
2. IPCC. Climate Change 2013: The physical science basis. Contribution of working group I to the fifth assessment report of the intergovernmental panel on climate change. In: Stocker, TF, Qin D, Plattner G-K, Tignor M, Allen SK, Boschung J, Nauels A, Xia Y, Bex V, Midgley PM (eds) Cambridge University Press, Cambridge. 2013. p 1535.
3. Randerson JT, Chapin FS, Harden JW, Neff JC, Harmon ME. Net ecosystem production: a comprehensive measure of net carbon accumulation by ecosystems. Ecol Appl. 2002;12:937–47.
4. Ryan MG, Harmon ME, Birdsey RA, Giardina CP, Heath LS, Houghton RA, Jackson RB, McKinley DC, Morrison JG, Murray BC, Pataki DE, Skog KE. A synthesis of the science on forests and carbon for US Forests. Issues in Ecology, Ecological Society of America, Report Number 13, 2010.
5. EISA, Energy Independence and Security Act. Public Law 110-140, United States Congress. 2007. http://www.gpo.gov/fdsys/pkg/PLAW-110publ140/pdf/PLAW-110publ140.pdf.
6. NACP, North American Carbon Program. http://nacarbon.org.
7. Federal Register. National Forest System Land Management Planning. Department of Agriculture, Forest Service. 36 CFR Part 219. 2012. http://www.fs.usda.gov/internet/fse_documents/stelprdb5362536.pdf.
8. Alaback PB. Comparative ecology of temperate rainforests of the Americas along analogous climatic gradients. Revista Chilena Historia Naturel. 1991;64:399–412.
9. USDA Forest Service. Tongass Young-Growth Management Strategy. Tongass National Forest, Region 10. 2014.
10. Harris AS, Farr WA. The forest ecosystem of southeast Alaska. 7: Forest ecology and timber management. 1974;Gen. Tech. Rep. PNW-25. Portland: USDA Forest Service, Pacific Northwest Forest and Range Experiment Station.
11. Lowell E, Dykstra C, Monserud R. Evaluating effects of thinning on wood quality in southeast Alaska. West J Appl For. 2012;27:72–83.
12. Hanley TA, Robbins CT, Spalinger DE. Forest habitats and the nutritional ecology of Sitka black-tailed deer: a research synthesis with implications for forest management. 1989;Gen. Tech. Rep. PNW-GTR-230. Portland: US Department of Agriculture, Forest Service, Pacific Northwest Research Station. p 52.

13. Deal RL, Farr WA. Composition and development of conifer regeneration in thinned and unthinned natural stands of western hemlock and Sitka spruce in southeast Alaska. Can J For Res. 1994;24:976–84.

14. McClellan MH. Recent research on the management of hemlock-spruce forest in southeast Alaska for multiple values. Landscape Urban Planning. 2005;72:65–78.

15. Carey AB. Biocomplexity and restoration of biodiversity in temperate coniferous forest: inducing spatial heterogeneity with variable-density thinning. Forestry. 2003;76:127–36.

16. Jenkins JC, Chojnacky DC, Heath LS, Birdsey RA. National scale biomass estimators for United States tree species. For Sci. 2003;49:12–35.

17. Standish JT, GH Manning, JP Demaerschalk. Development of biomass equations for British Columbia tree species. Info. Rep. BC-X-264. Victoria: Canadian Forest Service, Pacific Forest Resource Center. 1985, p 47.

18. Woodall CW, Heath LS, Domke GM, Nichols MC. Methods and equations for estimating aboveground volume, biomass, and carbon for trees in the US forest inventory. Gen. Tech. Rep. NRS-88. Newtown Square: US Department of Agriculture, Forest Service, Northern Research Station. 2011, p 30.

19. DeMars DJ. Stand-density study of spruce-hemlock stands in southeastern Alaska. 2000; General Technical Report PNW-GTR-496. USDA Forest Service, Pacific Northwest Research Station, Portland, Oregon.

20. McGuire AD, Melillo JM, Kicklighter DW, Joyce LA. Equilibrium responses of soil carbon to climate change: empirical and process-based estimates. J Biogeograph. 1995;22:785–96.

21. Harmon ME, Krankina ON, Yatskov M, Matthews E. Predicting broadscale carbon stores of woody detritus from plot-level data. In: Lai R, Kimble J, Stweart BA, editors. Assessment methods for soil carbon. New York: CRC Press; 2001. p. 533–52.

22. Lyssaert S, Schulze ED, Borner A, Knohl A, Hessenmooler D, Law BE, Ciais P, Grace J. Old-growth forests as global carbon sinks. Nature. 2008;455:213–5. doi:10.1038/nature07276.

23. Hudiburg T, Law B, Turner DP, Campbell J, Donato D, Duane M. Carbon dynamics of Oregon and Northern California forests and potential land-based carbon storage. Ecol Appl. 2009;19:163–80.

24. Hember RA, Kurz WA, Metsaranta JM, Black TA, Guy RD, Coops NC. Accelerating regrowth of temperate-maritime forests due to environmental change. Glob Change Biol. 2012;18:2026–40. doi:10.1111/j.1365-2486.2012.02669.x.

25. Malmsheimer RW, Bowyer JL, Fried JS, Gee E, Izlar RL, Miner RA, Munn IA, Oneil E, Stewart WC. Managing forests because carbon matters: integrating energy, products, and land management policy. J For. 2011;109: Number 7S.

26. Hanley TA, McClellan MH, Barnard JC, Friberg MA. Precommercial thinning: Implications of early results from the Tongass-Wide Young-Growth Studies experiments for deer habitat in southeast Alaska. 2013; Res. Pap. PNW-RP-593. Portland: U.S. Department of Agriculture, Forest Service, Pacific Northwest Research Station. p 64.

27. Southeast Alaska Integrated Resource Plan (SEIRP). Alaska Energy Authority project, Black and Veatch report no. 172744. 2011.

28. Saunders M, Tobin B, Black K, Gioria M, Nieuwenhuis M, Osborne BA. Thinning effects on the net ecosystem exchange of a Sitka spruce forest are temperature dependent. Agric For Meteor. 2012;157:1–10. doi:10.1016/j.agrformet.2012.01.008.

29. Eriksson E. Thinning operations and their impact on biomass production in stands of Norway spruce and Scots pine. Biomass Bioenergy. 2006;30:848–54.

30. Clark J, Sessions J, Krankina O, Maness T. Impacts of thinning on carbon stores in the PNW: a plot level analysis. Corvallis: Oregon State University, College of Forestry. 2011, p 61.

31. Hahm WJ, Riebe CS, Lukens CE, Araki S. Bedrock composition regulates mountain ecosystems and landscape evolution. PNAS. 2014;111:3338–43.

32. Neiland BJ. The forest-bog complex of southeast Alaska. Vegetatio. 1971;22:1–64.

33. Sidle RC, Shaw CG. III. Evaluation of planting sites common to a southeast Alaska clear-cut. I. Nutrient status. Can J For Res. 1983;13:1–8.

34. Wittwer RF, Lynch TB, Huebschmann MM. Thinning improves growth of crop tree in natural shortleaf pine stands. South J Appl For. 1996;4:182–7.

35. Peterson JA, Seiler JR, Nowak J, Ginn SE, Kreh RE. Growth and physiological responses of young loblolly pine stands to thinning. For Sci. 1997;43:529–34.

36. Ritchie MW, Zhang J, Hamilton TA. Aboveground tree biomass for *Pinus ponderosa* in Northeastern California. Forests. 2013;4:179–96.

37. Gyawali N. Aboveground biomass partitioning due to thinning in naturally regenerated even-aged shortleaf pine (Pinus echinata Mill.) stands in southeast Oklahoma. 2003; M.S. thesis, Oklahoma State University, p 76.

38. Poage NJ, Marshall DD, McClellan MH. Maximum stand-density index of 40 western hemlock-sitka spruce stands in southeast Alaska. West J Appl For. 2007;22:99–104.

39. Poage NJ. Long-term basal area and diameter growth responses of western hemlock-sitka spruce stands in southeast Alaska to a range of thinning intensities. In: Deal, R.L, tech. editors. Integrated restoration of forested ecosystems to achieve multiresource benefits: proceedings of the 2007 national silviculture workshop. 2008; Gen.Tech. Rep. PNW-GTR-733. Portland: US Department of Agriculture, Forest Service, Pacific Northwest Research Station: 271–280.

40. McClellan MH, Hennon PE, Heuer PG, Coffin KW. Conditions and deterioration rate of precommercial thinning slash at False Island, Alaska. 2013; Res. Pap. PNW-RP-594. Portland: U.S. Department of Agriculture, Forest Service, Pacific Northwest Research Station. p 29.

41. Lamlon S, Savidge R. A reassessment of carbon content in wood: variation within and between 41 North American species. Biomass Bioenergy. 2003;25:381–8.

42. Oliver CD, Larson BC. Forest stand dynamics. John Wiley and Sons Inc., New York. 1996 (ISSN:0471138339).

43. Leighty WW, Hamburg SP, Caouette J. Effects of management on carbon sequestration in forest biomass in southeast Alaska. Ecosystems. 2006;9:1051–65.

44. Pinheiro J, Bates D, DebRoy S, Sarkar D, R Core Team. nlme: linear and Nonlinear Mixed Effects Models. 2012; R package version 3.1-105.

45. Richards FJ. A flexible growth function for empirical use. J Exp Bot. 1959;10:290–301.

6

Evaluating revised biomass equations: are some forest types more equivalent than others?

Coeli M. Hoover[*] and James E. Smith

Abstract

Background: In 2014, Chojnacky et al. published a revised set of biomass equations for trees of temperate US forests, expanding on an existing equation set (published in 2003 by Jenkins et al.), both of which were developed from published equations using a meta-analytical approach. Given the similarities in the approach to developing the equations, an examination of similarities or differences in carbon stock estimates generated with both sets of equations benefits investigators using the Jenkins et al. (For Sci 49:12–34, 2003) equations or the software tools into which they are incorporated. We provide a roadmap for applying the newer set to the tree species of the US, present results of equivalence testing for carbon stock estimates, and provide some general guidance on circumstances when equation choice is likely to have an effect on the carbon stock estimate.

Results: Total carbon stocks in live trees, as predicted by the two sets, differed by less than one percent at a national level. Greater differences, sometimes exceeding 10–15 %, were found for individual regions or forest type groups. Differences varied in magnitude and direction; one equation set did not consistently produce a higher or lower estimate than the other.

Conclusions: Biomass estimates for a few forest type groups are clearly not equivalent between the two equation sets—southern pines, northern spruce-fir, and lower productivity arid western forests—while estimates for the majority of forest type groups are generally equivalent at the scales presented. Overall, the possibility of very different results between the Chojnacky and Jenkins sets decreases with aggregate summaries of those 'equivalent' type groups.

Keywords: Biomass estimation, Allometry, Forest carbon stocks, Tests of equivalence, Individual-tree estimates by species group

Background

Nationally consistent biomass equations can be important to forest carbon research and reporting activities. In general, the consistency is based on an assumption that allometric relationships within forest species do not vary by region. Essentially, nearly identical trees even in distant locations should have nearly identical carbon mass. In 2003, Jenkins et al. published a set of 10 equations for estimating live tree biomass, developed from existing equations using a meta-analytical approach, which were intended to be applicable over temperate forests of the United States [1]. These equations were developed to support US forest carbon inventory and reporting, and had several key elements: (1) a national scale, so that regional variations in biomass estimates due to the use of local biomass equations was eliminated, (2) the exclusion of height as a predictor variable, and (3) in addition to equations to estimate aboveground biomass, a set of component equations allowing the separate estimation of biomass in coarse roots, stem bark, stem wood, and foliage. Since their introduction, these equations have been incorporated into the Fire and Fuels Extension of the Forest Vegetation Simulator as a calculation option [2],

*Correspondence: choover@fs.fed.us
USDA Forest Service, Northern Research Station, Durham, NH, USA

utilized in NED-2 [3], and have provided the basis for calculating the forest carbon contribution to the US annual greenhouse gas inventories for submission years 2004–2011 (e.g., see [4]). Researchers in Canada [5, 6] and the US (e.g. [7–9]) have also employed the equations while other investigators have adopted the component ratios to estimate biomass in coarse roots or other components (e.g. [10, 11]).

In 2014, Chojnacky et al. [12] introduced a revised set of generalized biomass equations for estimating aboveground biomass. These equations were developed using the same underlying data compilations and general approaches to developing the individual tree biomass estimates as for Jenkins et al. [1], but with greater differentiation among species groups, resulting in a set of 35 generalized equations: 13 for conifers, 18 for hardwoods, and 4 for woodland species. Important distinctions are: the database used to generate the revised equations was updated to include an additional 838 equations that appeared in the literature since the publication of the 2003 work or were not included at that time, taxonomic groupings were employed to account for differences in allometry, and taxa were further subdivided in cases where wood density varied considerably within a taxon. The only component equation revised by Chojnacky et al. [12] was for roots; equations were fitted for fine and coarse roots, in contrast to Jenkins et al. [1] where fine roots were not considered separately.

Based on the similarity of the equation development approach, it is likely that applications using the Jenkins et al. [1] set would have essentially the same basis for employing the revised equations. Since the primary objective of Chojnacky et al. [12] was to present the updated equations and describe the nature of the changes, only a brief discussion of the behavior of the updated equations vs. the Jenkins et al. [1] equation set was included. The authors noted that at a national level results were similar, while differences occurred in some species groups, for example, western pines, spruce/fir types, and woodland species. Given the limited information provided in Chojnacky et al. [12] we felt that a more thorough investigation of the differences in carbon stock estimates as generated with both sets of equations was needed.

One potentially practical result from a comparison of the two approaches is to identify where one set effectively substitutes for the other, which then suggests that revising or updating estimates would change little from previous analyses. For this reason we applied equivalence tests to determine the effective difference of the Chojnacky-based estimates relative to the Jenkins values. Note that hereafter we label the respective equations and species groups as Chojnacky and Jenkins (i.e., in reference to their products not the publications, per se).

In this paper, we: (1) provide a roadmap for applying the Chojnacky equations to the tree species of the US Forest Service's forest inventory [13], (2) present results of equivalence testing for carbon stock estimates computed using both sets of equations, and (3) provide general guidance on the circumstances when the choice of equation is likely to have an important effect on the carbon stock estimate. Note that we do not attempt any evaluation of relative accuracy or the relative merit of one approach relative to the other.

Results and discussion

We conducted multiple equivalence tests on data aggregated at various levels of resolution. As noted by Chojnacky et al. [12], at a national level the carbon density predicted by both equations was the same when grouped by just hardwoods and softwoods, while some type groups showed differences (though no statistical comparisons were conducted). Relative differences emerged as four regions (Fig. 1) relative to the entire United States were used to summarize total carbon stocks in the aboveground portion of live trees as shown in Fig. 2. Totals for the US as well as separate summaries according to either softwood or hardwood forest type groups (not shown) are about 1 % different. This similarity in aggregate values between the two approaches holds for the Rocky Mountain and North regions, where there is less than a 1 % difference between the two. There are more sizeable differences in the Pacific Coast and South regions, notably differing in direction and magnitude. The largest difference is in the South. Note that our results are presented in terms of carbon mass rather than biomass.

To examine the drivers of those differences, we carried out equivalence tests by forest type group at both the national and regional levels on the mean density of carbon in aboveground live trees; a summary of the results

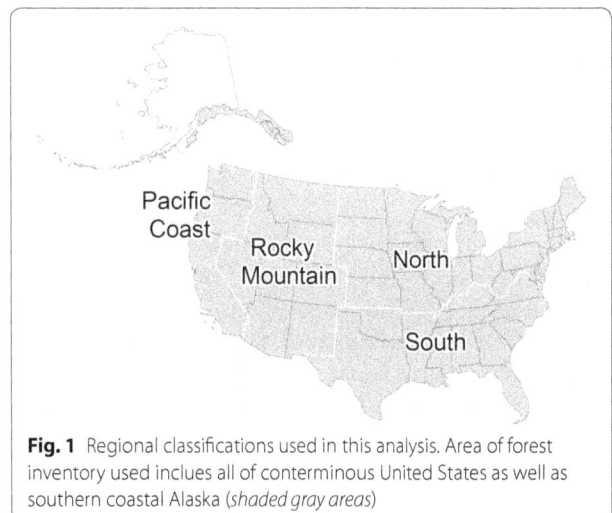

Fig. 1 Regional classifications used in this analysis. Area of forest inventory used inclues all of conterminous United States as well as southern coastal Alaska (*shaded gray areas*)

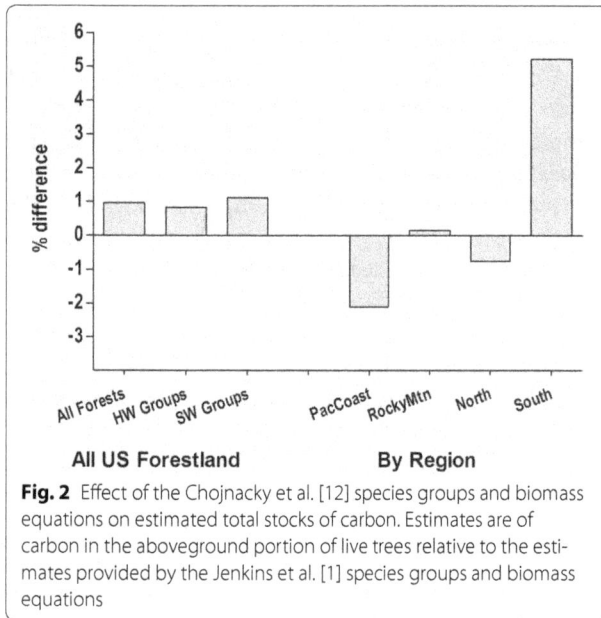

Fig. 2 Effect of the Chojnacky et al. [12] species groups and biomass equations on estimated total stocks of carbon. Estimates are of carbon in the aboveground portion of live trees relative to the estimates provided by the Jenkins et al. [1] species groups and biomass equations

is given in Table 1. The quantity tested is mean difference (Chojnacky − Jenkins) in plot level tonnes carbon per hectare; the test for equivalence was based on the percentage difference relative to the Jenkins based estimate (i.e. $100 \times ((Chojnacky − Jenkins)/Jenkins))$. The 5 (or 10) % of Jenkins, which was set as the equivalence interval, was put in units of tonnes per hectare for comparison with the 95 % confidence interval for the $\alpha = 0.05$ (or $\alpha = 0.1$) two one-sided tests (TOST) of equivalence. Of the 26 forest type groups included in the analysis, 20 are equivalent (at 5 or 10 %) at the national level, with most equivalent at 5 %. The exceptions are: spruce/fir, longleaf/slash pine, loblolly/shortleaf pine, pinyon/juniper, other western softwoods, and woodland hardwoods. At a regional level, differences emerge; in the North, only spruce/fir and loblolly/shortleaf pine are not equivalent (too few plots were available in pinyon/juniper for a reliable test statistic) while in the South, the pine types lacked equivalence, as did pinyon/juniper. This is very likely a reflection of the fact that the Chojnacky equations divide some taxa by specific gravity, while the Jenkins equations do not; softwoods generally display a larger range of specific gravity values within a species group than do hardwoods [14]. Researchers have noted considerable variability in the estimates produced by different southern pine biomass equations [15], even between different sets of local equations. Specific gravity, as mentioned above, is a factor, (southern pines exhibit considerable variability in specific gravity), as well as stand origin, and the mathematical form of the equation itself. Melson et al. [16], in their investigation of the effects of model selection on carbon stock estimates in northwest

Oregon, noted that the national level Jenkins [1] equations produced biomass estimates for *Picea* that were consistently lower than from approaches developed by the investigators, and hypothesized that differences in form between *Picea* species introduced bias into the generalized equation.

Pinyon/juniper was not equivalent in any region in which it was tested. While fir/spruce/mountain hemlock was not equivalent in the Rocky Mountains, the stock estimates were equivalent to 5 % in the Pacific Coast region, likely a function of the species and size classes that dominate the groups in each of these regions. The elm/ash/cottonwood category is represented in each region, and was equivalent to 5 % in all areas except the Pacific Coast. The woodland class has been less well studied than the others, and so less data and fewer equations are available to construct generalized equations like those in Jenkins et al. [1] and Chojnacky et al. [12]. Consequently, the woodland equations are not equivalent at the national level or in any region.

We also explored the effect of size class on equation performance, testing each combination of forest type group and stand size class and found notable differences among size classes, though no evidence of a systematic pattern. A summary of the results is given in Fig. 3a and 3b; the error bars represent the 95 % confidence interval transformed to percentage. Not every combination is shown; groups with results similar to another or comprising a very small proportion of plots are not included. While some groups such as ponderosa pine, oak/hickory, lodgepole pine, and white/red/jack pine show small differences between size classes and are equivalent (or nearly so), others such as loblolly/shortleaf pine, longleaf/slash pine (data not shown), woodland hardwoods, and spruce/fir show a strong pattern of increasing differences with increasing stand size, with a lack of equivalence between the small and large sawtimber classes. Note that both the direction and magnitude of the differences were variable across the forest type groups. Hemlock/Sitka spruce displayed a strong trend in the opposite direction, with large differences between the two approaches for the small and medium size classes, and a very small difference in the large sawtimber class. The difference between the two sets of estimates for the woodland group that is shown in Table 1 is readily apparent in Fig. 3a, with a large increase in the percent difference as the stand size class increases. This may be due to the lack of woodland biomass equations based on diameter at root collar (drc) and the difficulty of obtaining accurate drc measurements. Bragg [17] and Bragg and McElligott [15] have discussed the importance of diameter at breast height (dbh) in some detail, comparing the performance of local, regional, and national equations for southern

Table 1 **Mean stock of carbon in aboveground live tree biomass as computed using the equations from Jenkins et al. [1] and Chojnacky et al. [12]**

Forest type group	All US[a]		North		South		Rocky Mountain		Pacific Coast	
	Jenkins	Chojnacky	Jenkins	Chojnacky	Jenkins	Chojnacky	Jenkins	Chojnacky	Jenkins	Chojnacky
White/red/jack pine	68.7**	67.2**	67.7**	66.2**	92.4**	93.5**				
Spruce/fir	45.8	40.1	47.5	41.6					20.5*	18.9*
Longleaf/slash pine	35.4	40.6			35.4	40.6				
Loblolly/shortleaf pine	47.0	54	59.0	67.1	47.2	54.1				
Pinyon/juniper	18.4	22.5	◊15.5	◊17.2	11.5	13.3	19.6	24.1	21.4	23.4
Douglas-fir	114.5*	108.0*					71.4*	66.5*	148.6*	140.9*
Ponderosa pine	50.0**	50.7**	37.3**	37.9**			46.3**	47.1**	53.5**	54.2**
Western white pine	66.2**	67.6**							◊74.6	◊76.7
Fir/spruce/mtn hemlock	92.2*	87.1*					71.8	64.4	119.4**	117.4**
Lodgepole pine	48.6**	48.2**					48.2**	47.2**	49.5**	49.7**
Hemlock/sitka spruce	155.1**	151.0**					108.8*	101.4*	159.7**	155.9**
Western larch	62.6**	65.2**					55.4**	57.5**	69.6	72.6
Redwood	236.2**	235.3**							236.2**	235.3**
Other western softwoods	27.0	35.3					43.2*	45.8*	19.5	30.4
California mixed conifer	134.7**	132.8**							134.7**	132.8**
Oak/pine	54.1**	56.6**	64.4**	65.5**	50.9*	53.9*				
Oak/hickory	72.7**	72.8**	78.7**	78.8**	65.2**	65.3**				
Oak/gum/cypress	78.1**	79.7**	86.9**	85.2**	78.5**	80.3**				
Elm/ash/cottonwood	56.6**	56.6**	60.6**	59.8**	50.4**	52.2**	48.8**	48.2**	82.3	71.8
Maple/beech/birch	80.7**	80.3**	80.1**	79.7**	82.1**	83.3**				
Aspen/birch	45.3**	43.2**	43.9**	41.8**			52.8**	50.4**	38.0**	36.5**
Alder/maple	98.5**	100.1**							99.4**	101.0**
Western oak	64.7*	61.1*							64.7**	61.1**
Tanoak/laurel	131.2**	134.6**							131.2**	134.6**
Other hardwoods	49.6**	51.2**	43.0*	45.8*	43.2*	45.9*			67.5**	66.3**
Woodland hardwoods	8.6	11.1			5.0	7.0	12.7	15.7	22.1	29.5

Values followed by a double asterisk (**) are equivalent at 5 %; values followed by a single asterisk (*) are equivalent at 10 %. Regions are as shown in Fig. 1. A diamond preceding a value indicates that the sample size was too small for a reliable test of equivalence. Data not shown for categories represented by fewer than 10 plots

[a] As shown in Fig. 1

pines across a range of diameters. While most equations returned fairly similar estimates for trees up to 50 cm dbh, equation behavior diverged at larger diameters, in some cases returning estimates that were considerably different. In these examples, the national level Jenkins equations [1] did not produce extreme estimates, they were intermediate to those returned by local and regional equations. Melson et al. [16] also noted that considerable error could be introduced when applying equations to trees with a dbh value outside the range on which the equations were developed.

Equivalence was not tested at the level of the individual tree, though a random subset of individual tree estimates were plotted for each species group to compare tree-level biomass estimates. These plots reflect the patterns demonstrated above, with one method producing values consistently higher or lower than the other, the differences becoming more apparent at larger diameters. Tree data were also classified by east and west to further explore equation behavior within species groups where there are considerable differences in the range of tree diameters, east versus west. In many cases, no trends were revealed, but there are some key differences; a notable example is shown in Fig. 4a, b, which show the results of tree-level carbon estimates by each set of equations, categorized as east and west. In Fig. 4a, the eastern US, the Jenkins estimates are larger than those produced from the Chojnacky equations, while in Fig. 4b, the western US, the Jenkins estimates are generally somewhat lower, with the exception of the "Abies; LoSG" group. Figure 5 shows similar data for the woodland taxa; again, there is a considerable difference between the estimates

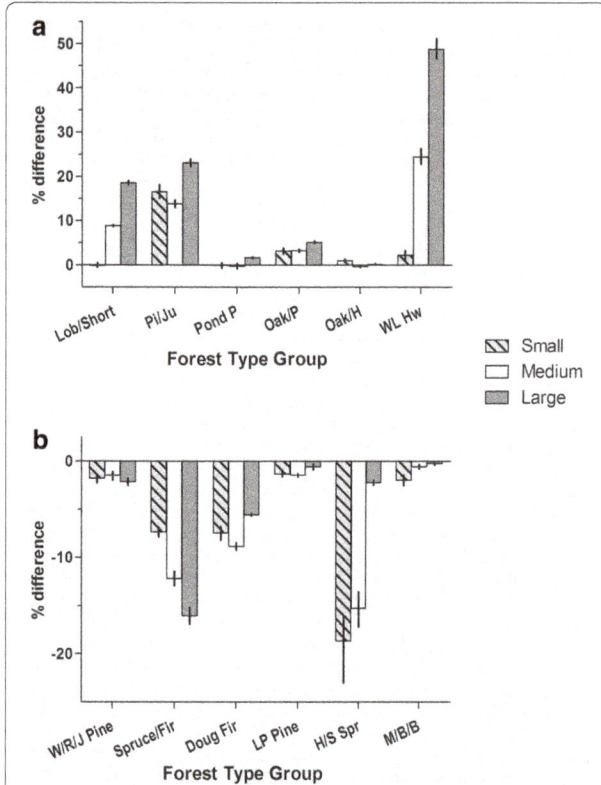

Fig. 3 Effect of the two alternate biomass equations as relative difference in stock (*panel* **a**, positive difference, *panel* **b**, negative). Estimates are classified by forest type group and stand size class. The *error bar* represents the confidence interval used in the equivalence tests. In general, small stands have at least 50 % of stocking in small diameter trees, large stands have at least 50 % of stocking in large and medium diameter trees, with large tree stocking ≥ medium tree. The 12 forest type groups included here are: loblolly/shortleaf pine, pinyon/juniper, ponderosa pine, oak/pine, oak/hickory, and woodland hardwoods in *panel* **a**, and white/red/jack pine, spruce/fir, Douglas-fir, lodgepole pine, hemlock/Sitka spruce, and maple/beech/birch in *panel* **b**

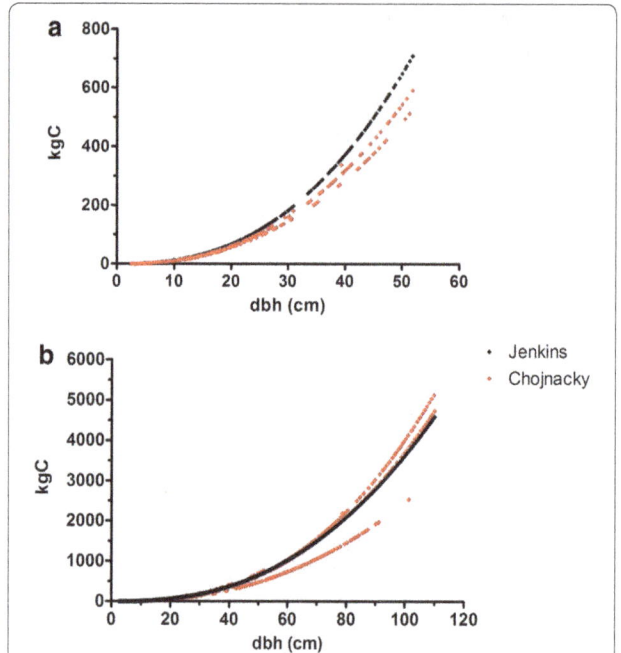

Fig. 4 Examples of the Chojnacky-based and Jenkins-based estimates for aboveground carbon mass (kg) of individual live trees (plotted by diameter at breast height, dbh). Separate panels show the East (**a**, North and South) and the West (**b**, Pacific Coast and Rocky Mountain). This example includes trees within the fir species group of Jenkins (*black*) and their mapping to Chojnacky (*red*) species groups, which are identified in Table 2. Data points include applicable live trees in the FIADB tree data table up to the 99th percentile of diameters in the east and west, respectively

computed with the two methods, with the Jenkins equations producing consistently lower estimates than the Chojnacky equations. In this case, we see no obvious differences between the predictions in the East or West.

As mentioned above, the belowground component equations were also revised in the 2014 publication, and while not divided according to hardwood and softwood, the revised root component equations are subdivided by coarse and fine roots. There are important differences in the shape of the root component curve between the two approaches (Fig. 6), and the Jenkins hardwood equation yields a consistently lower proportion than the Chojnacky equation. This suggests that adopting the Chojnacky estimates for full above- and belowground tree would add up to an additional 2–3 % of biomass for hardwoods but would also affect some softwood estimates.

A preliminary analysis did show an effect on the test for the 5 % equivalence for some categories. However, our emphases here are the various species groups/equations and not the components.

Conclusions

The revised approach to developing these biomass equations has the effect of providing better regional differentiation/representation at the plot/stand level summaries by allowing for separation within the taxonomic classes according to wood properties or growth habit. The emergence of Southern pines as distinctly different under the Chojnacky groups is one example. It is challenging to provide specific criteria for choosing one set of equations over the other, since validating any biomass equation requires the destructive sampling of multiple stems across a range of diameters. The Chojnacky groups appear to provide greater resolution across forest types and regions. From this, investigators working in southern pine, northern spruce-fir, pinyon-juniper, and woodland types may be advised to use the updated equations [12], which provide more taxonomic resolution. It should also be noted that estimates of change over time

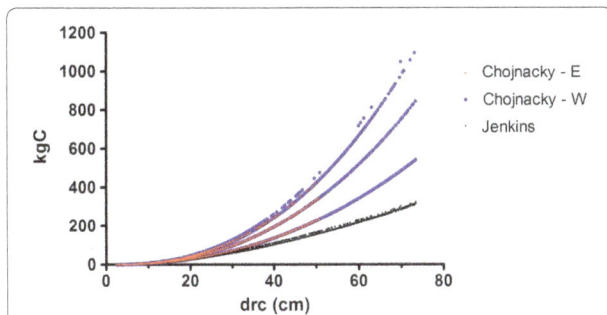

Fig. 5 Examples of the Chojnacky-based and Jenkins-based estimates for aboveground carbon mass (kg) of individual live trees by dbh. This example includes all trees within the woodland species group of Jenkins (*black*) and their mapping to Chojnacky species groups (not identified) in the East (*red*, North and South) and the West (*blue*, Pacific Coast and Rocky Mountain). Data points include all applicable live trees in the FIADB tree data table up to the 99th percentile of diameters in the East and West, respectively

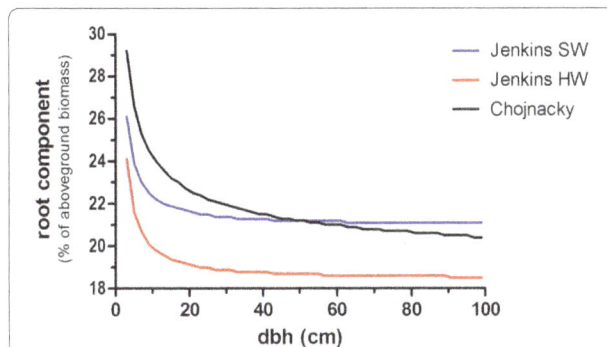

Fig. 6 Root component by diameter of the Chojnacky-based estimates (*black*) relative to the softwood (*blue*) and hardwood (*red*) root components of the Jenkins-based estimates. Root biomass is calculated as equal to a proportion of aboveground biomass

are somewhat less sensitive to equation choice than stock estimates, so if change is the primary variable of interest, the user can select either equation set, based on personal preference.

Individual large diameter trees can be very different—Chojnacky relative to Jenkins—given the general trends of the tree-level estimates (Figs. 4 and 5 in this manuscript as well as Figs. 2, 3, and 4 in Chojnacky et al. [12]). This effect of one or a very few larger trees can result in very different estimates even in an "equivalent" forest type group, and this potential for larger differences is reflected in plot-level data. For example, in some eastern hardwood type groups, which were consistently identified as equivalent, up to one-third of the plots were individually more than 5 % different. The oak/gum/cypress type group in the South had 8 % of the plots with greater carbon density by over 5 % with the Jenkins estimates, while 27 % of plots had over 5 % greater carbon.

The remaining 65 % of the individual plots are within the 5 % bounds (data not shown here). This is consistent with our observation about similarities between the two sets and scale (Fig. 2)—the sometimes obvious and large differences for some forest type groups (all scales) become obscured when summed to total live tree carbon for the US. Singling out the correct or most accurate equations is beyond the scope here; however, caution is always warranted when applying equations to trees that are considerably outside the range of diameters used to construct the equations [16].

Our results point to a few forest type groups that are clearly not equivalent—southern pines, northern spruce-fir, and lower productivity arid western forests—while the majority of forest type groups are generally equivalent at the scales presented. Overall, the possibility of very different results between the Chojnacky and Jenkins sets decreases with aggregate summaries of those 'equivalent' type groups.

Methods
Tree data source
In order to implement the revised biomass equations and identify applications where they are effectively interchangeable, or equivalent, we used the Forest Inventory and Analysis Data Base (FIADB) compiled by the Forest Inventory and Analysis (FIA) Program of the US Forest Service [13]. The data are based on continuous systematic annualized sampling of US forest lands, which are then compiled and made available by the FIA program of the US Forest Service [18]; the specific data in use here were downloaded from http://apps.fs.fed.us/fiadb-downloads/datamart.html on 02 June 2015. Surveys are organized and conducted on a large system of permanent plots over all land within individual states so that a portion of the survey data is collected each year on a continuous cycle, with remeasurement at 5 or 10 years depending on the state. The portion of the data used here include the conterminous United States (i.e., 48 states), and the portion of southern coastal Alaska that has the established permanent annual survey plots (the gray areas in Fig. 1).

Our focus here is on the tree data of the FIADB, and for this analysis we present the Chojnacky and Jenkins estimates in terms of carbon mass (i.e., kg carbon per tree or tonnes per hectare per plot). We use the entire tree data table to assure that all applicable species (the gray areas in Fig. 1) are represented. All other summaries are based on the most recent (most up-to-date) set of tree and plot data available per state, with the Chojnacky and Jenkins estimates expressed as tonnes of carbon per hectare in live trees on forest inventory plots. These plot-level values are expanded to population totals, that is, total carbon stock per state, as provided within the FIADB as the basis for the result presented in Fig. 2. A subset of the current forest plot level summaries where the entire plot is identified as forested (i.e., single condition forest plots) is the basis for the results provided in Table 1 and Fig. 3.

Application of Chojnacky et al. [12] to the FIADB

Chojnacky et al. [12] provided a revised and expanded set of biomass equations following the approach of Jenkins et al. [1]. The revised equations are based on an approach similar to that of Jenkins et al. [1] and with an expanded database of published biomass equations; see Chojnacky et al. [12] for details. The new set of 35 Chojnacky species groups are based on taxon (family or genera), growth habit, or average wood density. See Table 2 for the links between species in the FIADB and the Jenkins and Chojnacky classifications. This allocation to the newer categories is not a simple mapping of the 10 Jenkins groups to Chojnacky groups. That is, while Jenkins groups are split among Chojnacky groups, so also the Chojnacky groups are in some cases composed of species from different Jenkins groups. While Chojnacky et al. [12] developed the set of new groups based on the FIADB, similar to Jenkins et al. [1], a very small percentage of hardwood species were not explicitly named (i.e., families were not listed [12]). We assigned these to the "Cor/Eri/Lau/Etc" group (Table 2).

In order to systematically assign all the biomass estimates presented in Chojnacky et al. [12] to trees in the FIADB (as in this analysis), we present a short set of steps to make this link. Note that these include our interpretation of some of the assignments of species to groups that are not explicit such as some assignments to the woodland groups or allocation to deciduous versus evergreen. These seven steps, which also include application of the revised root component, are the basis for the biomass equation group assignments in Table 2. Note that tables and figures referenced in this list refer to those in Chojnacky et al. [12]:

1. Overall, follow the placement of taxa as suggested within the manuscript (i.e., as in Tables 2, 3, 4, and Figs. 2, 3, and 4).
2. If a tree record is one of the five families (of Table 4) and the tree diameter is measured as diameter at root collar then one of the Table 4 woodland equations applies. Otherwise, if one of the five (Table 4) families and diameter is dbh then use the appropriate equation from Tables 2 or 3. If not one of the five Table 4 families but tree diameter is provided as a root collar measurement, then convert drc to dbh following information provided in Fig. 1 before applying a Table 2 or 3 equation.

3. The calculations for the woodland (Table 4) Cupressaceae ("Cupre; WL") uses the "2nd juniper" equation from footnote #2 in Table 5.
4. The Fabaceae/Juglandaceae split into the two groups—"Fab/Jug/Carya" and "Fab/Jug"—is according to the genus *Carya* versus all others (i.e., not-*Carya*).
5. Fagaceae's deciduous/evergreen split—"Faga; Decid" and "Faga; Evergrn"—sets deciduous as the default. The Fagaceae allocated to evergreen are those five species explicitly listed as evergreen in Table 3 and those identified as evergreen from the USDA PLANTS database [19], which currently includes the addition of three live oak species.
6. The 6-family general equation at the middle of page 136 (in Table 3 of Chojnacky et al. [12])—"Cor/Eri/Lau/Etc"—is assigned trees by family from 3 sources: (a) the six families listed in Table 3; (b) the five additional families noted in the Fig. 3 caption, and (c) any additional formerly unassigned hardwood species.
7. Roots—the Chojnacky estimates use both of the belowground root equations of Table 6 (the sum of the two is generally equivalent to the original Jenkins root component). Note these are dbh-based, so a drc tree should first convert drc-to-dbh according to Fig. 1. Also note, all other (other than root) components of the original Jenkins et al. [1] are applicable here.

Identifying equivalence between the alternate biomass estimates

Tests of equivalence of the plot level (tonnes carbon per hectare) representation of the Jenkins and Chojnacky groups are included principally as guidance as to where the choice of biomass equations may matter. The analysis does not address relative accuracy of the two alternatives. Specifically, we focused on equivalence tests of the mean difference between the two estimates at the plot, or stand, level according to region and forest type groups. While these are species (group) level equations, any practical effect (of interest) is at plot to landscape to national (carbon reporting) levels. Equivalence tests are appropriate where the questions are more directly "are the groups similar, or effectively the same?" and not so much "are they different?" [20, 21]. This distinction follows from the idea that failure to reject a null hypothesis of no difference between populations does not necessarily indicate that the null hypothesis is true. The essential characteristic of an equivalence test is that the null hypothesis is stated such that the two populations are different [22, 23] which can be viewed as the reverse of the more common approach to hypothesis testing. The specific measure, or threshold, of where two populations can be considered

Table 2 Guide to applying Chojnacky species groups (as shown in Table 5, Chojnacky et al. [12]) to US species

Scientific name	Common name	Jenkins group	Chojnacky et al. parameters when diameter is measured at	
			Breast height	Root collar
Abies spp.	Fir spp.	T Fir/Hem	Abies; HiSG	Pinac; WL
A. amabilis	Pacific silver fir	T Fir/Hem	Abies; HiSG	Pinac; WL
A. balsamea	Balsam fir	T Fir/Hem	Abies; LoSG	Pinac; WL
A. bracteata	Bristlecone fir	T Fir/Hem	Abies; HiSG	Pinac; WL
A. concolor	White fir	T Fir/Hem	Abies; HiSG	Pinac; WL
A. fraseri	Fraser fir	T Fir/Hem	Abies; HiSG	Pinac; WL
A. grandis	Grand fir	T Fir/Hem	Abies; HiSG	Pinac; WL
A. lasiocarpa var. arizonica	Corkbark fir	T Fir/Hem	Abies; HiSG	Pinac; WL
A. lasiocarpa	Subalpine fir	T Fir/Hem	Abies; LoSG	Pinac; WL
A. magnifica	California red fir	T Fir/Hem	Abies; HiSG	Pinac; WL
A. shastensis	Shasta red fir	T Fir/Hem	Abies; HiSG	Pinac; WL
A. procera	Noble fir	T Fir/Hem	Abies; HiSG	Pinac; WL
Chamaecyparis spp.	White-cedar spp.	Cedar/Larch	Cupr; MedSG	Cupre; WL
C. lawsoniana	Port Orford cedar	Cedar/Larch	Cupr; MedSG	Cupre; WL
C. nootkatensi	Alaska yellow cedar	Cedar/Larch	Cupr; HiSG	Cupre; WL
C. thyoides	Atlantic white cedar	Cedar/Larch	Cupr; MedSG	Cupre; WL
Cupressus spp.	Cypress	Woodland	Cupr; HiSG	Cupre; WL
C. arizonica	Arizona cypress	Woodland	Cupr; HiSG	Cupre; WL
C. bakeri	Baker/Modoc cypress	Woodland	Cupr; HiSG	Cupre; WL
C. forbesii	Tecate cypress	Woodland	Cupr; HiSG	Cupre; WL
C. macrocarpa	Monterey cypress	Woodland	Cupr; HiSG	Cupre; WL
C. sargentii	Sargent's cypress	Woodland	Cupr; HiSG	Cupre; WL
C. macnabiana	MacNab's cypress	Woodland	Cupr; HiSG	Cupre; WL
Juniperus spp.	Redcedar/juniper spp.	Cedar/Larch	Cupr; HiSG	Cupre; WL
J. pinchotii	Pinchot juniper	Woodland	Cupr; HiSG	Cupre; WL
J. coahuilensis	Redberry juniper	Woodland	Cupr; HiSG	Cupre; WL
J. flaccida	Drooping juniper	Woodland	Cupr; HiSG	Cupre; WL
J. ashei	Ashe juniper	Woodland	Cupr; HiSG	Cupre; WL
J. californica	California juniper	Woodland	Cupr; HiSG	Cupre; WL
J. deppeana	Alligator juniper	Woodland	Cupr; HiSG	Cupre; WL
J. occidentalis	Western juniper	Woodland	Cupr; HiSG	Cupre; WL
J. osteosperma	Utah juniper	Woodland	Cupr; HiSG	Cupre; WL
J. scopulorum	Rocky Mtn. juniper	Woodland	Cupr; HiSG	Cupre; WL
J. virginiana var. silcicola	Southern redcedar	Cedar/Larch	Cupr; HiSG	Cupre; WL
J. virginiana	Easterm redcedar	Cedar/Larch	Cupr; HiSG	Cupre; WL
J. monosperma	Oneseed juniper	Woodland	Cupr; HiSG	Cupre; WL
Larix spp.	Larch spp.	Cedar/Larch	Larix	Pinac; WL
L. laricina	Tamarack	Cedar/Larch	Larix	Pinac; WL
L. lyallii	Subalpine larch	Cedar/Larch	Larix	Pinac; WL
L. occidentalis	Western larch	Cedar/Larch	Larix	Pinac; WL
Calocedrus decurrens	Incense-cedar	Cedar/Larch	Cupr; MedSG	Cupre; WL
Picea spp.	Spruce spp.	Spruce	Pice; HiSG	Pinac; WL
P. abies	Norway spruce	Spruce	Pice; HiSG	Pinac; WL
P. breweriana	Brewer spruce	Spruce	Pice; HiSG	Pinac; WL
Picea engelmannii	Englemann spruce	Spruce	Pice; LoSG	Pinac; WL
P. glauca	White spruce	Spruce	Pice; HiSG	Pinac; WL
P. mariana	Black spruce	Spruce	Pice; HiSG	Pinac; WL

Table 2 continued

Scientific name	Common name	Jenkins group	Chojnacky et al. parameters when diameter is measured at	
			Breast height	**Root collar**
P. pungens	Blue spruce	Spruce	Pice; HiSG	Pinac; WL
P. rubens	Red spruce	Spruce	Pice; HiSG	Pinac; WL
P. sitchensis	Sitka spruce	Spruce	Pice; LoSG	Pinac; WL
Pinus spp.	Pine spp.	Pine	Pinu; LoSG	Pinac; WL
P. albicaulis	Whitebark pine	Pine	Pinu; LoSG	Pinac; WL
P. aristata	Rocky Mtn. bristlecone pine	Pine	Pinu; LoSG	Pinac; WL
P. attenuata	Knobcone pine	Pine	Pinu; LoSG	Pinac; WL
P. balfouriana	Foxtail pine	Pine	Pinu; LoSG	Pinac; WL
P. banksiana	Jack pine	Pine	Pinu; LoSG	Pinac; WL
P. edulis	Common/two-needle pinyon	Pine	Pinu; HiSG	Pinac; WL
P. clausa	Sand pine	Pine	Pinu; HiSG	Pinac; WL
P. contorta	Lodgepole pine	Pine	Pinu; LoSG	Pinac; WL
P. coulteri	Coulter pine	Pine	Pinu; LoSG	Pinac; WL
P. echinata	Shortleaf pine	Pine	Pinu; HiSG	Pinac; WL
P. elliottii	Slash pine	Pine	Pinu; HiSG	Pinac; WL
P. engelmannii	Apache pine	Pine	Pinu; LoSG	Pinac; WL
P. flexilis	Limber pine	Pine	Pinu; LoSG	Pinac; WL
P. strobiformis	Southwestern white pine	Pine	Pinu; LoSG	Pinac; WL
P. glabra	Spruce pine	Pine	Pinu; LoSG	Pinac; WL
P. jeffreyi	Jeffrey pine	Pine	Pinu; LoSG	Pinac; WL
P. lambertiana	Sugar pine	Pine	Pinu; LoSG	Pinac; WL
P. leiophylla	Chihauhua pine	Pine	Pinu; LoSG	Pinac; WL
P. monticola	Western white pine	Pine	Pinu; LoSG	Pinac; WL
P. muricata	Bishop pine	Pine	Pinu; HiSG	Pinac; WL
P. palustris	Longleaf pine	Pine	Pinu; HiSG	Pinac; WL
P. ponderosa	Ponderosa pine	Pine	Pinu; LoSG	Pinac; WL
P. pungens	Table Mountain pine	Pine	Pinu; HiSG	Pinac; WL
P. radiata	Monterey pine	Pine	Pinu; LoSG	Pinac; WL
P. resinosa	Red pine	Pine	Pinu; LoSG	Pinac; WL
P. rigida	Pitch pine	Pine	Pinu; HiSG	Pinac; WL
P. sabiniana	Gray pine	Pine	Pinu; LoSG	Pinac; WL
P. serotina	Pond pine	Pine	Pinu; HiSG	Pinac; WL
P. strobus	Eastern white pine	Pine	Pinu; LoSG	Pinac; WL
P. sylvestris	Scotch pine	Pine	Pinu; LoSG	Pinac; WL
P. taeda	Loblolly pine	Pine	Pinu; HiSG	Pinac; WL
P. virginiana	Viginia pine	Pine	Pinu; HiSG	Pinac; WL
P. monophylla	Singleleaf pinyon	Pine	Pinu; LoSG	Pinac; WL
P. discolor	Border pinyon	Pine	Pinu; LoSG	Pinac; WL
P. arizonica	Arizona pine	Pine	Pinu; LoSG	Pinac; WL
P. nigra	Austrian pine	Pine	Pinu; LoSG	Pinac; WL
P. washoensis	Washoe pine	Pine	Pinu; LoSG	Pinac; WL
P. quadrifolia	Four leaf pine	Pine	Pinu; LoSG	Pinac; WL
P. torreyana	Torrey pine	Pine	Pinu; LoSG	Pinac; WL
P. cembroides	Mexican pinyon pine	Pine	Pinu; LoSG	Pinac; WL
P. remota	Papershell pinyon pine	Pine	Pinu; LoSG	Pinac; WL
P. longaeva	Great Basin bristlecone pine	Pine	Pinu; LoSG	Pinac; WL
P. monophylla var. fallax	Arizona pinyon pine	Pine	Pinu; LoSG	Pinac; WL

Table 2 continued

Scientific name	Common name	Jenkins group	Chojnacky et al. parameters when diameter is measured at	
			Breast height	Root collar
P. elliottii var. elliottii	Honduras pine	Pine	Pinu; LoSG	Pinac; WL
Pseudotsuga spp.	Douglas-fir spp.	Doug Fir	Pseud	Pinac; WL
P. macrocarpa	Bigcone Douglas-fir	Doug Fir	Pseud	Pinac; WL
P. menziesii	Douglas-fir	Doug Fir	Pseud	Pinac; WL
Sequoia sempervirens	Redwood	Cedar/Larch	Cupr; MedSG	Cupre; WL
Sequoiadendron giganteum	Giant sequoia	Cedar/Larch	Cupr; MedSG	Cupre; WL
Taxodium spp.	Baldcypress spp.	Cedar/Larch	Cupr; HiSG	Cupre; WL
T. distichum	Baldcypress	Cedar/Larch	Cupr; HiSG	Cupre; WL
T. ascendens	Pondcypress	Cedar/Larch	Cupr; HiSG	Cupre; WL
T. mucronatum	Montezuma baldcypress	Cedar/Larch	Cupr; HiSG	Cupre; WL
Taxus spp.	Yew spp.	T Fir/Hem	Pseud	
T. brevifolia	Pacific yew	T Fir/Hem	Pseud	
T. floridana	Florida yew	T Fir/Hem	Pseud	
Thuja spp.	Thuja spp.	Cedar/Larch	Cupr; MedSG	Cupre; WL
T. occidentalis	Northern white-cedar	Cedar/Larch	Cupr; LoSG	Cupre; WL
T. plicata	Western redcedar	Cedar/Larch	Cupr; MedSG	Cupre; WL
Torreya spp.	Torreya (nutmeg) spp.	T Fir/Hem	Pseud	
T. californica	California torreya	T Fir/Hem	Pseud	
T. taxifolia	Florida torreya	T Fir/Hem	Pseud	
Tsuga spp.	Hemlock spp.	T Fir/Hem	Tsug; HiSG	Pinac; WL
T. canadensis	Eastern hemlock	T Fir/Hem	Tsug; LoSG	Pinac; WL
T. caroliniana	Carolina hemlock	T Fir/Hem	Tsug; HiSG	Pinac; WL
T. heterophylla	Western hemlock	T Fir/Hem	Tsug; HiSG	Pinac; WL
T. mertensiana	Mountain hemlock	T Fir/Hem	Tsug; HiSG	Pinac; WL
Dead conifer	Unknown dead conifer	Pine	Pinu; LoSG	
Acacia spp.	Acacia spp.	Woodland	Fab/Jug	Fab/Ros; WL
A. farnesiana	Sweet acacia	Woodland	Fab/Jug	Fab/Ros; WL
A. greggii	Catclaw acacia	Woodland	Fab/Jug	Fab/Ros; WL
Acer spp.	Maple spp.	S Maple/Bir	Acer; LoSG	
A. barbatum	Florida maple	S Maple/Bir	Acer; HiSG	
A. macrophyllum	Bigleaf maple	S Maple/Bir	Acer; LoSG	
A. negundo	Boxelder	S Maple/Bir	Acer; LoSG	
A. nigrum	Black maple	H Maple/Oak	Acer; HiSG	
A. pensylvanicum	Striped maple	S Maple/Bir	Acer; LoSG	
A. rubrum	Red maple	S Maple/Bir	Acer; LoSG	
A. saccharinum	Silver maple	S Maple/Bir	Acer; LoSG	
A. saccharum	Sugar maple	H Maple/Oak	Acer; HiSG	
A. spicatum	Mountain maple	S Maple/Bir	Acer; LoSG	
A. platanoides	Norway maple	S Maple/Bir	Acer; LoSG	
A. glabrum	Rocky Mtn. maple	Woodland	Acer; LoSG	
A. grandidentatum	Bigtooth maple	Woodland	Acer; LoSG	
A. leucoderme	Chalk maple	Mixed HW	Acer; LoSG	
Aesculus spp.	Buckeye spp.	Mixed HW	Hip/Til	
A. glabra	Ohio buckeye	Mixed HW	Hip/Til	
A. flava	Yellow buckeye	Mixed HW	Hip/Til	

Table 2 continued

| Scientific name | Common name | Jenkins group | Chojnacky et al. parameters when diameter is measured at | |
			Breast height	Root collar
A.californica	California buckeye	Mixed HW	Hip/Til	
A.glabra var. arguta	Texas buckeye	Mixed HW	Hip/Til	
A.pavia	Red buckeye	Mixed HW	Hip/Til	
A.sylvatica	Painted buckeye	Mixed HW	Hip/Til	
Ailanthus altissima	Ailanthus	Mixed HW	Cor/Eri/Lau/Etc	
Albizia julibrissin	Mimosa/silktree	Mixed HW	Fab/Jug	Fab/Ros; WL
Alnus spp.	Alder spp.	Aspen/Alder	Betu; LoSG	
A. rubra	Red alder	Aspen/Alder	Betu; LoSG	
A. rhombifolia	White alder	Aspen/Alder	Betu; LoSG	
A. oblongifolia	Arizona alder	Aspen/Alder	Betu; LoSG	
A. glutinosa	European alder	Aspen/Alder	Betu; LoSG	
Amelanchier spp.	Serviceberry spp.	Mixed HW	Cor/Eri/Lau/Etc	Fab/Ros; WL
A. arborea	Common serviceberry	Mixed HW	Cor/Eri/Lau/Etc	Fab/Ros; WL
A. sanguinea	Roundleaf serviceberry	Mixed HW	Cor/Eri/Lau/Etc	Fab/Ros; WL
Arbutus spp.	Madrone spp.	Mixed HW	Cor/Eri/Lau/Etc	
A. menziesii	Pacific madrone	Mixed HW	Cor/Eri/Lau/Etc	
A. arizonica	Arizona madrone	Mixed HW	Cor/Eri/Lau/Etc	
A. xalapensis	Texas madrone	Mixed HW	Cor/Eri/Lau/Etc	
Asimina triloba	Pawpaw	Mixed HW	Cor/Eri/Lau/Etc	
Betula spp.	Birch spp.	S Maple/Bir	Betu; Med1SG	
B. alleghaniensis	Yellow birch	S Maple/Bir	Betu; Med2SG	
B. lenta	Sweet birch	S Maple/Bir	Betu; HiSG	
B. nigra	River birch	S Maple/Bir	Betu; Med1SG	
B. occidentalis	Water birch	S Maple/Bir	Betu; Med2SG	
B. papyrifera	Paper birch	S Maple/Bir	Betu; Med1SG	
B. uber	Virginia roundleaf birch	S Maple/Bir	Betu; Med2SG	
B. utahensis	Northwestern paper birch	S Maple/Bir	Betu; Med2SG	
B. populifolia	Gray birch	S Maple/Bir	Betu; Med1SG	
Sideroxylon lanuginosum	Chittamwood/gum bumelia	Mixed HW	Cor/Eri/Lau/Etc	
Carpinus caroliniana	American hornbeam	Mixed HW	Betu; Med2SG	
Carya spp.	Hickory spp.	H Maple/Oak	Fab/Jug/Carya	
C. aquatica	Water hickory	H Maple/Oak	Fab/Jug/Carya	
C. cordiformis	Bitternut hickory	H Maple/Oak	Fab/Jug/Carya	
C. glabra	Pignut hickory	H Maple/Oak	Fab/Jug/Carya	
C. illinoinensis	Pecan	H Maple/Oak	Fab/Jug/Carya	
C. laciniosa	Shellbark hickory	H Maple/Oak	Fab/Jug/Carya	
C. myristiciformis	Nutmeg hickory	H Maple/Oak	Fab/Jug/Carya	
C. ovata	Shagbark hickory	H Maple/Oak	Fab/Jug/Carya	
C. texana	Black hickory	H Maple/Oak	Fab/Jug/Carya	
C. alba	Mockernut hickory	H Maple/Oak	Fab/Jug/Carya	
C. pallida	Sand hickory	H Maple/Oak	Fab/Jug/Carya	
C. floridana	Scrub hickory	H Maple/Oak	Fab/Jug/Carya	
C. ovalis	Red hickory	H Maple/Oak	Fab/Jug/Carya	
C. carolinae-septentrionalis	Southern shagbark hickory	H Maple/Oak	Fab/Jug/Carya	

Table 2 continued

Scientific name	Common name	Jenkins group	Chojnacky et al. parameters when diameter is measured at	
			Breast height	**Root collar**
Castanea spp.	Chestnut spp.	Mixed HW	Faga; Decid	Fagac; WL
C. dentata	American chestnut	Mixed HW	Faga; Decid	Fagac; WL
C. pumila	Allegheny chinkapin	Mixed HW	Faga; Decid	Fagac; WL
C. pumila var. ozarkensis	Ozark chinkapin	Mixed HW	Faga; Decid	Fagac; WL
C. mollissima	Chinese chestnut	Mixed HW	Faga; Decid	Fagac; WL
Chrysolepis chrysophylla	Giant/golden chinkapin	Mixed HW	Faga; Evergrn	Fagac; WL
Catalpa spp.	Catalpa spp.	Mixed HW	Cor/Eri/Lau/Etc	
C. bignonioide	Southern catalpa	Mixed HW	Cor/Eri/Lau/Etc	
C. speciosa	Northern catalpa	Mixed HW	Cor/Eri/Lau/Etc	
Celtis	Hackberry spp.	Mixed HW	Cor/Eri/Lau/Etc	
C. laevigata	Sugarberry	Mixed HW	Cor/Eri/Lau/Etc	
C. occidentalis	Hackberry	Mixed HW	Cor/Eri/Lau/Etc	
C. laevigata var. reticulata	Netleaf hackberry	Mixed HW	Cor/Eri/Lau/Etc	
Cercis canadensis	Eastern redbud	Mixed HW	Fab/Jug	Fab/Ros; WL
Cercocarpus ledifoliu	Curlleaf mountain-mahogany	Woodland	Cor/Eri/Lau/Etc	Fab/Ros; WL
Cladrastis kentukea	Yellowwood	Mixed HW	Fab/Jug	Fab/Ros; WL
Cornus spp.	Dogwood spp.	Mixed HW	Cor/Eri/Lau/Etc	
C. florida	Flowering dogwood	Mixed HW	Cor/Eri/Lau/Etc	
C. nuttallii	Pacific dogwood	Mixed HW	Cor/Eri/Lau/Etc	
Crataegus spp.	Hawthorn spp.	Mixed HW	Cor/Eri/Lau/Etc	Fab/Ros; WL
C. crusgalli	Cockspur hawthorn	Mixed HW	Cor/Eri/Lau/Etc	Fab/Ros; WL
C. mollis	Downy hawthorn	Mixed HW	Cor/Eri/Lau/Etc	Fab/Ros; WL
C. brainerdii	Brainerd's hawthorn	Mixed HW	Cor/Eri/Lau/Etc	Fab/Ros; WL
C. calpodendron	Pear hawthorn	Mixed HW	Cor/Eri/Lau/Etc	Fab/Ros; WL
C. chrysocarpa	Fireberry hawthorn	Mixed HW	Cor/Eri/Lau/Etc	Fab/Ros; WL
C. dilatata	Broadleaf hawthorn	Mixed HW	Cor/Eri/Lau/Etc	Fab/Ros; WL
C. flabellata	Fanleaf hawthorn	Mixed HW	Cor/Eri/Lau/Etc	Fab/Ros; WL
C. monogyna	Oneseed hawthorn	Mixed HW	Cor/Eri/Lau/Etc	Fab/Ros; WL
C. pedicellata	Scarlet hawthorn	Mixed HW	Cor/Eri/Lau/Etc	Fab/Ros; WL
Eucalyptus spp.	Eucalyptus spp.	Mixed HW	Cor/Eri/Lau/Etc	
E. globulus	Tasmanian bluegum	Mixed HW	Cor/Eri/Lau/Etc	
E. camaldulensi	River redgum	Mixed HW	Cor/Eri/Lau/Etc	
E. grandis	Grand eucalyptus	Mixed HW	Cor/Eri/Lau/Etc	
E. robusta	Swamp mahogany	Mixed HW	Cor/Eri/Lau/Etc	
Diospyros spp.	Persimmon spp.	Mixed HW	Cor/Eri/Lau/Etc	
D. virginiana	Common persimmon	Mixed HW	Cor/Eri/Lau/Etc	
D. texana	Texas persimmon	Mixed HW	Cor/Eri/Lau/Etc	
Ehretia anacua	Anacua knockaway	Mixed HW	Cor/Eri/Lau/Etc	

Table 2 continued

Scientific name	Common name	Jenkins group	Chojnacky et al. parameters when diameter is measured at	
			Breast height	**Root collar**
Fagus grandifolia	American beech	H Maple/Oak	Faga; Decid	Fagac; WL
Fraxinus spp.	Ash spp.	Mixed HW	Olea; LoSG	
F. americana	White ash	Mixed HW	Olea; HiSG	
F. latifolia	Oregon ash	Mixed HW	Olea; LoSG	
F. nigra	Black ash	Mixed HW	Olea; LoSG	
F. pennsylvanica	Green ash	Mixed HW	Olea; LoSG	
F. profunda	Pumpkin ash	Mixed HW	Olea; LoSG	
F. quadrangulata	Blue ash	Mixed HW	Olea; LoSG	
F. velutina	Velvet ash	Mixed HW	Olea; LoSG	
F. caroliniana	Carolina ash	Mixed HW	Olea; LoSG	
F. texensis	Texas ash	Mixed HW	Olea; LoSG	
Gleditsia spp.	Honeylocust spp.	Mixed HW	Fab/Jug	Fab/Ros; WL
G. aquatica	Waterlocust	Mixed HW	Fab/Jug	Fab/Ros; WL
G. triacanthos	Honeylocust	Mixed HW	Fab/Jug	Fab/Ros; WL
Gordonia lasianthus	Loblolly-bay	Mixed HW	Cor/Eri/Lau/Etc	
Ginkgo biloba	Ginkgo	Mixed HW	Cor/Eri/Lau/Etc	
Gymnocluadus diocicus	Kentucky coffeetree	Mixed HW	Fab/Jug	Fab/Ros; WL
Halesia spp.	Silverbell spp.	Mixed HW	Cor/Eri/Lau/Etc	
H. carolina	Carolina silverbell	Mixed HW	Cor/Eri/Lau/Etc	
H. diptera	Two-wing silverbell	Mixed HW	Cor/Eri/Lau/Etc	
H. parviflora	Little silverbell	Mixed HW	Cor/Eri/Lau/Etc	
Ilex opaca	American holly	Mixed HW	Cor/Eri/Lau/Etc	
Juglans spp.	Walnut spp.	Mixed HW	Fab/Jug	
J. cinerea	Butternut	Mixed HW	Fab/Jug	
J. nigra	Black walnut	Mixed HW	Fab/Jug	
J. hindsii	No. California black walnut	Mixed HW	Fab/Jug	
J. californica	So. California black walnut	Mixed HW	Fab/Jug	
J. microcarpa	Texas walnut	Mixed HW	Fab/Jug	
J. major	Arizona walnut	Mixed HW	Fab/Jug	
Liquidambar styraciflua	Sweetgum	Mixed HW	Hama	
Liriodendron tulipifera	Yellow poplar	Mixed HW	Magno	
Lithocarpus densiflorus	Tanoak	Mixed HW	Faga; Evergrn	Fagac; WL
Maclura pomifera	Osage orange	Mixed HW	Cor/Eri/Lau/Etc	
Magnolia spp.	Magnolia spp.	Mixed HW	Magno	
M. acuminata	Cucumbertree	Mixed HW	Magno	
M. grandiflora	Southern magnolia	Mixed HW	Magno	
M. virginiana	Sweeetbay	Mixed HW	Magno	
M. macrophylla	Bigleaf magnolia	Mixed HW	Magno	
M. fraseri	Mountain/Frasier magnolia	Mixed HW	Magno	
M. pyramidata	Pyramid magnolia	Mixed HW	Magno	
M. tripetala	Umbrella magnolia	Mixed HW	Magno	
Malus spp.	Apple spp.	Mixed HW	Cor/Eri/Lau/Etc	Fab/Ros; WL
M. fusca	Oregon crab apple	Mixed HW	Cor/Eri/Lau/Etc	Fab/Ros; WL

Table 2 continued

Scientific name	Common name	Jenkins group	Chojnacky et al. parameters when diameter is measured at	
			Breast height	Root collar
M. angustifolia	Southern crabapple	Mixed HW	Cor/Eri/Lau/Etc	Fab/Ros; WL
M. coronaria	Sweet crabapple	Mixed HW	Cor/Eri/Lau/Etc	Fab/Ros; WL
M. ioensi	Prairie crabapple	Mixed HW	Cor/Eri/Lau/Etc	Fab/Ros; WL
Morus spp.	Mulberry spp.	Mixed HW	Cor/Eri/Lau/Etc	
M. alba	White mulberry	Mixed HW	Cor/Eri/Lau/Etc	
M. rubra	Red mulberry	Mixed HW	Cor/Eri/Lau/Etc	
M. microphyll	Texas mulberry	Mixed HW	Cor/Eri/Lau/Etc	
M. nigra	Black mulberry	Mixed HW	Cor/Eri/Lau/Etc	
Nyssa spp.	Tupelo spp.	Mixed HW	Cor/Eri/Lau/Etc	
N. aquatica	Water tupelo	Mixed HW	Cor/Eri/Lau/Etc	
N. ogeche	Ogeechee tupelo	Mixed HW	Cor/Eri/Lau/Etc	
N. sylvatica	Blackgum	Mixed HW	Cor/Eri/Lau/Etc	
N. biflora	Swamp tupelo	Mixed HW	Cor/Eri/Lau/Etc	
Ostrya virginiana	Eastern hophornbeam	Mixed HW	Betu; HiSG	
Oxydendrum arboreum	Sourwood	Mixed HW	Cor/Eri/Lau/Etc	
Paulownia tomentosa	Paulownia/empress tree	Mixed HW	Cor/Eri/Lau/Etc	
Persea spp.	Bay spp.	Mixed HW	Cor/Eri/Lau/Etc	
Persea borbonia	Redbay	Mixed HW	Cor/Eri/Lau/Etc	
Planera aquatica	Water elm/planetree	Mixed HW	Cor/Eri/Lau/Etc	
Platanus spp.	Sycamore spp.	Mixed HW	Cor/Eri/Lau/Etc	
P. racemosa	California sycamore	Mixed HW	Cor/Eri/Lau/Etc	
P. occidentalis	American sycamore	Mixed HW	Cor/Eri/Lau/Etc	
P. wrightii	Arizona sycamore	Mixed HW	Cor/Eri/Lau/Etc	
Populus spp.	Cottonwood/poplar spp.	Aspen/Alder	Sali; HiSG	
P. balsamifera	Balsam poplar	Aspen/Alder	Sali; LoSG	
P. deltoides	Eastern cottonwood	Aspen/Alder	Sali; HiSG	
P. grandidentata	Bigtooth aspen	Aspen/Alder	Sali; HiSG	
P. heterophylla	Swamp cottonwood	Aspen/Alder	Sali; HiSG	
P. deltoides	Plains cottonwood	Aspen/Alder	Sali; HiSG	
P. tremuloides	Quaking aspen	Aspen/Alder	Sali; HiSG	
P. balsamifera	Black cottonwood	Aspen/Alder	Sali; LoSG	
P. fremontii	Fremont cottonwood	Aspen/Alder	Sali; HiSG	
P. angustifolia	Narrlowleaf cottonwood	Aspen/Alder	Sali; HiSG	
P. alba	Silver poplar	Aspen/Alder	Sali; HiSG	
P. nigra	Lombardy poplar	Aspen/Alder	Sali; HiSG	
Prosopis spp.	Mesquite spp.	Woodland	Fab/Jug	Fab/Ros; WL
P. glandulosa	Honey mesquite	Woodland	Fab/Jug	Fab/Ros; WL
P. velutina	Velvet mesquite	Woodland	Fab/Jug	Fab/Ros; WL
P. pubescens	Screwbean mesquite	Woodland	Fab/Jug	Fab/Ros; WL
Prunus spp.	Cherry/plum spp.	Mixed HW	Cor/Eri/Lau/Etc	Fab/Ros; WL
P. pensylvanica	Pin cherry	Mixed HW	Cor/Eri/Lau/Etc	Fab/Ros; WL

Table 2 continued

Scientific name	Common name	Jenkins group	Chojnacky et al. parameters when diameter is measured at	
			Breast height	Root collar
P. serotina	Black cherry	Mixed HW	Cor/Eri/Lau/Etc	Fab/Ros; WL
P. virginiana	Chokecherry	Mixed HW	Cor/Eri/Lau/Etc	Fab/Ros; WL
P. persica	Peach	Mixed HW	Cor/Eri/Lau/Etc	Fab/Ros; WL
P. nigra	Canada plum	Mixed HW	Cor/Eri/Lau/Etc	Fab/Ros; WL
P. americana	American plum	Mixed HW	Cor/Eri/Lau/Etc	Fab/Ros; WL
P. emarginata	Bitter cherry	Woodland	Cor/Eri/Lau/Etc	Fab/Ros; WL
P. alleghaniensis	Allegheny plum	Mixed HW	Cor/Eri/Lau/Etc	Fab/Ros; WL
P. angustifolia	Chickasaw plum	Mixed HW	Cor/Eri/Lau/Etc	Fab/Ros; WL
P. avium	Sweet cherry (domestic)	Mixed HW	Cor/Eri/Lau/Etc	Fab/Ros; WL
P. cerasus	Sour cherry (domestic)	Mixed HW	Cor/Eri/Lau/Etc	Fab/Ros; WL
P. domestica	European plum (domestic)	Mixed HW	Cor/Eri/Lau/Etc	Fab/Ros; WL
P. mahaleb	Mahaleb cherry (domestic)	Mixed HW	Cor/Eri/Lau/Etc	Fab/Ros; WL
Quercus spp.	Oak spp.	H Maple/Oak	Faga; Decid	Fagac; WL
Q. agrifolia	California live oak	H Maple/Oak	Faga; Evergrn	Fagac; WL
Q. alba	White oak	H Maple/Oak	Faga; Decid	Fagac; WL
Q. arizonica	Arizona white oak	Woodland	Faga; Decid	Fagac; WL
Q. bicolor	Swamp white oak	H Maple/Oak	Faga; Decid	Fagac; WL
Q. chrysolepis	Canyon live oak	H Maple/Oak	Faga; Decid	Fagac; WL
Q. coccinea	Scarlet oak	H Maple/Oak	Faga; Decid	Fagac; WL
Q. douglasii	Blue oak	H Maple/Oak	Faga; Evergrn	Fagac; WL
Q. sinuata var. sinuata	Durand oak	H Maple/Oak	Faga; Decid	Fagac; WL
Q. ellipsoidalis	Northern pin oak	H Maple/Oak	Faga; Decid	Fagac; WL
Q. emoryi	Emory oak	Woodland	Faga; Decid	Fagac; WL
Q. engelmannii	Englemann oak	H Maple/Oak	Faga; Decid	Fagac; WL
Q. falcata	Southern red oak	H Maple/Oak	Faga; Decid	Fagac; WL
Q. pagoda	Cherrybark oak	H Maple/Oak	Faga; Decid	Fagac; WL
Q. gambelii	Gambel oak	Woodland	Faga; Decid	Fagac; WL
Q. garryana	Oregon white oak	H Maple/Oak	Faga; Decid	Fagac; WL
Q. ilicifolia	Scrub oak	H Maple/Oak	Faga; Decid	Fagac; WL
Q. imbricaria	Shingle oak	H Maple/Oak	Faga; Decid	Fagac; WL
Q. kelloggii	California black oak	H Maple/Oak	Faga; Decid	Fagac; WL
Q. laevis	Turkey oak	H Maple/Oak	Faga; Decid	Fagac; WL
Q. laurifolia	Laurel oak	H Maple/Oak	Faga; Evergrn	Fagac; WL
Q. lobata	California white oak	H Maple/Oak	Faga; Decid	Fagac; WL
Q. lyrata	Overcup oak	H Maple/Oak	Faga; Decid	Fagac; WL
Q. macrocarpa	Bur oak	H Maple/Oak	Faga; Decid	Fagac; WL
Q. marilandica	Blackjack oak	H Maple/Oak	Faga; Decid	Fagac; WL
Q. michauxi	Swamp chestnut oak	H Maple/Oak	Faga; Decid	Fagac; WL

Table 2 continued

Scientific name	Common name	Jenkins group	Chojnacky et al. parameters when diameter is measured at	
			Breast height	Root collar
Q. muehlenbergii	Chinkapin oak	H Maple/Oak	Faga; Decid	Fagac; WL
Q. nigra	Water oak	H Maple/Oak	Faga; Decid	Fagac; WL
Q. texana	Texas red oak	H Maple/Oak	Faga; Decid	Fagac; WL
Q. oblongifolia	Mexican blue oak	Woodland	Faga; Decid	Fagac; WL
Q. palustris	Pin oak	H Maple/Oak	Faga; Decid	Fagac; WL
Q. phellos	Willow oak	H Maple/Oak	Faga; Decid	Fagac; WL
Q. prinus	Chestnut oak	H Maple/Oak	Faga; Decid	Fagac; WL
Q. rubra	Northern red oak	H Maple/Oak	Faga; Decid	Fagac; WL
Q. shumardii	Shumard oak	H Maple/Oak	Faga; Decid	Fagac; WL
Q. stellata	Post oak	H Maple/Oak	Faga; Decid	Fagac; WL
Q. simili	Delta post oak	H Maple/Oak	Faga; Decid	Fagac; WL
Q. velutina	Black oak	H Maple/Oak	Faga; Decid	Fagac; WL
Q. virginiana	Live oak	H Maple/Oak	Faga; Evergrn	Fagac; WL
Q. wislizeni	Interier live oak	H Maple/Oak	Faga; Evergrn	Fagac; WL
Q. margarettiae	Dwarf post oak	H Maple/Oak	Faga; Evergrn	Fagac; WL
Q. minima	Dwarf live oak	H Maple/Oak	Faga; Evergrn	Fagac; WL
Q. incana	Bluejack oak	H Maple/Oak	Faga; Decid	Fagac; WL
Q. hypoleucoides	Silverleaf oak	Woodland	Faga; Decid	Fagac; WL
Q. oglethorpensis	Oglethorpe oak	H Maple/Oak	Faga; Decid	Fagac; WL
Q. prinoides	Dwarf chinkapin oak	H Maple/Oak	Faga; Decid	Fagac; WL
Q. grisea	Gray oak	Woodland	Faga; Decid	Fagac; WL
Q. rugosa	Netleaf oak	H Maple/Oak	Faga; Decid	Fagac; WL
Q. gracilliformis	Chisos oak	Woodland	Faga; Decid	Fagac; WL
Amyris elemifera	Sea torchwood	Mixed HW	Cor/Eri/Lau/Etc	
Annona glabra	Pond apple	Mixed HW	Cor/Eri/Lau/Etc	
Bursera simaruba	Gumbo limbo	Mixed HW	Cor/Eri/Lau/Etc	
Casuarina spp.	Sheoak spp.	Mixed HW	Cor/Eri/Lau/Etc	
C. glauca	Gray sheoak	Mixed HW	Cor/Eri/Lau/Etc	
C. lepidophloia	Belah	Mixed HW	Cor/Eri/Lau/Etc	
Cinnamomum camphora	Camphortree	Mixed HW	Cor/Eri/Lau/Etc	
Citharexylum fruticosum	Florida fiddlewood	Mixed HW	Cor/Eri/Lau/Etc	
Citrus spp.	Citrus spp.	Mixed HW	Cor/Eri/Lau/Etc	
Coccoloba diversifolia	Tietongue/pigeon plum	Mixed HW	Cor/Eri/Lau/Etc	
Colubrina elliptica	Soldierwood	Mixed HW	Cor/Eri/Lau/Etc	
Cordia sebestena	Longleaf geigertree	Mixed HW	Cor/Eri/Lau/Etc	
Cupaniopsis anacardioides	Carrotwood	Mixed HW	Cor/Eri/Lau/Etc	
Condalia hookeri	Bluewood	Woodland	Cor/Eri/Lau/Etc	
Ebenopsis ebano	Blackbead ebony	Woodland	Fab/Jug	Fab/Ros; WL
Leucaena pulverulenta	Great leadtree	Woodland	Fab/Jug	Fab/Ros; WL
Sophora affinis	Texas sophora	Woodland	Fab/Jug	Fab/Ros; WL
Eugenia rhombea	Red stopper	Mixed HW	Cor/Eri/Lau/Etc	
Exothea paniculata	Butterbough/inkwood	Mixed HW	Cor/Eri/Lau/Etc	
Ficus aurea	Florida strangler fig	Mixed HW	Cor/Eri/Lau/Etc	
Ficus citrifolia	Banyantree/shortleaf fig	Mixed HW	Cor/Eri/Lau/Etc	
Guapira discolo	Beeftree/longleaf blolly	Mixed HW	Cor/Eri/Lau/Etc	

Table 2 continued

Scientific name	Common name	Jenkins group	Chojnacky et al. parameters when diameter is measured at	
			Breast height	Root collar
Hippomane mancinella	Manchineel	Mixed HW	Cor/Eri/Lau/Etc	
Lysiloma latisiliquum	False tamarind	Mixed HW	Fab/Jug	Fab/Ros; WL
Mangifera indica	Mango	Mixed HW	Cor/Eri/Lau/Etc	
Metopium toxiferum	Florida poisontree	Mixed HW	Cor/Eri/Lau/Etc	
Piscidia piscipula	Fishpoison tree	Mixed HW	Fab/Jug	Fab/Ros; WL
Schefflera actinophylla	Octopus tree/schefflera	Mixed HW	Cor/Eri/Lau/Etc	
Sideroxylon foetidissimum	False mastic	Mixed HW	Cor/Eri/Lau/Etc	
Sideroxylon salicifolium	White bully/willow bustic	Mixed HW	Cor/Eri/Lau/Etc	
Simarouba glauca	Paradisetree	Mixed HW	Cor/Eri/Lau/Etc	
Syzygium cumini	Java plum	Mixed HW	Cor/Eri/Lau/Etc	
Tamarindus indica	Tamarind	Mixed HW	Fab/Jug	Fab/Ros; WL
Robinia pseudoacacia	Black locust	Mixed HW	Fab/Jug	Fab/Ros; WL
Robinia neomexicana	New Mexico locust	Woodland	Fab/Jug	Fab/Ros; WL
Acoelorraphe wrightii	Everglades palm	Mixed HW	Cor/Eri/Lau/Etc	
Coccothrinax argentata	Florida silver palm	Mixed HW	Cor/Eri/Lau/Etc	
Cocos nucifera	Coconut palm	Mixed HW	Cor/Eri/Lau/Etc	
Roystonea spp.	Royal palm spp.	Mixed HW	Cor/Eri/Lau/Etc	
Sabal Mexicana	Mexican palmetto	Mixed HW	Cor/Eri/Lau/Etc	
Sabal palmetto	Cabbage palmetto	Mixed HW	Cor/Eri/Lau/Etc	
Thrinax morrisii	Key thatch palm	Mixed HW	Cor/Eri/Lau/Etc	
Thrinax radiata	Florida thatch palm	Mixed HW	Cor/Eri/Lau/Etc	
Arecaceae	Other palms	Mixed HW	Cor/Eri/Lau/Etc	
Sapindus saponaria	Western soapberry	Mixed HW	Cor/Eri/Lau/Etc	
Salix spp.	Willow spp.	Aspen/Alder	Sali; HiSG	
S. amygdaloides	Peachleaf willow	Aspen/Alder	Sali; HiSG	
S. nigra	Black willow	Aspen/Alder	Sali; HiSG	
S. bebbiana	Bebb willow	Aspen/Alder	Sali; HiSG	
S. bonplandiana	Bonpland willow	Aspen/Alder	Sali; HiSG	
S. caroliniana	Coastal plain willow	Aspen/Alder	Sali; HiSG	
S. pyrifolia	Balsam willow	Aspen/Alder	Sali; HiSG	
S. alba	White willow	Aspen/Alder	Sali; HiSG	
S. scouleriana	Scouder's willow	Aspen/Alder	Sali; HiSG	
S. sepulcralis	Weeping willow	Aspen/Alder	Sali; HiSG	
Sassafras albidum	Sassafrass	Mixed HW	Cor/Eri/Lau/Etc	
Sorbus spp.	Mountain ash spp.	Mixed HW	Cor/Eri/Lau/Etc	Fab/Ros; WL
S. americana	American mountain ash	Mixed HW	Cor/Eri/Lau/Etc	Fab/Ros; WL
S. aucuparia	European mountain ash	Mixed HW	Cor/Eri/Lau/Etc	Fab/Ros; WL
S. decora	Northern mountain ash	Mixed HW	Cor/Eri/Lau/Etc	Fab/Ros; WL
Swietenia mahagoni	West Indian mahogany	Mixed HW	Cor/Eri/Lau/Etc	
Tilia spp.	Basswood spp.	Mixed HW	Hip/Til	
T. americana	American basswood	Mixed HW	Hip/Til	

Table 2 continued

Scientific name	Common name	Jenkins group	Chojnacky et al. parameters when diameter is measured at	
			Breast height	Root collar
T. americana var. heterophylla	White basswood	Mixed HW	Hip/Til	
T. americana var. caroliniana	Carolina basswood	Mixed HW	Hip/Til	
Ulmus spp.	Elm spp.	Mixed HW	Cor/Eri/Lau/Etc	
U. alata	Winged elm	Mixed HW	Cor/Eri/Lau/Etc	
U. americana	American elm	Mixed HW	Cor/Eri/Lau/Etc	
U. crassifolia	Cedar elm	Mixed HW	Cor/Eri/Lau/Etc	
U. pumila	Siberian elm	Mixed HW	Cor/Eri/Lau/Etc	
U. rubra	Slippery elm	Mixed HW	Cor/Eri/Lau/Etc	
U. serotina	September elm	Mixed HW	Cor/Eri/Lau/Etc	
U. thomasii	Rock elm	Mixed HW	Cor/Eri/Lau/Etc	
Umbellularia californica	California laurel	Mixed HW	Cor/Eri/Lau/Etc	
Yucca brevifolia	Joshua tree	Mixed HW	Cor/Eri/Lau/Etc	
Avicennia germinan	Black mangrove	Mixed HW	Cor/Eri/Lau/Etc	
Conocarpus erectus	Button mangrove	Mixed HW	Cor/Eri/Lau/Etc	
Laguncularia racemosa	White mangrove	Mixed HW	Cor/Eri/Lau/Etc	
Rhizophora mangle	American mangrove	Mixed HW	Cor/Eri/Lau/Etc	
Olneya tesota	Desert ironwood	Woodland	Fab/Jug	Fab/Ros; WL
Tamarix spp.	Saltcedar	Mixed HW	Cor/Eri/Lau/Etc	
Melaleuca quinquenervia	Melaleuca	Mixed HW	Cor/Eri/Lau/Etc	
Melia azedarach	Chinaberry	Mixed HW	Cor/Eri/Lau/Etc	
Triadica sebifera	Chinese tallowtree	Mixed HW	Cor/Eri/Lau/Etc	
Vernicia fordii	Tungoil tree	Mixed HW	Cor/Eri/Lau/Etc	
Cotinus obovatus	Smoketree	Mixed HW	Cor/Eri/Lau/Etc	
Elaeagnus angustifolia	Russian olive	Mixed HW	Cor/Eri/Lau/Etc	
Tree broadleaf	Unknown dead hardwood	Mixed HW	Cor/Eri/Lau/Etc	
Tree unknown	Unknown live tree	Mixed HW	Cor/Eri/Lau/Etc	
C. phaenopyrum	Washington hawthorn	Mixed HW	Cor/Eri/Lau/Etc	Fab/Ros; WL
C. succulenta	Fleshy hawthorn	Mixed HW	Cor/Eri/Lau/Etc	Fab/Ros; WL
C. uniflora	Dwarf hawthorn	Mixed HW	Cor/Eri/Lau/Etc	Fab/Ros; WL
F. berlandieriana	Berlandier ash	Mixed HW	Olea; LoSG	
Persea americana	Avocado	Mixed HW	Cor/Eri/Lau/Etc	
Ligustrum sinense	Chinese privet	Mixed HW	Olea; HiSG	
Q. gravesii	Graves oak	H Maple/Oak	Faga; Decid	Fagac; WL
Q. polymorpha	Mexican white oak	H Maple/Oak	Faga; Decid	Fagac; WL
Q. buckleyi	Buckley oak	H Maple/Oak	Faga; Decid	Fagac; WL
Q. laceyi	Lacey oak	H Maple/Oak	Faga; Decid	Fagac; WL
Cordia boissieri	Anacahuita Texas olive	Mixed HW	Cor/Eri/Lau/Etc	
Tamarix aphylla	Athel tamarisk	Mixed HW	Cor/Eri/Lau/Etc	

The first part of the Chojnacky parameter designator is the species group; text after a semicolon indicates the relevant category when more than one set of coefficients is given for a group

HiSG the coefficients given for the highest specific gravity in the designated species group, *LoSG* the lowest specific gravity given for a species group, *MedSG* select the coefficients given for the mid-range specific gravity. *WL* select the set of coefficients given for the woodland type. For example, Fagac; WL indicates that the second to the last line of Table 5, Woodland, Fagaceae should be used rather than the coefficients provided for Hardwood; Fagaceae

equivalent versus different is set by researchers and a conclusion of not-different, or equivalent, results from rejecting the null hypothesis (that the two are different).

Equivalence tests presented here are paired-sample tests [24, 25] because each sample is based on estimates from each of the Chojnacky and Jenkins groups. Our test statistic is the difference between estimates (Chojnacky minus Jenkins), and we set "equivalence" as a mean difference less than 5 % of the Jenkins-based estimate. Putting our test in terms of the null and alternative hypotheses following the format of publications describing this approach [22, 24], we have:

Null, H_0: (Chojnacky-Jenkins) <-5 % Jenkins or (Chojnacky-Jenkins) >5 Jenkins

and

Alternative, H_1: -5 % Jenkins $\leq$ (Chojnacky-Jenkins) ≤ 5 % Jenkins

We use the two one-sided tests (TOST) of our two-part null hypothesis that the plot-level difference was greater than 5 % of the Jenkins value and set $\alpha = 0.05$—one test that the mean difference is less than minus 5 % of the Jenkins estimate, and one test that the mean difference is greater than 5 % of the Jenkins estimate. Within an application of the TOST where α is set to 0.05, a one-step approach to accomplish the TOST result is establish a 2-sided 90 % confidence interval for the test statistic; if this falls entirely within the prescribed interval then the two populations can be considered equivalent [26]. We also extended the level of "equivalence" to within 10 % of the Jenkins-based estimates for some analyses in order to look for more general trends, or broad agreement between the two approaches.

Our equivalence tests are based on the paired estimates of carbon tonnes per hectare on the single-condition forested plots variously classified according to regions described in Fig. 1, forest type-groups listed in Table 1, or stand size class as in Fig. 3 (see [13] for additional details about these classifications). The distribution of the test statistic (mean difference) was obtained from resampling with replacement [27] ten thousand times, with a mean value determined for each sample. The number of plots available varied depending on the classification (Table 1; Fig. 3). We did not test for equivalence if fewer than 30 plots were available, and if over 2000 plots were available we randomly selected 2000 for resampling. The choice of 2000 is based on preliminary analysis of these data that showed the confidence interval from resampling converge with percentiles obtained directly from the distribution of the large number of sample plots, usually well below 1000; the 2000 is simply a round number well beyond this convergence without getting too computationally intense. The 90 % confidence interval (the same as the 95 % interval of TOST) obtained for the

distribution of the mean difference is according to a bias corrected and accelerated percentile method [28, 29]. Note that our tests for equivalence are based on comparing this confidence interval to the ±5 % of the corresponding Jenkins based estimate. Table 1 provides the estimates from the two approaches, with the equivalence test results indicated with asterisks. Similarly, the equivalence test results in Fig. 3 are not in the tonnes per hectare of the resampled values and the confidence interval, they are represented as percentage of Jenkins estimates—for this, equivalence is established if the entire confidence interval is within the zero side of the respective 5 %.

Authors' contributions

Design and analysis was split equally between JS and CH; JS was responsible for coding and calculations, CH developed the figures and tables, and writing was equally divided between JS and CH. All authors read and approved the final manuscript.

Acknowledgements

The authors gratefully acknowledge the helpful feedback from Linda Heath and William Leak on the draft manuscript. We would also like to thank the anonymous reviewers for their time and comments.

Competing interests

The authors declare that they have no competing interests.

References

1. Jenkins JC, Chojnacky DC, Heath LS, Birdsey RA. National-scale biomass estimators for United States tree species. For Sci. 2003;49:12–34.
2. Hoover CM, Rebain SA. Forest carbon estimation using the forest vegetation simulator: seven things you need to know. Gen. Tech. Rep. NRS-77. Newtown Square, PA: U.S. Department of Agriculture, Forest Service, Northern Research Station; 2011.
3. Twery MJ, Knopp, PD, Thomasma, SA, Nute, DE. NED-2 reference guide. Gen. Tech. Rep. NRS-86. Newtown Square, PA: U.S. Department of Agriculture, Forest Service, Northern Research Station; 2012.
4. US EPA. Inventory of U.S. Greenhouse gas emissions and sinks: 1990–2009. EPA 430-R-11-005. U.S. Environmental Protection Agency, Office of Atmospheric Programs, Washington, DC; 2011. http://www.epa.gov/climatechange/ghgemissions/usinventoryreport/archive.html.
5. Liénard JF, Gravel D, Strigul NS. Data-intensive modeling of forest dynamics. Environ Model Softw. 2015;67:138–48.
6. Ziter C, Bennett EM, Gonzalez A. Temperate forest fragments maintain aboveground carbon stocks out to the forest edge despite changes in community composition. Oecologia. 2014;176:893–902.
7. Carter DR, Tahey RT, Dreisilker K, Bialecki MB, Bowles ML. Assessing patterns of oak regeneration and C storage in relation to restoration-focused management, historical land use, and potential trade-offs. For Ecol Manage. 2015;343:53–62.
8. Reinikainen M, D'Amato AW, Bradford JB, Fraver S. Influence of stocking, site quality, stand age, low-severity canopy disturbance, and forest composition on sub-boreal aspen mixedwood carbon stocks. Can J For Res. 2014;44:230–42.
9. DeSiervo MH, Jules ES, Safford HD. Disturbance response across a productivity gradient: postfire vegetation in serpentine and nonserpentine forests. Ecosphere. 2015;6(4):60. doi:10.1890/ES14-00431.1.
10. Dore S, Kolb TE, Montes-Helu M, et al. Carbon and water fluxes from ponderosa pine forests disturbed by wildfire and thinning. Ecol Appl. 2010;20:663–83.

11. Magruder M, Chhin S, Palik B, Bradford JB. Thinning increases climatic resilience of red pine. Can J For Res. 2013;43:878–89.

12. Chojnacky DC, Heath LS, Jenkins JC. Updated generalized biomass equations for North American tree species. Forestry. 2014;87:129–51.

13. USDA Forest Service. Forest Inventory and Analysis National Program, FIA library: Database Documentation. U.S. Department of Agriculture, Forest Service, Washington Office; 2015. http://www.fia.fs.fed.us/library/database-documentation/.

14. Jenkins JC, Chojnacky DC, Heath LS, Birdsey RA. Comprehensive database of diameter-based biomass regressions for North American tree species. Gen. Tech. Rep. NE-319. Newtown Square, PA: U.S. Department of Agriculture, Forest Service, Northeastern Research Station; 2004.

15. Bragg DC, McElligott KM. Comparing aboveground biomass predictions for an uneven-aged pine-dominated stand using local, regional, and national models. J Ark Acad Sci. 2013;67:34–41.

16. Melson SL, Harmon ME, Fried JS, Domingo JB. Estimates of live-tree carbon stores in the Pacific Northwest are sensitive to model selection. Carbon Balance Manag. 2011;6:2.

17. Bragg DC. Modeling loblolly pine aboveground live biomass in a mature pine-hardwood stand: a cautionary tale. J. Ark. Acad. Sci. 2011;65:31–8.

18. USDA Forest Service. Forest Inventory and Analysis National Program: FIA Data Mart. U.S. Department of Agriculture Forest Service. Washington, DC; 2015. http://apps.fs.fed.us/fiadb-downloads/datamart.html. Accessed 2 June 2015.

19. USDA, NRCS. The PLANTS Database. National Plant Data Team, Greensboro, NC, USA; 2015. http://plants.usda.gov. Accessed 23 September 2015.

20. Robinson AP, Duursma RA, Marshall JD. A regression-based equivalence test for model validation: shifting the burden of proof. Tree Phys. 2005;25:903–13.

21. MacLean RG, Ducey MJ, Hoover CM. A comparison of carbon stock estimates and projections for the northeastern United States. For Sci. 2014;60(2):206–13.

22. Parkhurst DF. Statistical significance tests: equivalence and reverse tests should reduce misinterpretation. BioSci. 2001;51:1051–7.

23. Brosi BJ, Biber EG. Statistical inference, Type II error, and decision making under the US Endangered Species Act. Front Ecol Environ. 2009;7(9):487–94.

24. Feng S, Liang Q, Kinser RD, Newland K, Guilbaud R. Testing equivalence between two laboratories or two methods using paired-sample analysis and interval hypothesis testing. Anal Bioanal Chem. 2006;385:975–81.

25. Mara CA, Cribbie RA. Paired-samples tests of equivalence. Commun Stat Simulat. 2012;41:1928–43.

26. Berger RL, Hsu JC. Bioequivalence trials, intersection-union tests and equivalence confidence sets. Stat Sci. 1996;11(4):283–319.

27. Efron B, Tibshirani RJ. An introduction to the bootstrap. New York: Chapman and Hall; 1993

28. Carpenter J, Bithell J. Bootstrap confidence intervals: when, which, what? A practical guide for medical statisticians. Statis Med. 2000;19:1141–64.

29. Fox J. Bootstrapping regression models. In: Applied Regression Analysis and Generalized Linear Models, 2nd edition. Thousand Oaks: Sage, Inc; 2008. pp 587–606.

7

Rapid forest carbon assessments of oceanic islands: a case study of the Hawaiian archipelago

Gregory P. Asner[1]* ◉, Sinan Sousan[1], David E. Knapp[1], Paul C. Selmants[2], Roberta E. Martin[1], R. Flint Hughes[3] and Christian P. Giardina[3]

Abstract

Background: Spatially explicit forest carbon (C) monitoring aids conservation and climate change mitigation efforts, yet few approaches have been developed specifically for the highly heterogeneous landscapes of oceanic island chains that continue to undergo rapid and extensive forest C change. We developed an approach for rapid mapping of aboveground C density (ACD; units = Mg or metric tons C ha^{-1}) on islands at a spatial resolution of 30 m (0.09 ha) using a combination of cost-effective airborne LiDAR data and full-coverage satellite data. We used the approach to map forest ACD across the main Hawaiian Islands, comparing C stocks within and among islands, in protected and unprotected areas, and among forests dominated by native and invasive species.

Results: Total forest aboveground C stock of the Hawaiian Islands was 36 Tg, and ACD distributions were extremely heterogeneous both within and across islands. Remotely sensed ACD was validated against U.S. Forest Service FIA plot inventory data (R^2 = 0.67; RMSE = 30.4 Mg C ha^{-1}). Geospatial analyses indicated the critical importance of forest type and canopy cover as predictors of mapped ACD patterns. Protection status was a strong determinant of forest C stock and density, but we found complex environmentally mediated responses of forest ACD to alien plant invasion.

Conclusions: A combination of one-time airborne LiDAR data acquisition and satellite monitoring provides effective forest C mapping in the highly heterogeneous landscapes of the Hawaiian Islands. Our statistical approach yielded key insights into the drivers of ACD variation, and also makes possible future assessments of C storage change, derived on a repeat basis from free satellite data, without the need for additional LiDAR data. Changes in C stocks and densities of oceanic islands can thus be continually assessed in the face of rapid environmental changes such as biological invasions, drought, fire and land use. Such forest monitoring information can be used to promote sustainable forest use and conservation on islands in the future.

Keywords: Carbon stocks, Carnegie Airborne Observatory, Forest inventory, Invasive species, LiDAR, Random Forest Machine Learning

Background

Aboveground carbon (C) stock assessments have become a mainstay of forest management [1]. In the past decade, the importance of such assessments has also grown in the climate change mitigation arena [2]. In step with these efforts, there has been increasing focus on developing quantitative methods to monitor forest C stocks over time, as a means to support policies that reduce emissions from deforestation and forest degradation, and increase C storage in existing forests (REDD+) [3]. C storage has also become an important metric for assessing forest habitat and condition in the broader conservation arena [4, 5].

Based on the increasing value in understanding the geography of forest C stocks, both field-based and remote sensing-assisted C assessments have been

*Correspondence: gpa@carnegiescience.edu
[1] Department of Global Ecology, Carnegie Institution for Science, 260 Panama St, Stanford, CA 94305, USA
Full list of author information is available at the end of the article

undertaken over larger and larger geographic areas [6, 7]. Far less attention, however, has been given to oceanic islands, likely due to their relatively small land area. Oceanic islands provide model socio-ecological systems with which to examine spatial patterns in forest C stocks, because islands are often comprised of highly heterogeneous ecosystems, where many of the drivers of C storage (e.g., vegetation types, climate, fire, and land use) vary strongly over short distances [8, 9]. While C stocks on oceanic islands may be small in a global context, they provide unique opportunities to test fundamental concepts on the landscape ecology, sociology, economics and management of forest C sequestration. Further, forests on oceanic islands are quite important to the provisioning of ecosystem goods and services, including fresh water supply, prevention and mitigation of soil erosion that can deplete upland soil resources and pollute aquatic ecosystems including coral reefs [10], and both timber and non-timber forest products. Island forests also play a strong cultural role as a locus of subsistence and recreational activities [11, 12]. However, relative to continental ecosystems, forests on oceanic islands continue to undergo a much greater proportional extent and rate of change in cover and composition, which threatens the sustainability of forest-based good and services including C stocks [13, 14]. Not only have islands been heavily deforested in some regions of the world, they have also undergone enormous change via introduced disturbance regimes, such as fire, and alien invasive species [15, 16]. The effects of these and other changes on forest C stocks remain poorly understood, despite numerous local- to landscape-scale assessments [17]. Without continuous and spatially extensive forest monitoring, patterns of change and/or opportunities for recovery of island forests will remain a challenge to incorporate into conservation, management and resource policy initiatives.

Like most oceanic islands, aboveground forest C stocks within and across the Hawaiian Islands are poorly known, owing to extreme environmental heterogeneity combined with local inaccessibility and complex terrain. This has greatly limited efforts to develop and maintain operational, repeat forest inventory on the ground. Global remote sensing-based carbon mapping approaches generally yield lower spatial resolutions and C stock sensitivities [18–21], which are difficult to apply in regions of high ecological heterogeneity like islands. While high-resolution remote sensing methods, such as airborne Light Detection and Ranging (LiDAR) [22], are suitable for such settings [23], mapping remote or difficult-to-access areas with aircraft can be expensive. In particular, cloud cover is often persistent over higher-elevation forests of key interest in forest C and watershed assessments. As a result, airborne campaigns can

be prolonged and accumulate costs. An added challenge is that island forest assessments are needed on a repeat basis in response to the inherent vulnerability of many island landscapes to rapid change driven by land use, fire, storms (e.g., hurricanes), biological invasions and sea level rise. The issue of rapid change calls for the development of a low-cost, repeatable forest monitoring method for island forests. Such rapid, high-resolution assessment capabilities must be sensitive to the drivers of forest C change, not only as a metric for climate change mitigation, but also as a measure of forest health and provisioning services.

While mapping of forest C stocks has been challenged by uncertainty and cost [7], recent progress at subnational to national levels indicates that significant methodological hurdles can be overcome at larger scales, especially through the fusion of ground, aircraft and satellite based measurements [21, 24]. These approaches can simultaneously increase map resolution in ways that benefit forest managers, while reducing uncertainty to levels acceptable to policy makers. Despite these advances, important methodological questions remain regarding how to provide high resolution, low uncertainty monitoring at low cost in heterogeneous landscapes. A further need is the simultaneous assessment of the drivers of spatial variation in C storage.

We developed an approach for monitoring forest aboveground carbon density (ACD; units = Mg or metric tons C ha^{-1}) across island archipelagos at a spatial resolution of 30 m (0.09 ha) using a combination of airborne LiDAR and freely available satellite data (Fig. 1). The approach involves initial use of high-resolution LiDAR sampling of a selected island within an archipelago to derive vegetation canopy height data. These data from the sampled island are then used to train a geospatial model that incorporates maps of multiple environmental factors, as well as forest canopy structural metrics derived from Landsat or comparable satellite imagery [25]. The resulting model is applied to all islands within the archipelago using as input the same portfolio of environmental and satellite-based canopy structural maps used on the model-training island, thereby yielding a multi-island map of canopy height at 30-m spatial resolution. Finally a regionally-tuned equation is applied to relate mapped canopy height to ACD [26], resulting in a carbon density map at 30-m resolution for the entire island chain. Critically, once the model is built for an archipelago, subsequent changes in ACD can be detected using only Landsat imagery, thereby greatly reducing longer-term monitoring costs [24].

For this study, we first sampled Hawaii Island, by far the largest island in the Hawaiian archipelago, with airborne LiDAR to assess forest top-of-canopy height (TCH) responses to natural environmental gradients

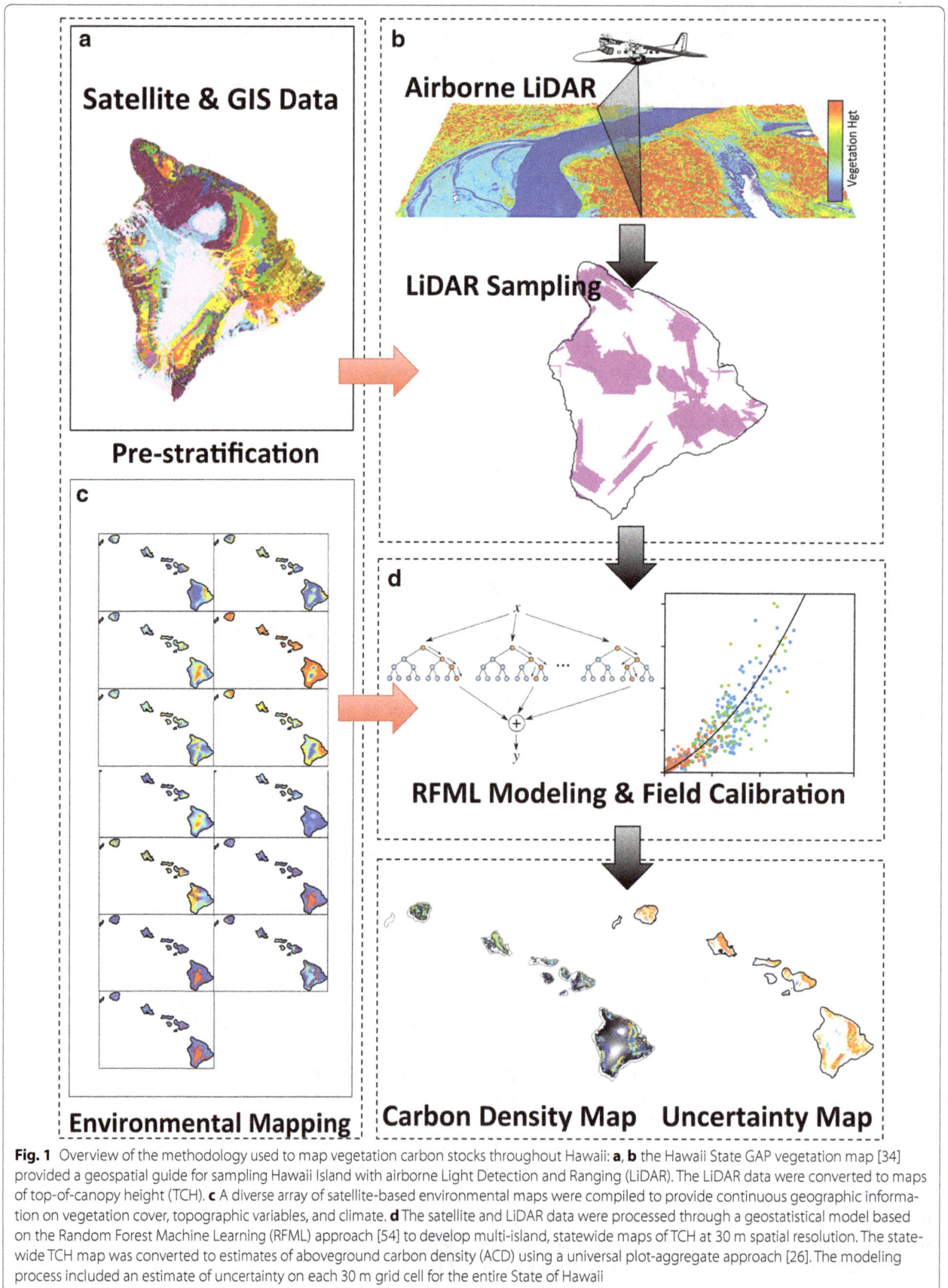

Fig. 1 Overview of the methodology used to map vegetation carbon stocks throughout Hawaii: **a**, **b** the Hawaii State GAP vegetation map [34] provided a geospatial guide for sampling Hawaii Island with airborne Light Detection and Ranging (LiDAR). The LiDAR data were converted to maps of top-of-canopy height (TCH). **c** A diverse array of satellite-based environmental maps were compiled to provide continuous geographic information on vegetation cover, topographic variables, and climate. **d** The satellite and LiDAR data were processed through a geostatistical model based on the Random Forest Machine Learning (RFML) approach [54] to develop multi-island, statewide maps of TCH at 30 m spatial resolution. The statewide TCH map was converted to estimates of aboveground carbon density (ACD) using a universal plot-aggregate approach [26]. The modeling process included an estimate of uncertainty on each 30 m grid cell for the entire State of Hawaii

and land use (Additional file 1: Figure S1). These LiDAR TCH data from Hawaii Island were used to calibrate a Random Forest Machine Learning (RFML) model, which was subsequently used to predict TCH at 30 m resolution on all islands from a portfolio of spatially explicit predictor maps (Additional file 1: Figures S2-S4). The resulting statewide model of forest TCH was then used to estimate forest ACD via a conversion equation developed for the Hawaiian Islands (Additional file 1: Figure S5). The resulting map was compared to US Forest Service Forest Inventory and Analysis (FIA—http://www.fia.fs.fed.us/) plot data for evaluation of mapped ACD precision. Finally, we used the new ACD map to assess aboveground forest C stocks within and among islands, in protected and unprotected areas, and among forests dominated by native and invasive plant species.

Results and discussion

Island carbon stocks and distributions

Total forest cover and aboveground carbon stock for seven main Hawaiian Islands was estimated at 550,065 ha

and 36.0 Tg (million metric tons), respectively (Fig. 2; Table 1). A map of estimated uncertainty indicated greatest absolute uncertainties of 20–40 % in very high-biomass forests, with much lower uncertainties in low-to-moderate biomass conditions (Additional file 1: Figure S6). Forest ACD varied widely by island (Fig. 3). Hawaii Island contained 57 % of the total forest cover of the State, and almost 20 Tg of the State's forest carbon. Kauai, Maui and Oahu islands collectively accounted for 36 % of the total forest cover and 14.7 Tg of aboveground C. Molokai, Lanai, and Kahoolawe together accounted for only 7 % of the State's forest cover and less than 1.4 Tg C. The small northwest-most island of Niihau was not considered in this study.

The highest forest ACDs were found on Hawaii Island, reaching 537 Mg C ha^{-1}. Maui supported the next highest ACDs, reaching 294 Mg C ha^{-1}. We also found extremely variable C stocks on each island (Additional file 1: Figures S7-S10). Aboveground forest C density varied up to three fold among State Districts, which are the minimum State-level political units of civil governance

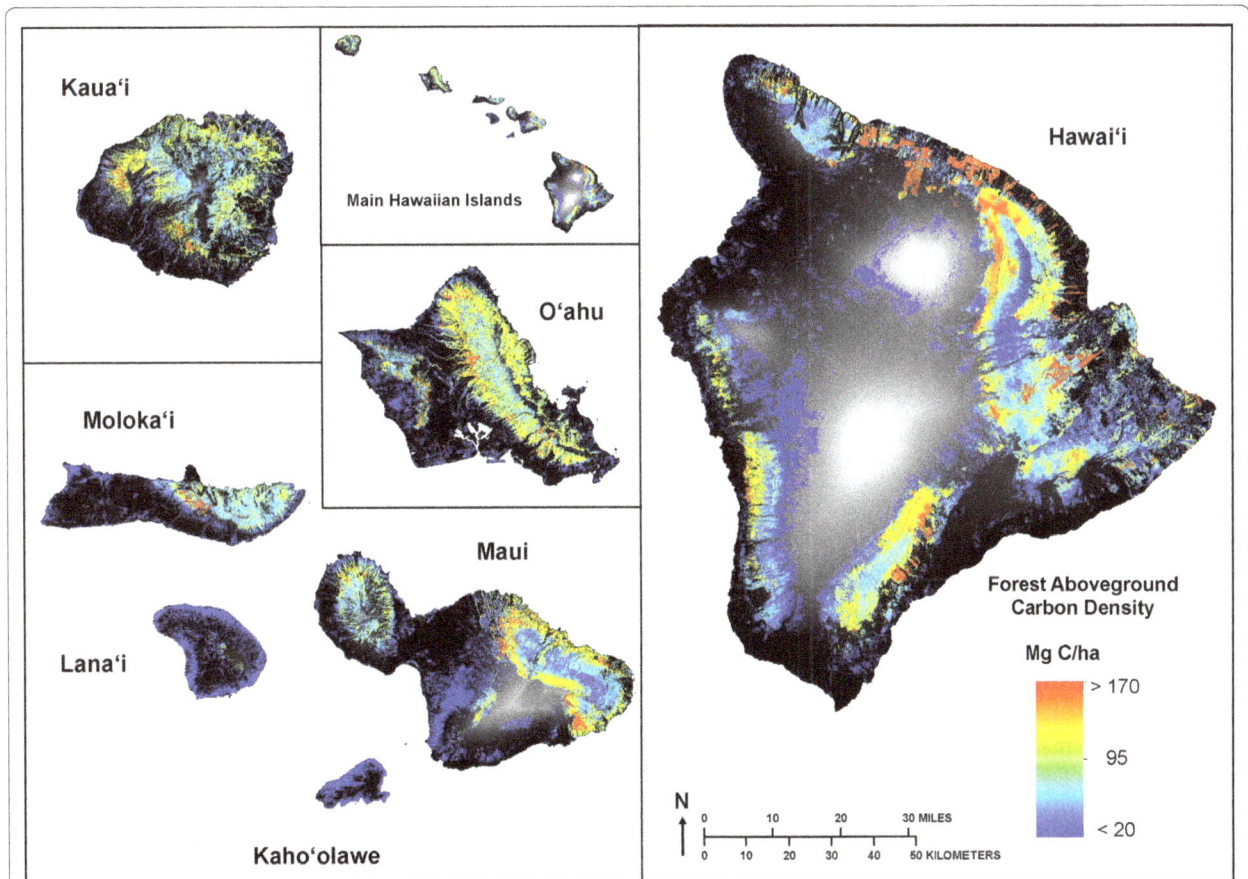

Fig. 2 Spatial distribution of forest aboveground carbon density (ACD; Mg C ha^{-1}) for the State of Hawaii at 30-m mapping resolution. A map of estimated uncertainty is provided in Additional file 1: Figure S6. The islands are displayed so that their relative sizes are preserved

Table 1 Forest cover and aboveground carbon stock and density for each island and the State's Districts

Island	Counties and districts	Forest cover (ha)	Aboveground carbon density (Mg C ha^{-1})	Aboveground carbon stock (Tg C)
Hawaii		311,977.0	64.0 + 43.7	20.0
	Hawaii County			
	Hamakua	23,391.8	51.4 + 47.4	1.2
	Kau	63,204.2	67.0 + 43.3	4.2
	North Hilo	18,598.8	93.3 + 49.3	1.7
	South Hilo	67,056.8	72.8 + 41.7	4.9
	North Kohala	8341.1	65.9 + 47.9	0.6
	South Kohala	7057.3	47.7 + 31.9	0.3
	North Kona	28,391.2	30.0 + 35.7	0.9
	South Kona	30,635.5	60.8 + 40.8	1.9
	Puna	65,302.4	66.3 + 36.8	4.3
Maui		75,532.9	67.1 + 47.0	5.1
	Maui County			
	Hana	29,763.6	78.7 + 41.4	2.3
	Lahaina	22,113.8	33.5 + 40.7	0.7
	Makawao	34,542.2	47.5 + 52.7	1.6
	Wailuku	7543.5	61.1 + 39.5	0.5
	Molokai	23,018.2	54.9 + 37.3	1.3
Molokai		23,018.2	54.9 + 37.3	1.3
Lanai		13,048.5	7.6 + 15.1	0.1
Kahoolawe		5391.1	3.0 + 2.5	0.02
Oahu		64,673.4	78.3 + 40.2	5.1
	Honolulu County			
	I	3562.3	77.4 + 39.2	0.3
	II	1648.4	92.4 + 33.3	0.2
	III	2803.9	63.4 + 47.0	0.2
	IV	6669.2	87.9 + 32.9	0.6
	V	10,988.6	95.5 + 32.0	1.1
	VI	14,449.8	80.2 + 38.9	1.2
	VII	4812.6	85.3 + 36.3	0.4
	VIII	5047.9	34.3 + 37.3	0.2
	IX	14,407.3	74.0 + 38.8	1.1
Kauai		56,424.0	80.4 + 35.5	4.5
	Kauai County			
	Hanalei	16,033.8	82.8 + 32.0	1.3
	Kawaihau	9020.1	83.0 + 32.1	0.8
	Koloa	3024.7	78.5 + 43.2	0.2
	Lihue	8886.3	79.2 + 32.7	0.7
	Waimea	19,449.3	78.1 + 39.2	1.5

(Table 1). On Hawaii Island, for example, forest ACD values varied from means of 30–93 Mg C ha^{-1} across Districts, yet within Districts, spatial variation in forest ACD ranged from 50 to 111 % of their District means. Moreover, three of nine Districts on Hawaii Island contained two-thirds of the entire island's forest C stock. The island with the most variable inter-District forest C stocks was Maui.

Model comparison to FIA plots

Comparison of modeled ACD to values estimated from FIA plot inventory indicated good precision ($R^2 = 0.67$) and accuracy (average root mean squared error or RMSE = 30.4 Mg C ha^{-1}) (Fig. 4). Bias was just 11.2 Mg C ha^{-1}, and heteroscedasticity was similar to that derived in plot-inventory comparison studies [27]. These map performances were particularly strong relative

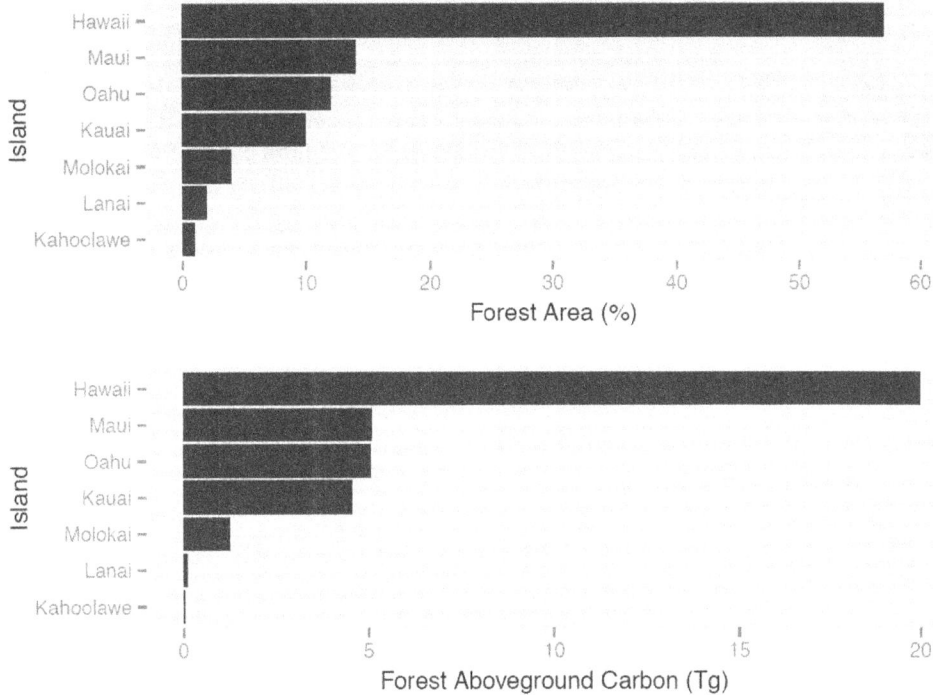

Fig. 3 Distribution of forest area and total aboveground carbon stock (Tg = million metric tons) for the main Hawaiian Islands. Percentages are given in terms of the entire State of Hawaii

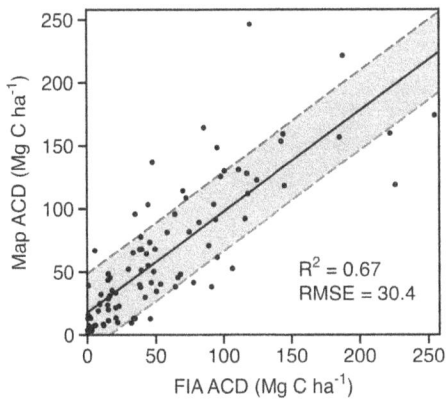

Fig. 4 Comparison of Hawaii statewide map of forest aboveground carbon density (ACD) against plot inventory-based estimates of ACD from the US Forest Service FIA plot-inventory data

to the accuracy of the equation used for estimating ACD from canopy height (Additional file 1: Figure S5).

Here we note the challenges involved in comparing the FIA plot data to mapped C densities based on remote sensing. First, there was an offset of about 6 years between the time the LiDAR flights were completed and the time the FIA measurements were taken in the field. Second, the FIA data in Hawaii were geo-located using

non-differentially corrected global positioning system (GPS) instruments. This leads to plot location uncertainties of up to 30 m. The combination of relatively small size (18 m radius), circular shape, and non-contiguity of the FIA plots (see "Methods"), explains higher uncertainty when comparing to ACD estimates in 30 m × 30 m mapping cells. Asner et al. [28] found that mismatches in location and plot shape alone account for up to 15 % uncertainty in field validation studies. Additionally, the allometric scaling applied to the FIA field measurements can result in additional uncertainties of up to 50 % of the plot mean value [29, 30].

Given these, and other sources of uncertainty, we contend that the verification step undertaken here was successful in validating the map results. Nonetheless, validation with FIA or other plots could be significantly improved by more accurate GPS measurements of plot locations, and by employing plot and sampling design that is better suited to validating remotely-sensed estimates of ACD. Specifically, plots should be similar in area to the final grid size and all trees >5 cm dbh should have height and diameter measured in each plot. Better allometry would also decrease uncertainty. Currently, we employ species-specific allometric equations only for the two most dominant native woody tree species (*Metrosideros polymorpha* and *Acacia koa*) and for four non-native tree species. Aboveground biomass for the remaining 114 tree species encountered in FIA plots was estimated using a general model for tropical trees that incorporates diameter, height and wood density [31]. Species-specific allometry for large, widespread non-native tree species, such as *Falcataria moluccana*, would almost certainly reduce uncertainty in estimates of their aboveground biomass.

Factors affecting carbon stocks

The geospatial analysis indicated that fractional canopy cover (FC) was the principal driver of spatial variation in forest carbon stocks throughout the Hawaiian archipelago, accounting for 27 % of the total variance in ACD (Fig. 5). Forest cover was closely followed by forest type, as defined using the vegetation-cover classification, which accounted for an additional 24 % of variation in ACD. Other important factors included mean annual precipitation, vegetation structure, and cloudiness, which individually explained 6–8 % of the ACD variation throughout the islands. Finally, fire return factors, elevation and additional climate variables individually explained 1–4 % of the variability in carbon density.

Note that while the results presented in Fig. 5 account for co-variation in explanatory factors, many of them are ecologically and/or geospatially convolved with one another. For example, forest FC is broadly related to elevation and topographic aspect, with less forest cover

often observed at high elevations and on leeward aspects, although low forest FC was also observed in deforested zones at lower elevations on windward aspects. Thus the factor rankings presented here indicate an additional effect of elevation and aspect not already explained by FC alone. Similar inter-factor co-variances occur among the model rankings in Fig. 5. Nonetheless, it is clear that FC and vegetation type explain much of the geographic variation in forest carbon stocks.

Effects of biological invasion on forest carbon

Although this study is limited to a single time step, the current Hawaii vegetation map allowed us to conduct the first statewide assessment of the large-scale effects of alien plant species on forest C stocks. Numerous plot- to landscape-scale studies have reported on this issue, with highly variable outcomes ranging from no effect of invasion on carbon densities, to increases and decreases in ACD following invasion [17, 23, 28, 32, 33]. Such wide-ranging results stem from underlying variability in the mediating factors, such as time-since-introduction, rates of invasion, relative changes in plant functional and structure types, and environmental filters such as soils and climate. There is thus a general need for large-scale, high-resolution assessments that go beyond local contextual results.

The Hawaii State vegetation map was generated using manual and automated classification of Landsat imagery

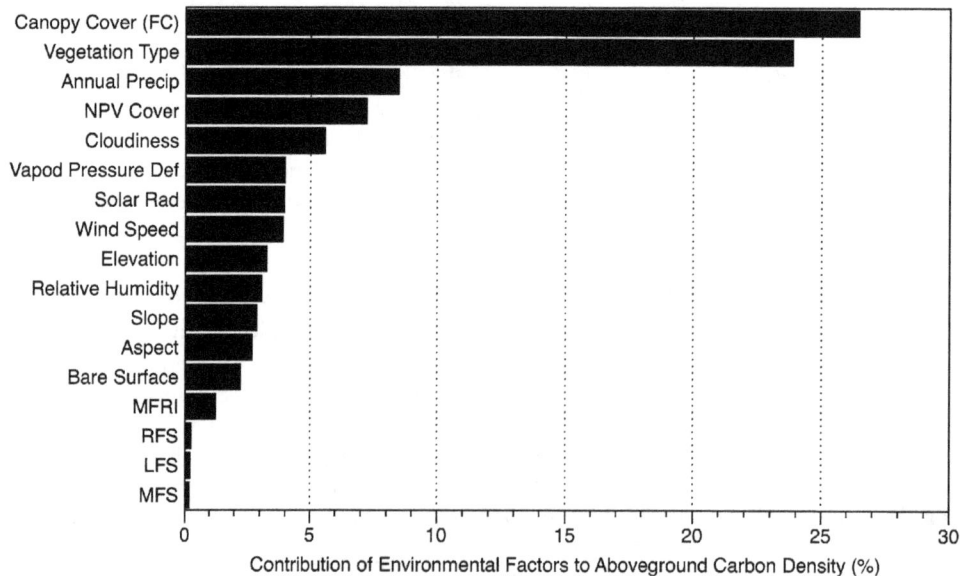

Fig. 5 Contribution of each potential explanatory factor determining aboveground carbon density (ACD) in the Hawaiian Islands. Fractional canopy cover (FC), non-photosynthetic vegetation (NPV) cover, and bare surface cover (soils, rock, infrastructure) were derived from sub-30 m resolution Landsat-based satellite mapping of the islands (see [25]). Vegetation type was provided by the Hawaii State GAP vegetation map [34]. *MFRI* mean fire return interval; *RFS* replacement fire severity; *LFS* low fire severity; *MFS* mixed fire severity

against aerial photography [34]. Experience with this map in field studies indicates that the "alien-dominated" classes are comprised of mature stands of non-native species, while "native-dominated" classes are comprised of mature stands of native species, particularly dominated by the keystone canopy species *Metrosideros polymorpha* and *Acacia koa*. We focused our analysis on these two groups because the Hawaii State vegetation map alone does not provide sufficient detail to partition the mapped C results into finer levels of invasion, particularly since the invasion process is ongoing and highly dynamic (in favor of alien invasive species dominance). We further partitioned the native- and alien-dominated groups by three major environmental filters known to mediate C stocks: annual precipitation, elevation and substrate age (from volcanic activity dating back to the early Pliocene) (Additional file 1: Table S1).

Our results show that, on medium-to-older substrates in both drier and wetter conditions, the total area of alien-dominated forest exceeds that of native-dominated forest in lower-elevation zones (Fig. 6a). In contrast, the majority of wetter, higher-elevation and/or older-substrate conditions remain dominated by native forest cover. Critically, however, we found that ACD is greater in native-dominated forests in low-to-medium elevation, dry-to-mesic regions of the islands, whereas alien-dominated forests tend to have slightly higher ACD levels in wetter environments across the board (Fig. 6b). At these broad multi-island scales, substrate age played only a small role in determining the *relative* difference in alien- and native-dominated forest ACD. This suggests strong limiting effects of nutrient-poor soils on growth and biomass accumulation for all species, independent of origin [35]. In contrast, higher biomass of native forest canopies in drier zones on older substrates may reflect evolutionary adaptation to these environments, as well as a lack of analog tree taxa in the current alien species pool on the islands.

Our results are also suggestive of how native biological diversity intersects with C storage, and how alien invasive species alter those relationships. For example, higher-elevation, drier forests on older substrates may be dominated by alien forest cover (smallest solid green dot; Fig. 6a), but native-dominated forests in similar environments support twice the stored C on a per-area basis (Fig. 6b). Thus actions to conserve and restore high-elevation native ecosystems yield a co-benefit of increased C storage. On the other hand, higher-elevation, drier conditions on younger substrates are areas currently dominated by native forest cover (open small red dot; Fig. 6a), but alien species can double the ACD levels in these environments (Fig. 6b). Forest managers and conservationists can use these landscape-scale relationships as trade-offs

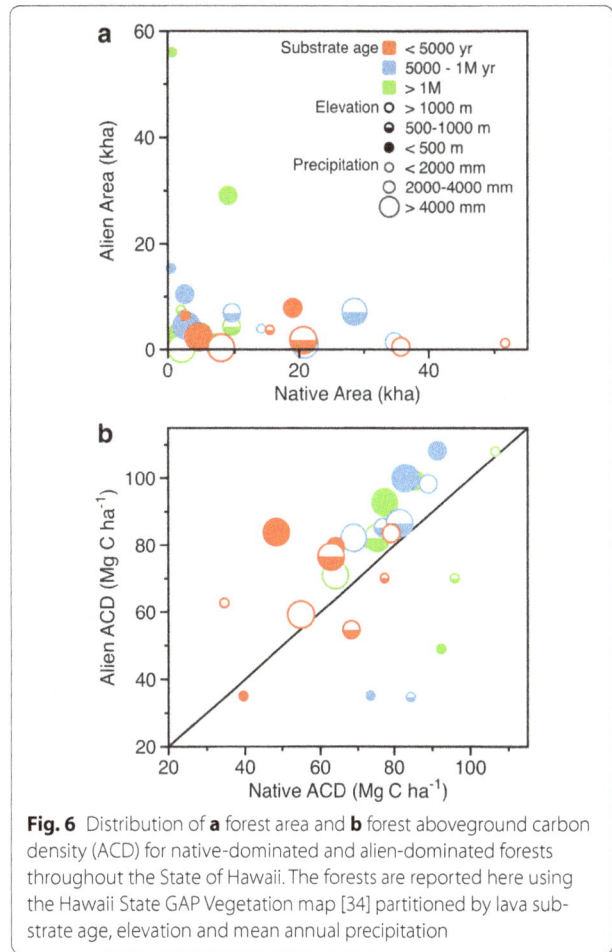

Fig. 6 Distribution of **a** forest area and **b** forest aboveground carbon density (ACD) for native-dominated and alien-dominated forests throughout the State of Hawaii. The forests are reported here using the Hawaii State GAP Vegetation map [34] partitioned by lava substrate age, elevation and mean annual precipitation

in planning efforts to increase C storage while managing for biological diversity [36, 37].

Forest carbon protections and opportunities

High-resolution C mapping also affords a way to assess current protections, threats and opportunities for sequestered carbon and generating healthy forests via land-use allocation and management [21]. Using land tenure data provided by the State of Hawaii, we quantified C stocks and densities on State, federal and private reserves. Of the total aboveground forest C stock found on the islands (36 Tg C), about 18.5 Tg C or 51 % is officially protected on State (e.g., Natural Area Reserves; Forest Reserves), federal (National Parks; Wildlife Refuges) and private (The Nature Conservancy; Kamehameha Schools lands) lands covering 257,691 ha (Fig. 7a, Additional file 1: Table S2). This is almost equally matched by forests outside of protected reserves, which in total cover more land area at 292,374 ha, but which contain 17.5 Tg of aboveground C. This finding indicates that a large amount of forest C could be incorporated into more formal reserve

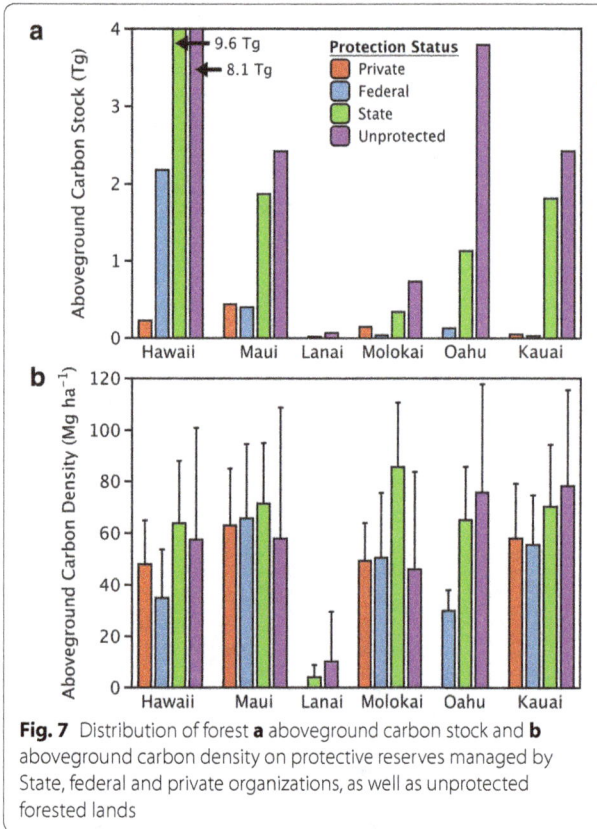

Fig. 7 Distribution of forest **a** aboveground carbon stock and **b** aboveground carbon density on protective reserves managed by State, federal and private organizations, as well as unprotected forested lands

protections. Moreover, we found that reserve ACD averages 61.8 ± 22.3 Mg C ha^{-1}, whereas non-reserve forests have carbon densities of 59.6 ± 34.2 3 Mg C ha^{-1} (Fig. 7b). Combined, these results underscore the C-storage benefit of adding long-term protection status to remaining island forests; Total forest aboveground C stock increases linearly with increasing reserve area (Additional file 1: Figure S11).

On all islands, 189 State-managed reserves hold the vast majority of protected carbon stocks—14.8 Tg C, while 25 federal and 14 private reserves contain just 2.8 and 0.9 Tg C, respectively (Additional file 1: Table S2). Carbon densities are highest in State reserves (66.3 ± 23.2 Mg C ha^{-1}), followed by private (56.1 ± 19.3 Mg C ha^{-1}) and federal reserves (41.4 ± 17.6 Mg C ha^{-1}). Differences in forest carbon densities are reflective of the location of the reserves (lowland vs. montane, wet windward vs. dry leeward) as well as species composition and management. A desired outcome of this work is to provide forest managers and the public with information to compare, for example, carbon stocks on a reserve-by-reserve basis against environmental maps, to identify opportunities for increasing C densities through conservation and management actions.

Replication on oceanic island chains

The approach we have developed and tested here for high-resolution mapping of aboveground forest carbon density is intended for replication on oceanic islands worldwide, but also any set of highly heterogeneous landscapes. The methodology is based on a previously established strategy that relies on airborne LiDAR sampling of forests found across a range of ecological conditions, but limited to one island [23]. Here we greatly advanced the approach by extending the initial LiDAR sampling of a single island, via a machine-learning algorithm [38, 39], to the multi-island or archipelago scale using a suite of environmental maps and satellite data that is, in combination, sufficiently sensitive to variation in the LiDAR-based estimates of canopy height. Shared environmental characteristics among neighboring islands usually include geology, climatic zones, and dominant vegetation types. Satellite-based metrics of forest structure, derived from Landsat-based spectral mixture analysis, are time-variant and key to the linkage with the LiDAR data. Strategically, these Landsat-based metrics can be updated through time using the fully automated CLASlite software [25].

The conversion of either LiDAR-scale or modeled canopy height to estimates of ACD requires plot-aggregate allometric equations [40]. This worked well in Hawaii, relative to plot-estimated ACD from U.S. Forest Service inventory data. The universal or regional plot-aggregate allometric equations proposed by Asner et al. [40] have also worked reasonably well in other regions [17, 41, 42], and they tend to result in mismatches between LiDAR-based and field-based estimates of ACD of 10–15 % when applied at 1-ha spatial resolution [26]. Nonetheless, application of these conversion equations to oceanic islands requires further validation, particularly for isolated islands in which vegetation types (and thus allometrics) may diverge from general databases.

There is an initial cost for installing a forest C monitoring program on any given island chain or archipelago. It includes an initial airborne LiDAR survey of one island or part of the archipelago, which varies widely in cost depending upon whether the data are sourced from non-profit, government, or commercial organizations. Our LiDAR data collection and processing cost was approximately $150,000 for the Island of Hawaii, but costs have greatly declined since the data acquisition was made for this study [43]. The LiDAR component was followed by personnel and computing costs required to link the LiDAR data to the satellite imagery and for validation work. However, the satellite imagery was free of charge, and CLASlite is also currently available at no charge [44], thereby providing us with a low-cost way to complete the initial carbon map. Moreover, the free imagery and software makes updates to the map extremely cost-efficient, likely requiring the effort of a single geospatial technician for the State of Hawaii. Even if field inventories could be done at large geographic scales on a spatially contiguous basis, which is not possible, the recurring costs would be extremely high for each monitoring step through time.

Conclusions
We have shown that a combination of one-time airborne LiDAR data acquisition, and freely available satellite data with automated analysis, can provide effective forest C mapping and monitoring of oceanic islands. The method is highly replicable and cost-effective. From the first map generated, and with regular updates using satellite data over time, assessments of C storage can be derived by political entity (e.g., State Districts), land-use allocation (e.g., protected vs. unprotected areas), or any other unit of governance or management. Moreover, changes in C stocks and densities can be continually assessed in the face of rapid environmental changes, such as climate, fire and biological invasion. The resulting information is spatially explicit, allowing for actions that promote sustainability of forests and the services they provide to island biodiversity and societies. High-resolution monitoring approaches also provide a geography of forest C stock that facilitates the inclusion of multiple stakeholders ranging from individual landowners to national governments. The resulting empowerment afforded by this type of ecological information will be important to the protection, enhancement and/or restoration of island ecosystems in the future.

Methods
Our mapping approach is summarized in Fig. 1. The necessary technologies are airborne Light Detection and Ranging (LiDAR), which yields highly detailed measurements of forest canopy height and vertical canopy profile, and satellite-derived maps of environmental variables and forest canopy fractional cover. A second component relies on machine learning algorithms to scale airborne LiDAR samples of one island up to multi-island or archipelago maps. Several studies have employed a Random Forest Machine Learning (RFML) algorithm to model the relationship between LiDAR-based estimates of forest structure or biomass and a suite of satellite data sets [19, 21, 45, 46]. RFML fits multiple environmental datasets (predictors) to estimates of vegetation structure or biomass (response), as described later. In doing so, a direct scaling of LiDAR samples to full-coverage maps can be derived without artificial boundaries between ecosystems that often occur using traditional stratification approaches.

Random Forest Machine Learning also provides quantitative information on which predictors (e.g., satellite data) are most important in determining the response variable (LiDAR-derived canopy height) [47]. Here the importance of a predictor to the RFML model was assessed by randomly permuting the values of the factor within a validation dataset, and processing the validation data through the regression trees. In our implementation of RFML, a temporary validation dataset is created to build each regression tree, and is chosen as a randomly selected set of 250 samples left out of the full training dataset. To assess the importance of a single factor, we compared the mean square error (MSE) values of the validation data both before (MSE_i) and after (MSE^*_i) randomly permuting the values of the factor for each tree [48]. For each tree i, the difference between MSE_i and MSE^*_i, divided by MSE_i, was collected. The importance of the given factor was then taken to be the mean of these relative difference values across all trees. By repeating the above procedure for each explanatory factor, the relative importance of each factor could be compared.

LiDAR data acquisition and analysis
LiDAR data were collected using the Carnegie Airborne Observatory [49]. Flights covered 379,337 ha of Hawaii Island (Additional file 1: Figure S1) including all major forest types (Additional file 1: Figure S2b) [23]. LiDAR data were collected at 1000 or 2000 m above ground level, using two corresponding configurations: higher resolution with 0.56 m on-the-ground laser spot spacing, 24° field of view (FOV) and a 70 kHz pulse repetition frequency; low resolution with a 1.12 m spot spacing, 30° FOV and a 50 kHz pulse repetition frequency, respectively. Ground cover was sampled along parallel flight lines with 50 % overlap to ensure LiDAR coverage of no less than 4 laser shots m^{-2}.

Mean top-of-canopy height (TCH) was calculated for each 30 m × 30 m grid cell of LiDAR coverage on Hawaii Island (Additional file 1: Figure S1). To create this layer, the laser range measurements from the LiDAR were combined with the embedded high resolution Global Positioning System-Inertial Measurement Unit (GPS-IMU) data to determine the 3-D locations of the laser returns. This calculation produced a 'cloud' of LiDAR data. The LiDAR data cloud was processed to identify where the laser pulses penetrated the canopy volume, reaching the ground surface, from which a digital terrain model (DTM) was produced. This was achieved using a 10 m × 10 m filter kernel throughout the LiDAR coverage, and the lowest elevation in each kernel was deemed as possible ground detection. These filtered points were then evaluated by fitting a horizontal plane through each point. If the closest unclassified point was <1.5 m higher in elevation, the pre-filtered point was finalized as a ground-classified surface point. This process was repeated until all potential ground points within the LiDAR coverage were evaluated. A digital surface model (DSM), which is essentially the top-most surface (e.g., canopies, buildings, exposed ground), was also generated based on interpolations of all first-return points at 1.12 m spatial resolution. The DTM and DSM were combined as a tightly matched pair of data layers. The vertical difference between them resulted in a model of top-of-canopy height (TCH) at 1.12 m spatial resolution throughout the 379,337 ha LiDAR sampling coverage. Validation studies of this CAO LiDAR TCH estimation approach have shown it to be highly accurate across a wide range of forests including extremely densely foliated, tall tropical forests exceeding 60 m in height [28, 42].

Environmental predictor variables

We used 17 environmental predictor variables from co-aligned spatial datasets covering State of Hawaii to model canopy height based on the LiDAR TCH measurements made on Hawaii Island (Additional file 1: Figure S2-S4). All predictor variables were gridded at 30-m spatial resolution. Three predictor variables were fractional cover of forest canopy (FC), non-photosynthetic vegetation (NPV), and bare surfaces. These were determined from nine primary Landsat-8 images collected in 2013 and 2014. The mosaic of nine images included a few small cloud-covered areas, so those areas were backfilled with Landsat-7 and Landsat-8 data going back to 2010. The Landsat mosaic was run through a probabilistic spectral mixture analysis algorithm embedded in the CLASlite forest monitoring software package [25]. These fractional cover images have been validated and used in numerous studies in Hawaii and elsewhere [e.g., 23, 50].

An important additional predictor variable was the Hawaii State GAP vegetation map, which provides the highest resolution and most widely used vegetation cover type information for the State of Hawaii. The version used was based on Gon et al. [34], with improvements based on high-resolution satellite images and other more recent vegetation mapping information [51]. Three additional predictor variables were derived from 30-m Shuttle Radar Topography Mapping (SRTM) mission data: elevation, slope and aspect. In addition, mean annual precipitation (MAP), mean wind speed at 2 m above ground, vapor pressure deficit, total solar radiation, mean relative humidity, and cloud frequency data were acquired from http://climate.geography.   http://hawaii.edu/downloads. html. Finally, we used four fire-related predictor variables: low fire severity (LFS), mixed fire severity (MFS), replacement fire severity (RFS), and mean fire return

interval (MFRI) provide by http://www.landfire.gov/fireregime.php.

These 17 predictor maps and the 30-m LIDAR-derived TCH map were applied to the RFML model for Hawaii Island to develop the prediction-based regression trees. The regression trees were then used to predict TCH values across the entire State of Hawaii using the 17 predictor maps as input.

Estimating aboveground carbon density

We estimated ACD from the statewide TCH map using a plot-aggregate allometric scaling approach [26]. A biophysical link was previously developed to quantitatively link mapped TCH to field estimates of ACD by applying regional plot-aggregated estimates of vegetation wood density and diameter-to-height relationships. To develop a TCH-to-ACD calibration for Hawaiian forests and other vegetation types throughout the State, we used 209 field plots located on Hawaii Island for which ACD was measured using field plot-based inventory measurements as detailed by Asner et al. [23]. The resulting calibration between TCH and ACD is shown in Additional file 1: Figure S5, with in $R^2 = 0.82$ and RMSE $= 78.7$ Mg C ha^{-1}. The final calibration equation for relating TCH to ACD was: ACD $= 3.744 * \text{TCH}^{1.391}$.

Uncertainty map

The uncertainty of the mapped ACD estimates was estimated by developing a relationship between the mapped ACD values and the RMSE of ACD for those areas on Hawaii Island covered by the LiDAR data [21]. These RMSE values were partitioned into 30 bins across the range of RFML-modeled ACD values. A polynomial was fit to model the RMSE of an ACD estimate as a function of its predicted ACD value. The polynomial was then applied to the ACD map to produce an estimate of ACD uncertainty (Additional file 1: Figure S6).

Map validation

To evaluate the accuracy of the final carbon map, we compared data from the map to georeferenced plots surveyed across the Hawaiian Islands in 2011 and 2012 by the United States Department of Agriculture Forest inventory and Analysis (FIA) Program. The FIA Program is a national network of plots designed to represent all forest conditions across the United States [52]. Each FIA plot is a cluster of four circular 7.32-m radius subplots arranged in a fixed pattern. All trees and tree ferns ≥12.7 cm diameter at breast height (dbh; 1.37 m above the ground) had diameter, height, and species recorded in each subplot. Trees and tree ferns <12.7 cm dbh had diameter, height, and species were recorded in microplots, which are 2.07 m radius plots located within each subplot. Macroplots, which are 17.95 m radius and immediately surround each subplot, are usually reserved for destructive sampling. However, FIA plots sampled in Hawaii in 2011–2012 using the 'experimental forest' (EXPFOR) protocol (n = 96) had all trees ≥12.7 cm dbh measured in Macroplots as well, greatly enlarging the sample footprint of each plot. We used data from these 96 EXPFOR FIA plots to validate the accuracy of the final carbon map.

We estimated ACD for each tree measured in the 96 FIA plots using a combination of species-specific and general diameter-to-ACD and height-to-diameter models. We used locally derived, species-specific diameter to ACD models for eight species, including the two most common species in the FIA dataset: *Metrosideros polymorpha* and *Acacia koa* (Additional file 1: Table S3). For all other species, and for large trees that exceeded the diameter range of species-specific diameter-to-ACD models, we used a general allometric model for tropical trees developed by Chave et al. [31] that uses diameter, height, and wood density to estimate ACD (Additional file 1: Table S4). When the Chave model was employed, we used species-specific wood density values from Hawaii [23] and a global wood density database [53]. If a species-specific wood density value was unavailable, we used a mean value for the genus, and if this was not available we used a default value of 0.5 (Additional file 1: Table S5). We note here that wood densities are difficult to find for some commonly occurring oceanic island species, and thus we encourage research and measurement in this area. Occasionally, a height measurement was lacking for trees requiring the general Chave model. In these instances, we used locally derived, species-specific diameter to height models from Asner et al. [23]. When no species-specific diameter-to-height model was available, we used a general diameter-to-height model developed by Chave et al. for tropical trees that incorporates an environmental stress E parameter. Plot-level ACD was estimated by (1) estimating aboveground biomass (AGB) per unit area of microplots and macroplots within each FIA plot; (2) summing AGB per unit area within each FIA plot (n = 96); and (3) multiplying plot-level AGB per unit area by 0.48 to estimate ACD. The ACD of the 96 FIA plot locations were extracted from the statewide carbon map and averaged in a 3 × 3 pixel window (~1 ha) centered on each plot location.

Abbreviations
ACD: aboveground carbon density; AGB: aboveground biomass; C: carbon; FC: fractional cover; FIA: Forest Inventory and Analysis; LFS: low fire severity; LiDAR: Light Detection and Ranging; MFRI: mean fire return interval; MFS: mixed fire severity; NPV: non-photosynthetic vegetation; PV: photosynthetic vegetation; RFML: Random Forest Machine Learning; RFS: replacement fire severity; TCH: top of canopy height.

Authors' contributions
GA designed the study, led the airborne remote sensing data collection, analyzed data, and wrote the paper. SS analyzed satellite remote sensing data, and carried out the modeling analyses. DK analyzed field and airborne remote sensing data, and carried out the modeling analyses. PS analyzed field data, provided GIS data analyses, and contributed to the writing of the paper. RM, FH, and CG assisted with study design, acquisition of funding, data interpretation, and writing of the paper. All authors read and approved the final manuscript.

Author details
[1] Department of Global Ecology, Carnegie Institution for Science, 260 Panama St, Stanford, CA 94305, USA. [2] Department of Natural Resources and Environmental Management, University of Hawaii at Manoa, 1910 East–West Rd., Honolulu, HI 96822, USA. [3] USDA Forest Service, Pacific Southwest Research Station, Institute of Pacific Islands Forestry, 60 Nowelo Street, Hilo, HI 96720, USA.

Acknowledgements
We thank Lori Tango for assistance with interpreting FIA plot data, and Tom Thompson and Jane Reid with the USDA Forest Service FIA Program for access to Hawaii field plot data. We thank past and current Carnegie Airborne Observatory team members for assistance with data collection and processing. The USGS Biological Carbon Sequestration Program funded this project, a part of LandCarbon Carbon Assessment of Hawaii initiative. The Carnegie Airborne Observatory is made possible by the Avatar Alliance Foundation, John D. and Catherine T. MacArthur Foundation, Mary Anne Nyburg Baker and G. Leonard Baker Jr., and William R. Hearst III.

Competing interests
The authors declare that they have no competing interests.

References
1. Cannell M. Woody biomass of forest stands. For Ecol Manage. 1984;8(3–4):299–312.
2. Gibbs HK, Brown S, Niles JO, Foley JA. Monitoring and estimating tropical forest carbon stocks: making REDD a reality. Environ Res Lett. 2007;2:1–13.
3. Angelsen A. Moving Ahead with REDD: issues, options and implications. Bogor, Indonesia: Center for International Forestry Research (CIFOR); 2008.
4. Lindenmayer DB, Laurance WF, Franklin JF, Likens GE, Banks SC, Blanchard W, et al. New policies for old trees: averting a global crisis in a keystone ecological structure. Conserv Lett. 2013;7(1):61–9. doi:10.1111/conl.12013.
5. Berenguer E, Ferreira J, Gardner TA, Aragão LEOC, De Camargo PB, Cerri CE, et al. A large-scale field assessment of carbon stocks in human-modified tropical forests. Glob Change Biol. 2014;20(12):3713–26. doi:10.1111/gcb.12627.
6. Mitchard ETA, Feldpausch TR, Brienen RJW, Lopez-Gonzalez G, Monteagudo A, Baker TR, et al. Markedly divergent estimates of Amazon forest carbon density from ground plots and satellites. Glob Ecol Biogeogr. 2014;23(8):935–46. doi:10.1111/geb.12168.
7. Goetz S, Baccini A, Laporte N, Johns T, Walker W, Kellndorfer J, et al. Mapping and monitoring carbon stocks with satellite observations: a comparison of methods. Carbon Balance Manag. 2009;4(1):2. doi:10.1186/750-0680-4-2.
8. Vitousek PM. The Hawaiian Islands as a model system for ecosystem studies. Pac Sci. 1995;49:2–16.

9.   Loope LL, Hamman O, Stone CP. Comparative conservation biology of oceanic archipelagoes: Hawaii and the Galapagos. Bioscience. 1988;38:272–82.

10. Maina J, de Moel H, Zinke J, Madin J, McClanahan T, Vermaat JE. Human deforestation outweighs future climate change impacts of sedimentation on coral reefs. Nature Commun. 2013;4:1986. doi:10.1038/ncomms2986

11. Ticktin T, Whitehead AN, Fraiola HA. Traditional gathering of native hula plants in alien-invaded Hawaiian forests: adaptive practices, impacts on alien invasive species and conservation implications. Environ Conserv. 2006;33(03):185–94.

12. Berkes F, Colding J, Folke C. Rediscovery of traditional ecological knowledge as adaptive management. Ecol Appl. 2000;10(5):1251–62.

13. Fordham D, Brook B. Why tropical island endemics are acutely susceptible to global change. Biodivers Conserv. 2010;19(2):329–42.

14. D'Antonio CM, Dudley TL. Biological invasions as agents of change on islands versus mainlands. Ecol Stud. 1995;115:103–19.

15. D'Antonio CM, Vitousek PM. Biological invasions by exotic grasses, the grass/fire cycle, and global change. Annu Rev Ecol Syst. 1992;23:63–87.

16. Loope LL, Mueller-Dombois D. Characteristics of invaded islands, with special reference to Hawaii. In: Drake J, DiCastri F, Groves R, Kruger F, Mooney HA, Rejmanek M, et al., editors. Biological invasions: a global perspective. Chichester: Wiley and Sons; 1989. p. 257–80.

17. Hughes RF, Asner GP, Mascaro J, Uowolo A, Baldwin J. Carbon storage landscapes of lowland Hawaii: the role of native and invasive species through space and time. Ecol Appl. 2014;24(4):716–31.

18. Harris NL, Brown S, Hagen SC, Saatchi SS, Petrova S, Salas W, et al. Baseline map of carbon emissions from deforestation in tropical regions. Science. 2012;336(6088):1573–5. doi:10.1126/science.1217962.

19. Baccini A, Goetz SJ, Walker WS, Laporte NT, Sun M, Sulla-Menashe D, et al. Estimated carbon dioxide emissions from tropical deforestation improved by carbon-density maps. Nat Clim Change. 2012;. doi:10.1038/nclimate1354.

20. Mitchard E, Saatchi S, Baccini A, Asner G, Goetz S, Harris N, et al. Uncertainty in the spatial distribution of tropical forest biomass: a comparison of pan-tropical maps. Carbon Balance Manage. 2013;8(1):10.

21. Asner GP, Knapp DE, Martin RE, Tupayachi R, Anderson CB, Mascaro J, et al. Targeted carbon conservation at national scales with high-resolution monitoring. Proc Natl Acad Sci. 2014;111(47):E5016–22.

22. Lefsky MA, Cohen WB, Parker GG, Harding DJ. Lidar remote sensing for ecosystem studies. Bioscience. 2002;52(1):19–30.

23. Asner GP, Hughes RF, Mascaro J, Uowolo AL, Knapp DE, Jacobson J, et al. High-resolution carbon mapping on the million-hectare Island of Hawaii. Front Ecol Environ. 2011;9(8):434–9.

24. Asner GP. Tropical forest carbon assessment: integrating satellite and airborne mapping approaches. Environ Res Lett. 2009;3:1748–9326.

25. Asner GP, Knapp DE, Balaji A, Paez-Acosta G. Automated mapping of tropical deforestation and forest degradation: CLASlite. J Appl Remote Sens. 2009;3:033543.

26. Asner GP, Mascaro J. Mapping tropical forest carbon: Calibrating plot estimates to a simple LiDAR metric. Remote Sens Environ. 2014;140:614–24. doi:10.1016/j.rse.2013.09.023.

27. Brown S, Gillespie AJR, Lugo AE. Biomass estimation methods for tropical forests with application to forest inventory. Forest Sci. 1989;35:881–902.

28. Asner GP, Hughes RF, Varga TA, Knapp DE, Kennedy-Bowdoin T. Environmental and biotic controls over aboveground biomass throughout a tropical rain forest. Ecosystems. 2009;12:261–78.

29. Chave J, Andalo C, Brown S, Cairns MA, Chambers JQ, Eamus D, et al. Tree allometry and improved estimation of carbon stocks and balance in tropical forests. Oecologia. 2005;145:87–99. doi:10.1007/s00442-005-0100-x.

30. Keller M, Palace M, Hurtt G. Biomass estimation in the Tapajos National Forest, Brazil: examination of sampling and allometric uncertainties. For Ecol Manage. 2001;154:371–82.

31. Chave J, Réjou-Méchain M, Búrquez A, Chidumayo E, Colgan MS, Delitti WBC, et al. Improved allometric models to estimate the aboveground biomass of tropical trees. Glob Change Biol. 2014;20(10):3177–90. doi:10.1111/gcb.12629.

32. Asner GP, Martin RE, Knapp DE, Kennedy-Bowdoin T. Effects of *Morella faya* tree invasion on aboveground carbon storage in Hawaii. Biol Invasions. 2010;12:477–94. doi:10.1007/s10530-009-9452-1.

33. Mascaro J, Hughes RF, Schnitzer SA. Novel forests maintain ecosystem processes after the decline of native tree species. Ecol Monogr. 2011;82(2):221–8. doi:10.1890/11-1014.1.

34. Gon SM, Allison A, Cannarella RJ, Jacobi JD, Kaneshiro KY, Kido MH et al. A GAP analysis of Hawaii: Final report. US Department of the Interior. US Geological Survey, Washington, DC. 2006.

35. Funk JL, Cleland EE, Suding KN, Zavaleta ES. Restoration through reassembly: plant traits and invasion resistance. Trends Ecol Evol. 2008;23(12):695–703. doi:10.1016/j.tree.2008.07.013.

36. Stone CP, Cuddihy LW, Tunison JT. Responses of Hawaiian ecosystems to removal of feral pigs and goats. In: Stone CP, Smith CW, Tunison JT, editors. Alien Plant Invasions in Native Ecosystems of Hawaii: Management and Research. Honolulu: University of Hawaii Cooperative National Park Resources Study Unit; 1992. p. 666–704.

37. Loope LL, Scowcroft PG. Vegetation response within exclosures in Hawaii: A review. In: Stone CP, Scott JM, editors. Hawaii's terrestrial ecosystems: preservation and Management. Honolulu: University of Hawaii Cooperative National Park Resources Study Unit; 1985. p. 377–402.

38. Mascaro J, Asner GP, Knapp DE, Kennedy-Bowdoin T, Martin RE, Anderson C, et al. A tale of two "forests": Random Forest machine learning aids tropical forest carbon mapping. PLoS One. 2014;9(1):e85993. doi:10.1371/journal.pone.0085993

39. Breiman L. Random forests. Mach Learn. 2001;45:5–32.

40. Asner GP, Mascaro J, Muller-Landau HC, Vieilledent G, Vaudry R, Rasamoelina M, et al. A universal airborne LiDAR approach for tropical forest carbon mapping. Oecologia. 2012;168(4):1147–60. doi:10.1007/s00442-011-2165-z.

41. Mascaro J, Detto M, Asner GP, Muller-Landau HC. Evaluating uncertainty in mapping forest carbon with airborne LiDAR. Remote Sens Environ. 2011;115(12):3770–4. doi:10.1016/j.rse.2011.07.019.

42. Taylor PG, Asner GP, Dahlin K, Anderson CB, Knapp DE, Martin RE, et al. Landscape-scale controls on aboveground forest carbon stocks on the Osa Peninsula, Costa Rica. PLoS One. 2015;10(6):e0126748.

43. Mascaro J, Asner G, Davies S, Dehgan A, Saatchi S. These are the days of lasers in the jungle. Carbon Balance Manage. 2014;9(1):1–3. doi:10.1186/s13021-014-0007-0.

44. Asner GP. Satellites and psychology for improved forest monitoring. Proc Natl Acad Sci. 2014;111(2):567–8.

45. Baccini A, Asner GP. Improving pantropical forest carbon maps with airborne LiDAR sampling. Carbon Manag. 2013;4(6):591–600. doi:10.4155/cmt.13.66.

46. Mascaro J, Asner GP, Knapp DE, Kennedy-Bowdoin T, Martin RE, Anderson C et al. A tale of two "forests": Random Forest machine learning aids tropical forest carbon mapping. PLoS One. 2014;e85993.

47. Asner G, Mascaro J, Anderson C, Knapp D, Martin R, Kennedy-Bowdoin T, et al. High-fidelity national carbon mapping for resource management and REDD+. Carbon Balance Manage. 2013;8(1):7.

48. Gromping U. Variable importance assessment in regression: linear regression versus random forest. Am Stat. 2009;63(4):308–19. doi:10.2307/25652309.

49. Asner GP, Knapp DE, Kennedy-Bowdoin T, Jones MO, Martin RE, Boardman J, et al. Carnegie airborne observatory: in-flight fusion of hyperspectral imaging and waveform light detection and ranging for three-dimensional studies of ecosystems. J Appl Remote Sens. 2007;1:013536.

50. Reimer F, Asner GP, Joseph S. Advancing reference emission levels in subnational and national REDD + initiatives: a CLASlite approach. Carbon Balance Manag. 2015;10:5. doi:10.1186/s13021-015-0015-8

51. Jacobi JD, Price JP, Fortini LB, Berkowitz P. Baseline Land Cover. In: Z. Zhu e, U.S. Department of Interior, U.S. Geological Survey, editor. Baseline and projected future carbon storage and greenhouse-gas fluxes in ecosystems of Hawaii. 2015.

52. Woudenberg SW, Conkling BL, O'Connell BM, LaPoint EB, Turner JA, Waddell KL. The Forest Inventory and Analysis Database: Database description and users manual version 4.0 for Phase 2. 2010.

53. Chave J, Coomes D, Jansen S, Lewis SL, Swenson NG, Zanne AE. Towards a worldwide wood economics spectrum. Ecol Lett. 2009;12:351–66.

54. Liaw A, Wiener M. Classification and regression by randomForest. R News. 2002;2:18–22.

8

The global potential for carbon capture and storage from forestry

Yuanming Ni[1], Gunnar S. Eskeland[1,4], Jarl Giske[2] and Jan-Petter Hansen[1,3*]

Abstract

Background: Discussions about limiting anthropogenic emissions of CO_2 often focus on transition to renewable energy sources and on carbon capture and storage (CCS) of CO_2. The potential contributions from forests, forest products and other low-tech strategies are less frequently discussed. Here we develop a new simulation model to assess the global carbon content in forests and apply the model to study active annual carbon harvest 100 years into the future.

Results: The numerical experiments show that under a hypothetical scenario of globally sustainable forestry the world's forests could provide a large carbon sink, about one gigatonne per year, due to enhancement of carbon stock in tree biomass. In addition, a large amount of wood, 11.5 GT of carbon per year, could be extracted for reducing CO_2 emissions by substitution of wood for fossil fuels.

Conclusion: The results of this study indicate that carbon harvest from forests and carbon storage in living forests have a significant potential for CCS on a global scale.

Keywords: Carbon capture and storage, Simulations, Global forest

Background

According to the intergovernmental panel on climate change (IPCC), a reduction of the anthropogenic emissions of CO_2 to the atmosphere is necessary to avoid global warming beyond two degrees [1]. When also considering the projected population and consumption growth [2], the CO_2 reductions needed are daunting. It will require a transition to CO_2 free energy sources in many applications, and/or CCS from facilities such as fossil-based power plants. CO_2 free energy requires a significant build-up of nuclear and/or renewable power production, which involves large initial economic investments [3]. For CCS, a range of alternatives exist, each with its particular challenges. Industrial CCS requires energy and costly facilities and the captured gas has to be transported and stored in stable geological formations [4]. The cost of the capture process itself is estimated to

be in the range of 40–70 euros/tonne CO_2, depending on technology [5].

The scale of the problem should not be underestimated: to reach the less than two degree goal of IPCC, the annual CO_2 emissions must be reduced from the current level of 10 GT of carbon per year (GtC/year) to 5–6 GtC/year by 2050. Hence, the new energy sources need to deliver up to 5 terrawatts (TW) or above [5]. To contribute at this scale, the sequestrated amount of carbon by industrial CCS has to be of order 1 GtC/year or more, while the total amount of carbon stored so far is only a few tens of megatonnes, ie., a few per mille of the necessary amount. It is therefore important to consider alternative options. One alternative is to use the photosynthesis to increase the carbon sink by increased magnitude of the world's forests. Another option is to increase the carbon uptake by letting forest stands grow for longer periods [6]. Two other alternatives could be a large scale deployment of artificial photosynthesis [7], or to increase the carbon uptake of the oceans by adding active absorbers. However, the latter alternatives may involve high risk for unexpected drawbacks [8].

*Correspondence: Jan.Hansen@uib.no
[3] Department of Physics and Technology, University of Bergen, Allegt. 55, 5007 Bergen, Norway
Full list of author information is available at the end of the article

The annual sink of the world's forests has been estimated to be about 2.4 GtC [9]. Taking advantage of photosynthesis in forests requires global schemes for reducing deforestation in combination with planting and replanting programs. Also, a possibility is modified harvest and management. Recently, it was argued that this large-scale planting action may also have a negative total climate effect since the greening of open land areas will reduce the albedo of the earth [10]. An alternative would be to collect wood material at a constant rate and then store it. Independent strategies based on this idea were suggested a few years ago [11, 13]. In this context the storage problem would be small. Dry wood contains about 50 % carbon [14] and can be stored for example in decommissioned coal mines or in facilities near the forests in the long term. By using timber in fairly long-lasting applications (buildings, furniture), carbon storage could be even less expensive and more attractive. The global potential of this option was initially estimated to be as much as 5–15 GtC/year [11] and later estimated to be around 1–3 GtC/year [12] when land use, protection consideration and other factors were taken into account. This estimate is based on simulations using a model based on global carbon fluxes.

Storing carbon as standing forests or from harvested wood has long been recognized as a CCS option: for example, Schroder et al. estimated that 15–36 GtC could be stored in tropical plantations and 50–100 GtC sequestrated on a global scale [15, 16]. A detailed analysis of the Eastern US woodlands shows that 176 megatonnes of wood may be harvested annually without diverting current wood products, damaging habitat, or reducing terrestrial carbon sink [17]. A calculation by Lehmann [18] indicates that an equivalent of about 10 % of the US fossil fuel emissions may be harvested from biomass and stored as biochar. Based on simulating the growth of uneven-aged mixed beech-spruce stands in the temperate region, Kraxner et al. [19] found that 1–2 tonnes/year/ha can on average be extracted for storage. Several model studies also include economic aspects of forest management and carbon storage [20–22]. In terms of policies, our analysis goes in the same direction as for instance Hoel et al. [23]. They argue that forests and forest products may serve the climate better if valued not only as renewables but also for carbon storage purposes. While the literature is vast regarding regional and tree-specific studies, see e.g., [19, 24], to the best of our knowledge there has been no attempt to address the issue dynamically based on the nonlinear growth of trees to obtain numerical results with harvesting schemes for global forests as a whole. For example, by scaling up the simulation of Kraxner et al. [19], one attains an estimate of 0.7–1.4 GtC per year as the CCS potential of the temperate forests. However,

upscaling from detailed growth models needs to be performed with caution.

Here, we attempt to address this problem by modeling global potential for carbon storage in biomass based on a plant-harvest forestry program. It is assumed that a fraction of the forest is harvested and replanted every year. As a starting point we assume that forest growth in the boreal, temperate and tropical regions is latitude-dependent. Reasonable assumptions are also made regarding future reforestation and deforestation for different zones. The results are then directly based on non-linear forest growth and the instantaneous status of the world forests.

Results and discussion

We here demonstrate four applications of our model (described in the "Methods" section) in terms of strategies for harvesting, reforestation (replanting) and afforestation (new planting). The present forest area for each region is discretized to a sufficiently large number of initial forest areas. For example, when taking a discretized area size of 105 hectares, the simulation starts with 21111, 7394 and 11825 areas representing global forests in tropical, temperate and boreal zones respectively [25]. The carbon content in living biomass per ha is used [9] to reach an estimated implied initial average age of 94, 57 and 65 years for tropical, temperate and boreal zone trees.

In Fig. 1, we display the carbon content of world forests 100 years into the future, assuming constant (present) rates of deforestation and expansion without additional harvesting. Starting from 372 GtC, total carbon content is seen to increase to 530 GtC. This is a net increase of 158 GtC in

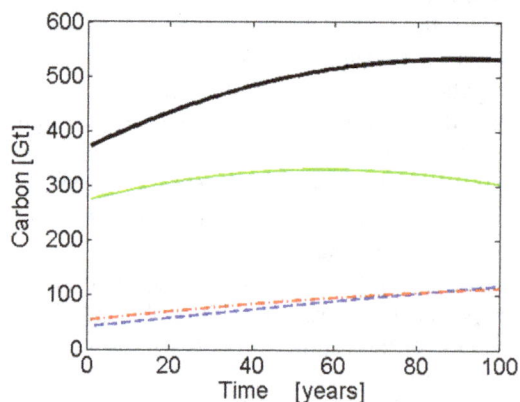

Fig. 1 Dynamic carbon content in tropical zone (*green full*), temperate zone (*blue dashed*), boreal zone (*red dashed-dot*) and world forest (*black thick full*) forecast with our model. Parameters: deforestation rate of 0.44 % in the tropical zone, current expansion rates of 0.1 % in the boreal zone and 0.264 % in the temperate zone [25]. The initial year is 2010

spite of a present deforestation rate of 0.44 % per year in the tropical region [25]. The increased carbon content comes from afforestation in the temperate and boreal zones, combined with increased carbon storage due to growth in maturing tropical trees. The assumption of constant growth rates for 100 years into the future may clearly be questioned. In the extreme case of a complete stop in tropical deforestation from today, the forest carbon amount would for example increase by up to 230 GtC within 100 years.

With non-harvest (status quo), as a background, we are now ready to model forest development under four alternative harvest-replanting scenarios. All harvesting strategies assume immediate replanting so that the land remains functioning as forest land. We also assume that tropical deforestation is alleviated from 0.44–0.3 % during the whole simulation period. The first two simulations assume harvesting parameters of 0.3 and 0.45 %, respectively, for all three zones. In the third strategy, we reduce the harvesting parameter to 0 in the tropical zone, and increase it to 0.8 % in the two other zones. The reduced harvesting rate in the tropics is motivated by the special soil feature in tropical areas. Due to relative high temperature all year round, the decomposition of forest residues happens fast, and this results in a rather thin layer of soil. Harvesting and storing wood away under these conditions may cause significant reduction in ground soil, motivating the non-harvest strategy. In the final simulation an extra plantation program is introduced so that every year a specified fraction of new forest area is added into the model. This is to explore the potential under a large-scale policy change of using forests as a CCS method.

In simulation 1 with 0.3 % harvest rate the total forest carbon increases from 372 to 447 GtC over 100 years (Fig. 2). Considering the harvested fraction of 0.3 %, it

seems that these 1.5–2 GtC can be extracted every year without harming forest production. The total carbon sequestration can thus be 2.25–2.75 GtC/year from both standing stock and harvested wood. The major contribution comes from the tropical zone, due to its stock size and productivity. At the end of the period, the constant deforestation rate finally brings down the total carbon stock in the tropical zone, while the other two zones still display growth.

The instantaneous status of the forest areas initially, after 50 and 100 years illustrates the development of the forest regions in this simulation (Fig. 3). Each bar represents 100 randomly merged simulation areas of each region, i.e., a forest area of 10500 ha. The height of each bar represents the relative carbon content in these areas normalized to the initial situation (red line). After 50 years, we observe some growth in areas without harvesting, and some regions of significantly less carbon content where harvest has taken place. After 100 years we observe that the carbon content of the untouched forest areas has increased even further while the number of harvested regions has increased as well. Finally, the initial number of forest areas has decreased in the tropical region due to deforestation while forest areas have increased in the temperate and boreal zones, due to afforestation.

The more aggressive harvest strategy in simulation 2 (without additional afforestation areas or actions to reduce tropical deforestation) may not be sustainable (Fig. 4). Here, more than 2 GtC is harvested, with a harvest rate of 0.45 % per year. The forest carbon stock will still, after 100 years, maintain its original state. Note, however, that the forest biomass will decrease towards the end of the period. Thus, the limit for what can be

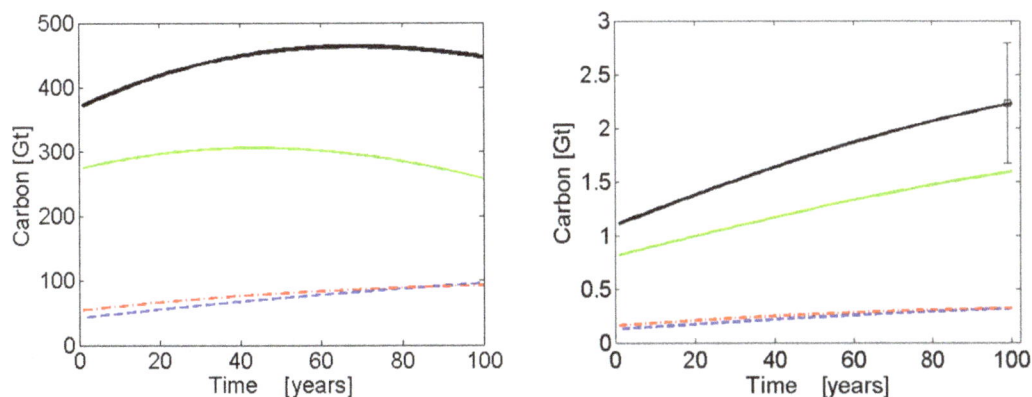

Fig. 2 *Left panel* Total forest carbon development in tropical zone (*green full*), temperate zone (*blue dashed*), boreal zone (*red dashed-dot*) and world forest (*black thick full*). *Right panel* Yearly harvested carbon development in tropical zone (*green full*), temperate zone (*blue dashed*), boreal zone (*red dashed-dot*) and world forest (*black thick full*) with indicative error bar (*square*) at year 100. Parameters: harvesting rate of 0.3 % in all zones; deforestation rate of 0.3 % in tropical zone; current expansion rates of 0.1 % in the boreal and 0.264 % in the temperate zones

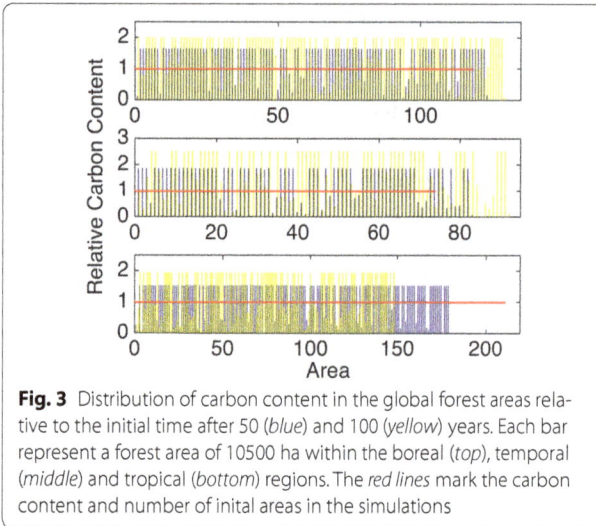

Fig. 3 Distribution of carbon content in the global forest areas relative to the initial time after 50 (*blue*) and 100 (*yellow*) years. Each bar represent a forest area of 10500 ha within the boreal (*top*), temporal (*middle*) and tropical (*bottom*) regions. The *red lines* mark the carbon content and number of inital areas in the simulations

extracted globally every year is about 2 GtC in a 100 year perspective.

In the final two simulations we extract carbon only from temperate and boreal zones to protect the tropical forest. After the simulation period, boreal and temperate zones retain almost the same amount of forest carbon as today, in the regime of annual harvest rate of 0.8 % (Fig. 5). In total, forest carbon grows from 372 to 496 GtC and the extracted average amount is about 1.25 GtC per year, with a total of 2.5 GtC sequestrated by both standing stocks and harvested wood every year.

In simulation 4, storage may be raised further if an additional planting program adds 0.05 % of the original area in boreal and temperate zones every year while in tropical zone the deforestation rate is reduced from 0.3 to 0.25 %.

The forest area then increases by 80 million ha. Under these assumptions the total harvested wood remains the same while the carbon content of the standing stocks increases up to 526 GtC, or 1.5 GtC per year. The total amount of sequestration is about 2.8 GtC annually.

The uncertainty in the estimates above stems partly from the model itself and partly from external factors. The model error bars are mainly determined by the deviation of the applied growth curves from the real average growth characteristics of global forests. Comparison with available growth data indicates that this uncertainty is in the order of 20 % or less. Additional caveats come with external factors: future technology improving the growth may occur on one side; deterioration of the soil or lack of nutrients or water supplies hindering growth may occur on the other. Regional catastrophic events such as wildfire, insect attacks, wind throw or volcano eruptions and the like may also happen. However, considering the annual global forest area change rate of −0.165 % from 1990 to 2010 [25], we believe that external factors in total lead to less than 5 % additional uncertainty. We thus conclude that the final figures above come with a total uncertainty of about 25 % reflected by the uncertainty of the growth model and indicated by black error bars at year 100 in the preceding figures.

Conclusions

We have developed a dynamic model for carbon storage in forests and harvested wood as an active CCS strategy. With harvest parameters limited by the need to conserve the forests, applications of the model show that in a 100 year perspective the capacity of harvested wood is 1–1.5 GtC per year, and the total sequestration including storage in living trees is on average more than 2–2.5 Gt

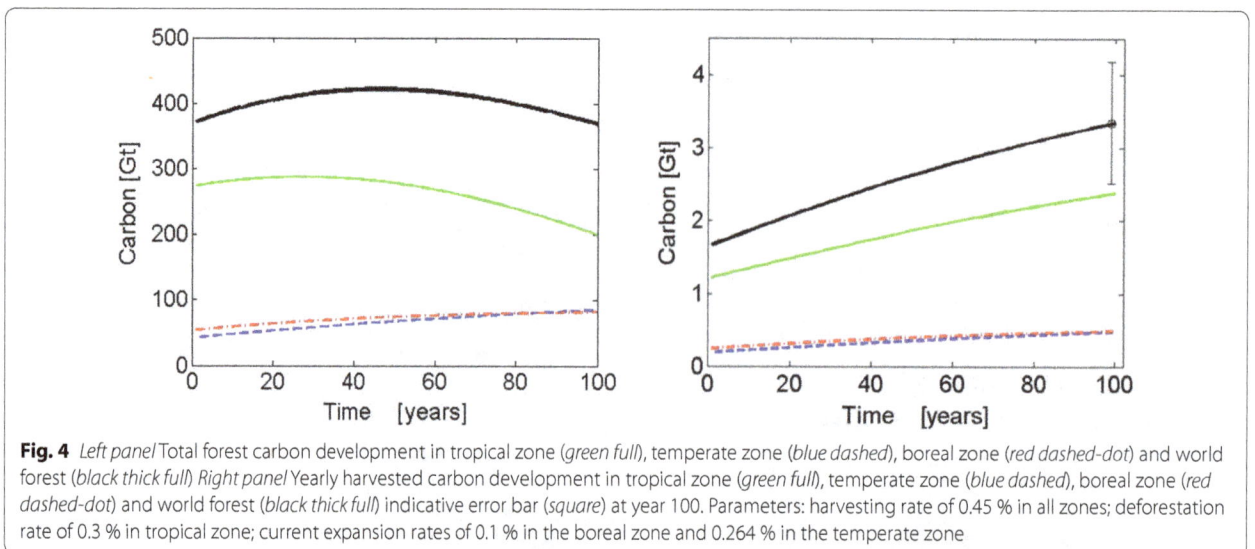

Fig. 4 *Left panel* Total forest carbon development in tropical zone (*green full*), temperate zone (*blue dashed*), boreal zone (*red dashed-dot*) and world forest (*black thick full*) *Right panel* Yearly harvested carbon development in tropical zone (*green full*), temperate zone (*blue dashed*), boreal zone (*red dashed-dot*) and world forest (*black thick full*) indicative error bar (*square*) at year 100. Parameters: harvesting rate of 0.45 % in all zones; deforestation rate of 0.3 % in tropical zone; current expansion rates of 0.1 % in the boreal zone and 0.264 % in the temperate zone

carbon per year. This is in fair agreement with the previous estimate of Zeng [12], but it also indicates that the harvest estimate is at the upper edge of what is possible without a serious reduction in total forest areas.

Our model also shows that the amount of stored carbon depends critically on the strategies for harvesting and planting. As compared to pure forest conservation with no harvest, the advantage of an active harvest strategy is that it may be applied at a constant rate on a timescale of several hundred years, not being constrained by maturing forests that cannot store more. Also, this proposal does not change the albedo since the small percentage areas of harvesting are rapidly becoming green again after the replanting. Our obtained numbers are a factor of five smaller than Zeng's [11] estimate of carbon storage potential from collection of dead and mature trees. The latter requires harvesting from the entire world forest areas, while the present plant and harvesting (PH, described in the Methods section) strategy is performed in concentrated regions of less than one percent of the global forest area.

A main advantage of implementing the proposed approach is storage at low costs. For industrial CCS, estimated costs are in the order of 100 USD per tonne of CO_2, which are about 150 % of the costs of electricity generation by fossil fuels [5]. Compared to this, harvesting and storing wood will very likely be competitive and the upper estimate is between 25 and 50 USD per tonne of CO_2 [13]. In conclusion, we believe that the present results are relevant in a medium-term future scenario of continued and even increased consumption of fossil fuels.

Methods

In this section we derive the plant-harvest (PH) model used in the simulations and assess its validity in comparison with tree growth data. The idea in our approach is to develop a characteristic average growth curve of wood at a latitude and to discretize the forest area at each latitude. The PH-model associates forest growth potential worldwide with different temperature and latitude characteristics in three zones: tropical, temperate and boreal. Within each zone the wood volume growth follows this characteristic curve after planting. Every year a certain area is harvested and immediately replanted. Additionally, new areas may be opened up for afforestation.

The global forest volume in cubic meters is computed as a function of time,

$$F_A(t) = F_A^0 + \sum_{i=1,2,3} \int_{t_0}^{t} \left[G_i(t') - H_i(t') \right] dt' \qquad (1)$$

Here F_A^0 is the initial global forest volume [m³] and G_i and H_i are the volume growth [m³] and harvest [m³ year⁻¹] in zone i.

The growth functions are non-linear and determined mainly by the incoming radiation, the CO_2 concentration and the availability of water and nutrients, in addition to the age of the tree stand within each area. Equations describing the growth characteristics of specific trees are in general empirical in their origins, such as the logistic equation or its generalization, the Richards equation [27]. Other applied growth curves are the Gompertz model and the modified Weibull model [28]. For a global approach, it is necessary to represent the forest growth within each area in terms of a characteristic growth function. This may be inappropriate when describing the growth of a single tree, a single stand, or a particular species, but quite accurate in terms of the expected large-scale carbon production stored in wood over time. We start out by deriving this growth model for that purpose.

Let T be the age at which planted trees within a large area start to spend all of their energy on maintaining the total mass, so that volume growth after T is effectively zero within the area. During the time interval $(0, T)$ the wood volume reaches its maximum size, i.e., $V(t) \in (0, V_{max})$. Growth is generated by a total area of leaves being exposed to incoming electromagnetic radiation. Thus, the volume growth can be assumed to be proportional to the exposed area $A(t)$ set up by the leaves of all trees within the given fixed region,

$$\frac{dV}{dt} = \epsilon(t)A(t) \qquad (2)$$

Note that both the exposed area A and the proportionality factor are time-dependent: $A(t)$ increases with the volume growth of the trees and is taken in the following to be proportional to the total wood volume, $A(t) = \alpha V(t)$. However, for the photosynthesis to be active, each plant uses energy internally, for instance for transport of water and other molecules up to its leaves. Some of these costs will increase with the size of the tree. At the time when the stand has reached its maximum size the proportionality factor is zero, $\epsilon(T) \simeq 0$. We will here assume a linear dependence from its initial value,

$$\epsilon(t) = \left(1 - \frac{V(t)}{V_{max}} \right) \qquad (3)$$

Thus we have obtained a logistic equation for wood volume growth within an entire region

$$\frac{dV}{dt} = \alpha V(t) \left(1 - \frac{V(t)}{V_{max}} \right), \qquad (4)$$

where the two proportionality constants have been merged into a single time independent parameter $\alpha = \epsilon_0 \epsilon_1$. (We here side-remark that by assuming a non-linear efficiency function for $\epsilon(t)$, the Richardson equation can be obtained). The solution of Eq. (4) is,

$$V(t) = \frac{V_{max}}{1 + e^{-\alpha(t-t_p)}} \qquad (5)$$

where t_p is the time at which the volume growth is at its largest. Next is to find a reasonable way to estimate V_{max}.

According to the World Energy Assessment 2000, the net energy yield (E_Y) for wood is in the range of 30–80 GJ/ha/year [29]. E_Y is what the forest has converted to bioenergy, in terms of wood and what can finally be harvested after a period of time. In this paper we apply 76, 62 and 38 GJ/ha/year for the tropical, temperate and boreal region respectively. Based on existing studies, we take T for each zone to be 200, 150 and 140 years [30]. It is reasonable to assume an average dry wood density ρ_w of 0.6 tonne/m^3. The calorific value of wood (C_w) depends on wood type, but in concordance with this approach we apply a value of $20 \cdot 10^9$ J/tonne for all three zones.

$$V_{max} \cdot \rho_w \cdot C_w = E_Y \cdot T \qquad (6)$$

As a consistency check of the present numbers we compare the computed V_{max} from Eq. (6) with the data for power density of tropical plantations and commercial boreal forestry in [31]. In the first case, taking a burn efficiency of 35 %, our computed V_{max} corresponds to an energy density of 0.7 and 0.34 W/m^2 in the tropical and boreal regions respectively. The corresponding numbers from Ref. [31] are in the range 0.6–1.1 W/m^2 in the tropical regions and around 0.3 W/m^2 in northern Europe wood forestry. Thus, V_{max} is pinpointed by a single parameter E_Y and the total growth time T taken from Ref. [30]. The other free parameter is the proportionality factor α of Eq. (4). The numerical values of input parameters and growth curve parameters are summarized in Table 1.

The resulting growth curves of each zone to be applied in the PH model (Fig. 6) may be compared to existing data sets, even if they cannot be expected to reproduce the growth of a single stand or the forest of a localized area very accurately. The data points refer to compiled data from the literature. We start out to discuss the comparison with the most extensive data set we could find covering the temperate zone (blue squares). These come

from US forests covering the West and East coast areas [32]. We here select representative temperate forests species from Northeast, Northern lake states, Northern prairie states, Pacific Northwest, Pacific Southwest and Rocky Mountain South respectively. The data consists of time series from year 5 to year 125 measured every 10 years of the forest stand yield (in (m^3 /ha) of a certain species or species combination after reforestation. The average growth data of these samples are in good agreement with our temperate growth curve. The error bars, calculated from measurement and sampling uncertainty are below 25 m^3 /ha at all times.

Similar extensive forest data could not be found in the two other regions. However, we may compare the regional growth curve with data for single stands. This is a smaller problem in boreal forests than in the other regions, because boreal forests are often dominated by only a few species. However, local environmental conditions will always have impact on growth. As an example, we compare stand growth data for Norwegian spruce with the boreal growth curve from [33]. Typical data points with error bars covering the spread of tree data have been extracted and multiplied again by a timber value of 0.2. In this case the boreal growth curve is in agreement with the data as well. For the tropical region (green), the open circles are average diameter at breast height data extracted from São Paulo State Park of Serra do Mar in Southeastern Brazil, 400 m above sea level [34]. The stored carbon volume per ha is calculated from 1500 stems ha^{-1}, an average tree volume of 6 m^2, a timber value of 0.2 and a linear relationship between the diameter at breast height (0–75 cm) and the age (0–200 years). Since these are measurements of individual trees, the transformation to wood volume as function of age gives an error bar along both axes. The indicated error bar covers the scattered data quite reasonably [35]. With these values our growth model is seen to fit relatively well with existing data.

In summary, the comparison with average wood growth from the regional USA data indicates an uncertainty in the growth model of less than 20 %. The model performs well with individual stands and tree data samples in the boreal and tropical region as well. However, we cannot deduce an absolute model error unless large data sets over a 100 years period from all regions had been available. The fact that the growth curves agree well with the data actually available supports the assumption of an accuracy of the order 20 % which induces a similar uncertainty in the simulation results.

With the growth functions at hand, the model propagates in time, year by year according to different assumptions of harvest rate, deforestation rate and expansion rate. In the start each discretized forest area has the same

Table 1 Extracted total growth time T, net energy yield E_Y, computed growth model parameter V_{max} (m^3/ha) and applied growth rate parameter α of each region

Region	T (years)	E_Y (GJ/ha/year)	V_{max} (m^3/ha)	α (1/year)
Tropical	200	76	1262	0.013
Temperate	150	62	775	0.015
Boreal	140	38	406	0.02

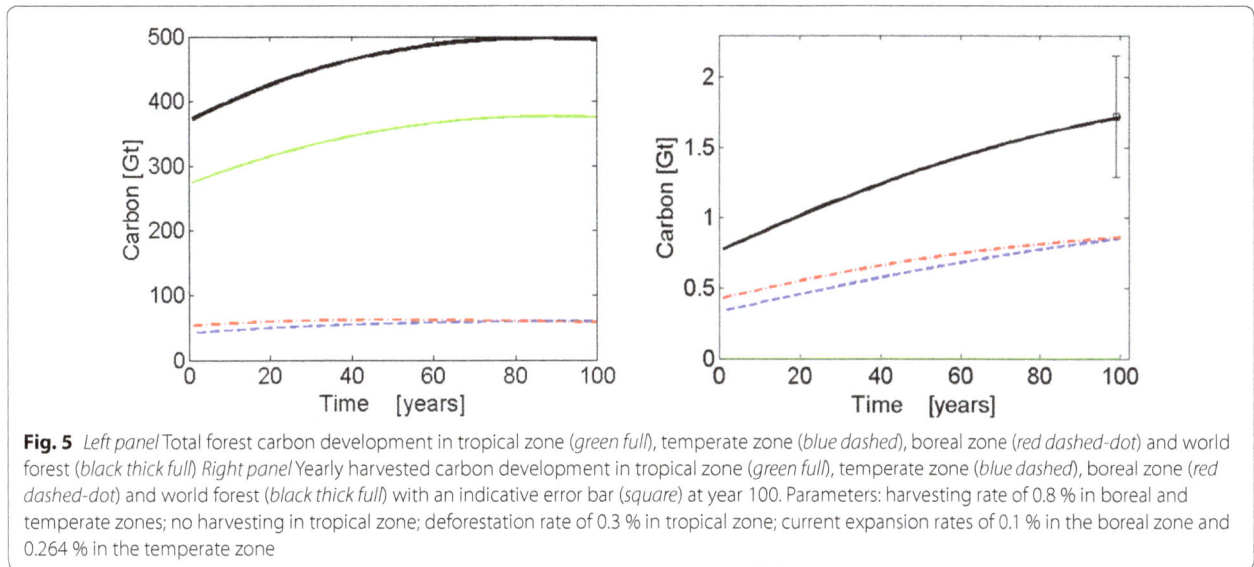

Fig. 5 *Left panel* Total forest carbon development in tropical zone (*green full*), temperate zone (*blue dashed*), boreal zone (*red dashed-dot*) and world forest (*black thick full*) *Right panel* Yearly harvested carbon development in tropical zone (*green full*), temperate zone (*blue dashed*), boreal zone (*red dashed-dot*) and world forest (*black thick full*) with an indicative error bar (*square*) at year 100. Parameters: harvesting rate of 0.8 % in boreal and temperate zones; no harvesting in tropical zone; deforestation rate of 0.3 % in tropical zone; current expansion rates of 0.1 % in the boreal zone and 0.264 % in the temperate zone

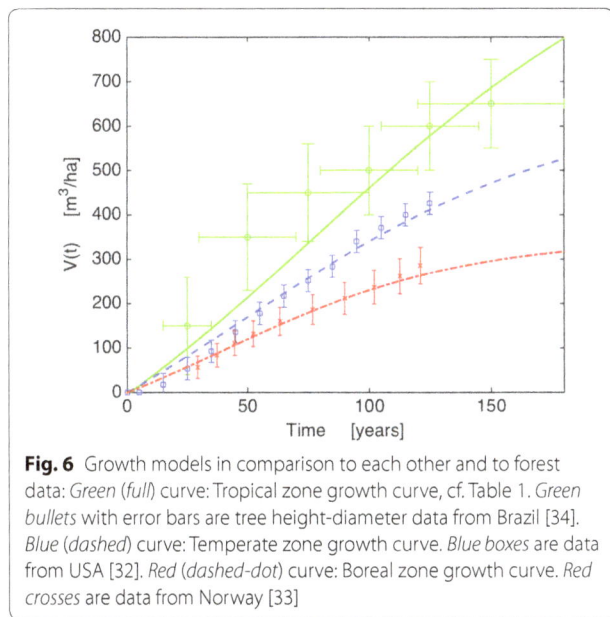

Fig. 6 Growth models in comparison to each other and to forest data: *Green* (*full*) curve: Tropical zone growth curve, cf. Table 1. *Green bullets* with error bars are tree height-diameter data from Brazil [34]. *Blue* (*dashed*) curve: Temperate zone growth curve. *Blue boxes* are data from USA [32]. *Red* (*dashed-dot*) curve: Boreal zone growth curve. *Red crosses* are data from Norway [33]

fraction of the total known carbon content of each of the three forest zones. This automatically sets the initial age where each area continues to grow from. The expansion rate of each forest zone decides the number of added area units being created every year. Once created, the new area will start accumulating carbon according to our growth function from zero time. Deforestation rates, only valid for tropical zone, determine how many areas will be deleted from the model each year. It is assumed that the carbon stored in these areas will be released immediately since most wood from deforested areas are burned as fuel directly. Harvest rates determine the areas in which the

carbon is harvested and stored as wood. After harvesting from one area unit it is assumed that the unit is replanted and continues to grow according to the growth curve from time zero. The total dynamic forest carbon is calculated by summing up the carbon content in all active growing areas of each region each year.

Authors' contributions

All authors have taken part in the development of the main idea and taken part in the writing of the manuscript. The model has been programmed by YN who also performed the series of simulations. The input data and the results of the simulation were analyzed and discussed among all four authors. All authors read and approved the final manuscript.

Author details

[1] Department of Business and Management Science, Norwegian School of Economics, Helleveien 30, 5045 Bergen, Norway. [2] Department of Biology, University of Bergen, 5020 Bergen, Norway. [3] Department of Physics and Technology, University of Bergen, Allegt. 55, 5007 Bergen, Norway. [4] Centre for Applied Research (SNF), Helleveien 30, 5045 Bergen, Norway.

Acknowledgements

The authors acknowledge Center for Sustainable Energy Studies (CenSES) for financial support.

Competing interests

The authors declare that they have no competing interests.

References

1. Smith JB, Schellnhuber HJ, Mirza MMQ. Vulnerability to climate change and reasons for concern: a synthesis. In: McCarthy JJ, Canziani OF, Leary NA, Dokken DJ, White KS, editors. Climate change 2001: impacts, adaptation, and vulnerability. Contribution of working group II to the third assessment report of the intergovernmental panel on climate change. Cambridge University Press; 2001.
2. Cohen JE. Human population: the next half century. Science. 2003;302:1172–5.

3. IPCC: climate change 2014 synthesis report. Geneva: IPCC; 2014.
4. Rubin ES. Understanding the pitfalls of ccscost estimates. IJGGC. 2012;10:181–90.
5. Narbel PA, Hansen JP, Lien JR. Energy technologies and economics. Springer Verlag; 2014.
6. Prato T. Natural resource and environmental economics. Wiley-Blackwell; 1999.
7. Meyer TJ. Chemical approaches to artificial photosynthesis. Acc Chem Res. 1989;22:163–70.
8. Royal society. Geoengineering the climate: science. London: Governance and Uncertainty: Royal Society; 2009.
9. Pan Y, Birdsey RA, Fang J, Houghton R, Kauppi PE, Kurz WA, et al. A large and persistent carbon sink in the worlds forests. Science. 2011;333:988–93.
10. Unger N. Human land-use-driven reduction of forest volatiles cools global climate. Nature Climate Change. 2014;4:907–10.
11. Zeng N. Carbon sequestration via wood burial. Carbon Balance Manag. 2008;3:1.
12. Zeng N, King AW, Zaitchik B, Wullschegger SD, Gregg J, Wang S, Kirk-Davidoff D. Carbon sequestration via wood harvest and storage: an assessment of its harvest potential. Climatic Change. 2013;118:245–57.
13. Scholz F, Hasse U. Permanent wood sequestration: the solution to the global carbon dioxide problem. Chemsuschem. 2008;1:381–4.
14. Lamlom SH, Savidge RA. A reassessment of carbon content in wood: variation within and between 41 North American species. Biomass Bioenergy. 2003;25:381–8.
15. Schroeder P. Carbon storage potential of short rotation tropical tree plantations. Forest Ecol Manag. 1992;50:31–41.
16. Winjum JK, Dixon RK, Schroeder PE. Estimating the global potential of forest and agroforest management practices to sequester carbon. In: Wisniewski J, Lugo AE, editors. Natural Sinks of CO_2. Netherlands: Springer Verlag; 1992. 213–27.
17. Sarah CD, Michael D, Evan D, Chris F, Steven PH, Scott L, et al. Harvesting carbon from Eastern US forests: Opportunities and impacts of an expanding bioenergy industry. Forests. 2012;3:370–97.
18. Lehmann J. A handful of carbon. Nature. 2007;447:143.
19. Kraxner F, Nilsson S, Obersteiner M. Negative emissions from bioenergy use, carbon capture and sequestration (BECS): The case of biomass production by sustainable forest management from semi-natural temperate forests. Biomass Bioenergy. 2003;24:285–96.
20. Sohngen B, Mendelsohn R. An optimal control model of forest carbon sequestration. Amer J Agr Econ. 2003;85:448–57.
21. Hennigar CR, MacLean DA, Amos-Binks LJ. A novel approach to optimize management strategies for carbon stored in both forests and wood products. Forest Ecol Manag. 2008;256:786–97.
22. Cunha-e-S MA, Rosa R, Costa-Duarte C. Natural carbon capture and storage (NCCS): forests, land use and carbon accounting. Res Energy Econ. 2003;35:148–70.
23. Hoel M, Holtsmark B, Holtsmark K. Faustmann and the climate. J Forest Econ. 2014;20:192–210.
24. Knauf M, Köhl M, Mues V, Olschofsky K, Fröhwald A. Modeling the CO_2 - effects of forest management and wood usage on a regional basis. Carbon Balance Manag. 2015;10:13.
25. FAO and JRC. Global forest land-use change 1990–2005. In: Lindquist EJ, Annunzio RD, Gerrand A, MacDicken K, Achard F, Beuchle R, Brink A, Eva HD, Mayaux P, San-Miguel-Ayanz J, Stibig HJ, editors. FAO Forestry Paper No. 169. Food and Agriculture Organization of the United Nations and European Commission Joint Research Centre. Rome: FAO; 2012.
26. Dasgupta B, Lall SV, Lozano-Gracia N. Urbanization and housing investment. World Bank Group. 2014.
27. Birch CSP. A new generalized logistic sigmoid growth equation compared with the Richards growth equation. Ann Botany. 1999;83:713–23.
28. Yan F, Wang S, Han N. Tree modeling based on GA-fitted growth function. International conference on computational intelligence and software engineering, CiSE, IEEE. 2009.
29. UNDP: World energy assessment: energy and the challenge of sustainability. United Nations Development Programme; 2000.
30. Lieberman D, Lieberman M, Hartshorn G, Peralta R. Growth rates and age-size relationships of tropical wet forest trees in Costa Rica. Trop Ecol. 1985;1:97–109.
31. MacKay D. Sustainable energy—without the hot air. UIT Cambridge; 2009. p. 43.
32. Smith JE, Heath LS, Skog KE, Birdsey RA. Methods for calculating forest ecosystem and harvested carbon with standard estimates for forest types of the United States. USDA Forest Service, General Technical Report, NE-343; 2006.
33. Pretzsch H, Biber P, Schütze G, Uhl E, Rötzer T. Forest stand growth dynamics in Central Europe have accelerated since 1870. Nat Comm. 2014;5:4967.
34. Marcos ASS, Luciana FA, Simone AV, Plinio BC, Carlos AJ, Luiz AM. Height-diameter relationships of tropical Atlantic moist forest trees in southeastern Brazil. Scientia Agricola. 2012;69:26–37.
35. Lieberman M, Lieberman D. Simulation of growth curves from periodic increment data. Ecology. 1985;66:632–5.

9

Carbon storage in Ghanaian cocoa ecosystems

Askia M. Mohammed[1]*, James S. Robinson[2], David Midmore[2] and Anne Verhoef[2]

Abstract

Background: The recent inclusion of the cocoa sector as an option for carbon storage necessitates the need to quantify the C stocks in cocoa systems of Ghana.

Results: Using farmers' fields, the carbon (C) stocks in shaded and unshaded cocoa systems selected from the Eastern (ER) and Western (WR) regions of Ghana were measured. Total ecosystem C (biomass C + soil C to 60 cm depth) ranged from 81.8 to 153.9 Mg C/ha. The bulk (~89 %) of the systems' C stock was stored in the soils. The total C stocks were higher in the WR (137.8 ± 8.6 Mg C/ha) than ER (95.7 ± 8.6 Mg C/ha).

Conclusion: Based on the cocoa cultivation area of 1.45 million hectares, the cocoa sector in Ghana potentially could store 118.6–223.2 Gg C in cocoa systems with cocoa systems aged within 30 years regardless of shade management. Thus, the decision to include the cocoa sector in the national carbon accounting emissions budget of Ghana is warranted.

Keywords: Carbon stocks, Cocoa ecosystem, Shaded and unshaded cocoa systems

Background

Cocoa is cultivated in the forest regions of Ghana where an estimated area of 1.45 million hectares of forest land has been displaced [1]. A substantial volume of literature is replete with evidence that the reductions in forest cover produced net sources of carbon dioxide (CO_2), the main greenhouse gas of the atmosphere [2, 3]. According to the Intergovernmental Panel on Climate Change (IPCC), global C stocks in terrestrial biomass have decreased by 25 % over the past century [3, 4]. This corresponds to an annual decline of 1.1 Gt of the global carbon stocks in forest biomass [5]. Stern [2] note that deforestation alone is responsible for 18 % of the world's greenhouse gas emissions.

Cocoa intensification for higher yields has led to a drastic reduction in shade tree density and, on many farms total elimination of the shade trees in cocoa ecosystems [6]. Essentially, cocoa expansion in Ghana has been closely linked to deforestation [7, 8]. One option to redress deforestation and create a carbon sink is to encourage the establishment of tree-crop farming or agroforestry systems [9–11]. Cocoa agroforestry is an age-old practice in the tropics [12]. Various recommendations have been made to farmers with regard to the number of non-cocoa trees to provide shade for cocoa during planting. However, the decision on how much shade is optimal often depends on the ecological system, social factors, biodiversity interests, ecological services and pod yields [7, 11].

With the recent inclusion of the cocoa sector in the national C emission accounting budgets of Ghana [13], the need to quantify the carbon sequestered in cocoa ecosystems is urgent. In addition to measuring the amounts of carbon stored in cocoa and shade tree biomass in the cocoa systems, the soil organic carbon content needs to be determined. Globally, the amount of C stored in soils is estimated to be 1.5–3 times more than in vegetation [9]. Thus, if Ghana is to include the C sequestered in the cocoa sector in its proposal for developing a national carbon accounting strategy, as outlined in its Readiness Plan Proposal [13], the C quantities stored both in the vegetation and the soils of the cocoa ecosystems must be included.

*Correspondence: mamusah@yahoo.com
[1] CSIR-Savanna Agricultural Research Institute, Nyankpala, PO Box 52, Tamale, Ghana
Full list of author information is available at the end of the article

This paper evaluates the C storage in cocoa ecosystems from two regions of Ghana under two shade management systems and two cocoa stand age categories. It was hypothesised that; (a) the distribution of the total C stocks in the cocoa ecosystem differs between vegetation and soils, and (b) the C stocks differ between regions and shade management. The objectives were: (1) to quantify the total carbon stocks and distribution in the cocoa ecosystem, and (2) to assess the influence of shade management and the region of cocoa production on the C stocks.

Results and discussion
Selected properties of the soils under the cocoa ecosystems
The present study showed a range of 1.1–1.9 Mg/m^3 as the bulk density of the soils under the cocoa ecosystems (Table 1). As expected, the bulk density increased with soil depth from the surface. The gravimetric moisture content of the soils under the cocoa ecosystems ranged from 12.6 to 17.9 % (w/w). The soil moisture only varied with soil depth with the topsoil, 0–20 cm, being the wettest (Table 1). The ranges of the particle size fractions were: clay, 6.6–13.6 %; sand, 49–53 %, and silt, 36–41 % (Table 1). The soils are characterised as having the texture of sandy silt throughout the 0–60 cm layer (Table 1).

Biomass C concentrations
The mean carbon concentrations in above-ground components for all of the ecosystems under evaluation are presented in Fig. 1. The measured litter carbon concentration of 36.1 ± 1.1 corroborates the value of 37 % C in forest litter by Smith and Heath [14] that is currently being used as a default C concentration for litter in agroecosystems by the Intergovernmental Panel on Climate Change [3]. Similarly, the current carbon concentration value of 42.0 ± 0.4 % for the cocoa trees is in agreement with 43.7 ± 2.1 % for cocoa carbon reported by Anglaaere [15].

With the exception of litter C (a proportion of which is lost through respiration as it decomposes), the other components had a narrow range of 42.0–45.6 % C, with cocoa trees having the least C and *Persea americana* (dominant shade species in the Western region) having the highest (Fig. 1). Although few studies on agroecosystem C stocks present direct measurements of carbon with the aid of a C-analyser [16, 17], several studies have used constant values ranging from 45 to 50 % as the proportion of C for all parts of tree biomass [18, 19]. The organic carbon levels in the shade trees in the current study are not markedly different from the constant 45 % C for forest species being used by other studies [20, 21].

Soil organic carbon concentration
The soil total organic carbon concentrations differed significantly (P < 0.05) between regions, systems and soil depths (Table 1). Soil C concentration decreased with soil depth from the surface. Similar trends with depth have been noted by Cifuentes-Jara [22] and Dawoe [23]. The topsoil, 0–20 cm, contained approximately 58.8 % of the soil organic C in the 0–60 cm soil profile. This undoubtedly reflects the great mass of litter fall in cocoa ecosystems. In addition, the high C concentration in the topsoil is in accordance with the presence of 80–85 % mat of lateral roots of cocoa trees being predominantly found within the top 0–30 cm [23–25], although visible roots were excluded in sampling for the current study. The soil C concentration range of 0.6–2.0 % lies within the soil C concentration range of 0.4–2.6 %, reported by Dawoe [23] for 15 and 30 year old cocoa ecosystems in the Ashanti region, Ghana.

Above-ground carbon stocks in cocoa ecosystems
The C contribution from different cocoa ecosystem components to the total above-ground biomass C varied among regions, system, and their interactions (Table 2). On a per hectare basis, the system's biomass C components ranged as follows: cocoa trees, 11.8–16.9 Mg C/ha;

Table 1 Grand mean ± standard error of selected properties of the soils in the cocoa ecosystems for region (n = 24), system (n = 24) and depth (n = 16)

Factor	Treatment	Bulk density (Mg/m³)	Clay (%)	Sand	Silt	Moisture	C
Region	Eastern	1.5 ± 0.1	8.3 ± 0.6	51 ± 2	40 ± 1	14.7 ± 0.7	0.7 ± 0.1
	Western	1.6 ± 0.1	11.7 ± 0.6	52 ± 2	38 ± 1	14.7 ± 0.7	1.5 ± 0.1
System	Shaded	1.6 ± 0.1	10.5 ± 0.6	53 ± 2	36 ± 1	14.8 ± 0.7	1.0 ± 0.1
	Unshaded	1.5 ± 0.1	9.6 ± 0.6	49 ± 2	41 ± 1	14.7 ± 0.7	1.3 ± 0.1
Soil depth	0–20 cm	1.1 ± 0.1	6.6 ± 0.7	53 ± 2	41 ± 1	17.9 ± 0.8	2.0 ± 0.1
	20–40 cm	1.6 ± 0.1	9.9 ± 0.7	51 ± 2	39 ± 1	12.6 ± 0.8	0.8 ± 0.1
	40–60 cm	1.9 ± 0.1	13.6 ± 0.7	50 ± 2	37 ± 1	13.7 ± 0.8	0.6 ± 0.1

Age of farms appearing as covariate

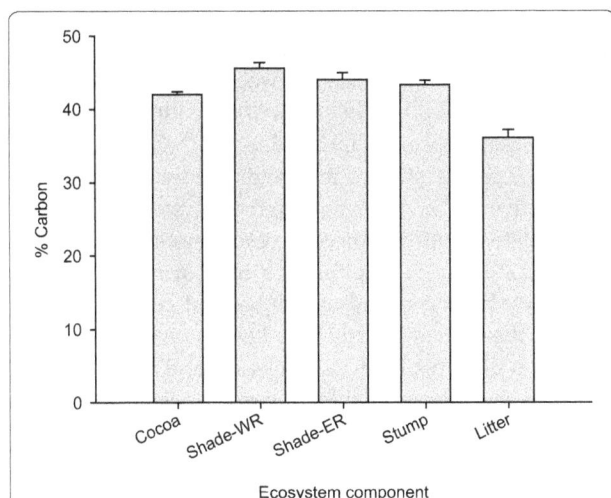

Fig. 1 Percentage carbon in cocoa ecosystem components ($n = 5$). *Persea americana* and *Newbouldia laevis* are dominant shade trees in Western and Eastern regions, respectively. *Error bars* represent standard errors

shade trees, 10.2–16.4 Mg C/ha; litter, 1.9–2.9 Mg C/ha, and stumps, 0.01–0.24 Mg C/ha (Table 2). These results compared well with the ranges of C stocks reported for cocoa trees in the literature [26, 27]. Similarly, the C stocks in shade trees of the current study agreed with the estimates for those in agro-ecosystems researched by Kürsten and Burschel [28] and Polzot [29] (3–25 and 1.9–31.8 Mg C/ha, respectively). Although the cocoa and shade trees' contributions were comparable, and together they contributed approximately 87.3–92.7 % of the total system's biomass C, only the biomass C contribution from the shade trees correlated significantly with the system's total biomass C (r = 0.9724, P < 0.001, Table 3). The lowest contribution to the system's C storage was obtained from stumps in shaded systems in the Western region (Table 2).

Overall, the mean carbon storage of cocoa trees was similar to that estimated for cocoa trees in a 30 year old cocoa system in Cameroon (14.4 Mg C/ha) reported by Norgrove and Hauser [27]. The present study estimated C stock in cocoa trees similar to those reported by Isaac et al. [30] as 10.3 Mg C/ha in an 8 year-old cocoa system in Ghana [30]. Isaac et al. [26] estimated the C storage of a 15 year-old cocoa system in Ghana as 16.8 and 15.9 Mg C/ha for a 25 year-old system, both of which agreed with the present finding that the average carbon storage of cocoa trees ranged between 11.8 and 16.9 Mg C/ha.

Soil organic carbon stocks
Understanding the effects of land use/land cover changes on ecosystem functions is often inferred from changes

in soil organic carbon. However, measurements of SOC have often been excluded in many studies on land-use change because of methodological uncertainties. Jones et al. [31] reported a measurement standard error of 1000 kg/ha for SOC, due largely to wide variation in the soil C estimation at deeper soil profiles. In the current study, uncertainty was reduced in the characterization of the soil C pools from the surface to 60 cm depth by measuring C stocks in different soil layers.

Table 4 presents the measured SOC contents for different layers to 60 cm depth. There were considerable variations in SOC contents between regions and systems. The soil organic C stocks ranges in the 0–20, 20–40 and 40–60 cm depths were 35.7–70.7, 15.0–46.7 and 11.5–31.3 Mg/ha, respectively. Clearly, the bulk of the SOC was concentrated in the topsoil, 0–20 cm depth. Moreover, SOC decreased with depth under all the factors. At all depths, soils in the W had the highest mean C stocks. Significantly (P < 0.05) higher SOC stocks were measured in E than W at all depths. The system of production affected SOC storage from the surface to 40 cm depth, but not between 40 and 60 cm (Table 4).

Total cocoa ecosystem carbon stocks and accumulation
Table 5 presents the mean C stocks distributed between the biomass and soil components of cocoa ecosystems. Total above-ground C stock in ecosystem biomass was estimated as the sum of the biomass C from cocoa trees, shade trees, stumps, and litter (Table 2). The total biomass C was highly variable in the cocoa ecosystems and ranged from a minimum mean value of 16.7 ± 2.2 Mg C/ha from unshaded cocoa systems in the Western region to a maximum mean value of 31.3 ± 2.2 Mg C/ha measured in shaded cocoa systems in the Eastern region (Table 5). Statistical analysis of the total system's biomass C showed significantly higher C stocks in E than W and in shaded than unshaded systems (Table 5).

Total SOC pools from the topsoil to 60 cm depth varied considerably from a minimum of 61.7 ± 7.7 Mg C/ha in unshaded cocoa system in the Eastern region to a maximum C stock of 137.8 Mg/ha in unshaded system in the Western region (Table 5). Results from this study estimated higher SOC stocks than the mean SOC value of 60.4 Mg/ha in Dawoe [23] for 0–60 cm depth of cocoa soils in the Ashanti region, Ghana. Cumulative (0–60 cm depth) SOC indicated significant (P < 0.05) variations between regions and also between management systems (Table 5).

The total ecosystem C stock of cocoa systems was estimated as the sum of soil C within 0–60 cm depth and above-ground biomass C (trees, stump and litter C). Total ecosystem C was higher in the Western region (137.7 ± 8.6 Mg C/ha) than in the Eastern region

Table 2 Mean C stocks ± standard error (Mg/ha) in cocoa trees, shade trees, stumps and litter components as influenced by region [Eastern (E), Western (W)] and system [shaded (S), unshaded (U)] and their interactions, (n = 12)

Factor	Treatment	Cocoa	Shade	Stumps	Litter
Region	E	15.2 ± 1.0	10.2 ± 6.4	0.16 ± 0.02	2.3 ± 0.2
	W	13.5 ± 1.0	16.4 ± 6.4	0.12 ± 0.02	2.4 ± 0.2
System	S	12.7 ± 1.1	13.3 ± 4.1	0.07 ± 0.02	2.6 ± 0.2
	U	16.1 ± 1.1	n.a.	0.21 ± 0.02	2.2 ± 0.2
Region * System	E * S	13.6 ± 1.5	10.2 ± 6.4	0.12 ± 0.03	2.3 ± 0.2
	E * U	16.9 ± 1.5	n.a.	0.19 ± 0.03	2.4 ± 0.2
	W * S	11.8 ± 1.6	16.4 ± 6.4	0.01 ± 0.03	2.9 ± 0.2
	W * U	15.2 ± 1.5	n.a.	0.24 ± 0.03	1.9 ± 0.2

Age of farms appearing as covariate in the statistical model used

Not applicable

Table 3 Pearson correlation coefficients (r) for linear relationships among biomass C components in cocoa ecosystems

	Ecosystem	Cocoa	Shade	Stumps
Cocoa	0.4936			
Shade	0.9724**	0.2801		
Stump	−0.1842	0.2948	−0.2536	
Litter	0.4945	0.5397	0.3703	−0.4871

Values with '**' are significant at P < 0.01, and without symbol are not significant, (2—tailed test)

(95.7 ± 8.6 Mg C/ha). These C estimates are very high when compared with data from Dawoe [23]. This is attributed to the low soil C stocks (35.5–80.4 Mg C/ha) from 0–60 cm depth reported by Dawoe [23], that were equivalent to the estimated C stocks in the current study's topsoil, 0–20 cm (35.7–70.7 Mg C/ha) (see Table 4). Notably, in the current study, the soils contributed between 3 and

5 times more C than the above-ground pools of the cocoa ecosystems. Given the age range (7–28 years) of farms used in the current studies, as well as the extensive cultivation of 1.45 million hectares of cocoa in Ghana [1], it appears that approximately 118.6–223.2 Gg C could be stored in cocoa systems with stands aged within 30 years, irrespective of the shade-management system.

The relative contribution of the cocoa systems (scatter) to the overall C stocks (line) in each component varied considerably when expressed on the basis of cocoa stand age (Fig. 2). The shaded and unshaded cocoa systems appeared to contain the same biomass stocks at stand age of 10 years in age. In the above-ground biomass C stocks, both shaded and unshaded cocoa systems increase with stand age but the contribution from the shaded systems to the overall biomass C trend was much higher than the unshaded system for cocoa stands older than 10 years (Fig. 2).

With respect to the effects of cocoa systems on soil C, there appears to be a general decline of the soil C stocks as time progressed. Whereas the shaded systems indicate a slight increase, the unshaded systems showed a slight decrease in soil C (Fig. 2). The two systems have similar soil C stocks at stand age of 25 years onwards.

The primary source of soil C is from litter and so the quantity and quality of the litter inputs affect the soil C dynamics [32]. Of the systems' contribution to the total C in the ecosystems, the trend follows that of the soil C since the bulk of C (>80 %) is stored in the soil (Table 5). The trends indicate that total carbon in shaded and unshaded systems are the same at age 17 years, but the shaded system thereafter, increased in the total C higher than that of the unshaded system (Fig. 2).

Conclusions

The need to quantify the carbon stocks in cocoa systems in Ghana is necessitated by the recent inclusion of the sector as an option that could result in a net increase in

Table 4 Mean soil organic C stocks ± standard error (Mg/ha) at 0–20, 20–40 and 40–60 cm layers as influenced by region [Eastern (E), Western (W)], and system [shaded (S), unshaded (U)], (n = 12)

Factor	Treatment	0–20 cm	20–40 cm	40–60 cm
Region	E	40.2 ± 3.4	16.6 ± 3.6	14.3 ± 1.6
	W	58.4 ± 3.4	33.3 ± 3.6	25.7 ± 1.6
System	S	45.4 ± 3.4	19.0 ± 3.6	18.5 ± 1.6
	U	53.2 ± 3.4	30.9 ± 3.6	21.4 ± 1.6
Region * System	E * S	44.7 ± 4.8	18.2 ± 5.0	17.1 ± 2.2
	E * U	35.7 ± 4.8	15.0 ± 5.0	11.5 ± 2.2
	W * S	46.1 ± 5.1	19.9 ± 5.4	20.0 ± 2.4
	W * U	70.7 ± 4.8	46.7 ± 5.1	31.3 ± 2.2

Age of farms appearing as covariate in the statistical model used

Table 5 Mean cocoa ecosystem carbon stocks ± standard error, distributed between the biomass and soil (0–60 cm depth) components according to region [Eastern (E), Western (W)], and system [shaded (S), unshaded (U)], (n = 12)

Factor	Treatment	Biomass C (Mg/ha)	Soil C (Mg/ha)	Total C (Mg/ha)
Region	E	25.2 ± 1.6	70.5 ± 5.4	95.7 ± 8.6
	W	18.4 ± 1.6	113.0 ± 5.4	137.7 ± 8.6
System	S	25.8 ± 1.6	83.7 ± 5.5	115.5 ± 8.6
	U	17.8 ± 1.6	99.8 ± 5.5	117.9 ± 8.6
Region * System	E * S	31.3 ± 2.2	79.3 ± 7.7	109.5 ± 12.0
	E * U	19.0 ± 2.2	61.7 ± 7.7	81.8 ± 12.0
	W * S	20.2 ± 2.3	88.1 ± 8.2	121.5 ± 12.0
	W * U	16.7 ± 2.2	137.8 ± 7.7	153.9 ± 12.1

Age of farms appearing as covariate in the statistical model used

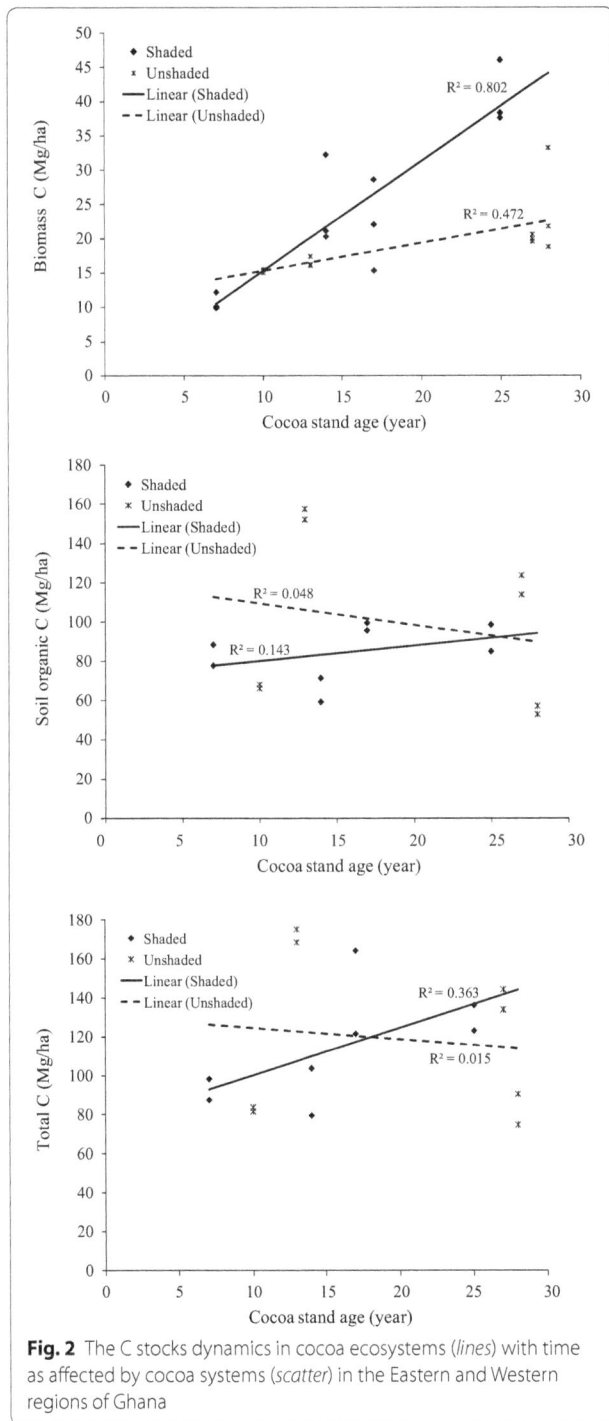

Fig. 2 The C stocks dynamics in cocoa ecosystems (*lines*) with time as affected by cocoa systems (*scatter*) in the Eastern and Western regions of Ghana

was twice that in unshaded systems, the two systems did not differ significantly with respect to total ecosystem C stocks. The bulk of the C stock was in the soil. The estimated high C stocks suggest that the cocoa sector holds a large amount of carbon and should be included in the national carbon accounting emission budget of Ghana.

Methods
Physiology of the study area
The field studies were carried out between July and October, 2011 in two regions of Ghana; the Eastern region at Duodukrom community in the Suhum district (6°2′N, 0°27′W), and the Western region at Anyinabrim in the Sefwi-Wiawso district (6°57′N, 2°35′W). Figure 3 presents the map of Ghana showing the regions and districts where the field studies were conducted.

The Eastern region covers a land area of 19,323 km^2 representing 8.1 % of the total land area of Ghana. It is located between latitude 6° and 7°N and longitude 1°30′W and 0°30′E. The region lies within the wet semi-equatorial zone which is characterized by double-maxima rainfall in June and October. The natural vegetation of the region is humid deciduous forest. Temperatures in the region are high and range between 26 °C in August and 30 °C in March. The relative humidity which is high throughout the year varies between 70 and 80 %.

The Western region occupies a land area of 23,921 km^2 which is approximately 10 % of the total land area of Ghana. The region lies in the equatorial climatic zone that is characterized by a double maxima rainfall occurring in May–July and September/October. Its vegetation is that of humid deciduous forest. The region is the wettest part of Ghana with an average rainfall of 1600 mm per annum and harbours about 24 forest reserves that account for about 40 % of the forest reserves in Ghana. The climate creates much moisture culminating in high relative humidity, ranging from 70 to 90 % in most part of the region. Temperatures range between 22 °C at nightfall and 34 °C during the day.

Thus, the two regions experience similar climate and vegetation. The major soils found in both regions are mostly well drained *Ochrosols* or *Oxisols* suitable for the production of industrial crops such as cocoa, pineapple, pawpaw cola nut and oil palm. However, the Eastern region has been producing cocoa long before cultivations started in the Western region.

Selection of farms
Eight farms, comprising four from the Duodukrom community in the Suhum district of the Eastern region, and four from the Anyinabrim community in the Sefwi-Wiawso district of Western region were selected for sampling cocoa stands on the basis of shade management

terrestrial carbon stocks. Hence, this paper estimated the carbon stocks in shaded and unshaded cocoa systems at different age categories; the fields were selected from the Eastern region (E) and Western region (W) of Ghana. Total ecosystem carbon was higher in the W than E. While the biomass C stock from shaded systems

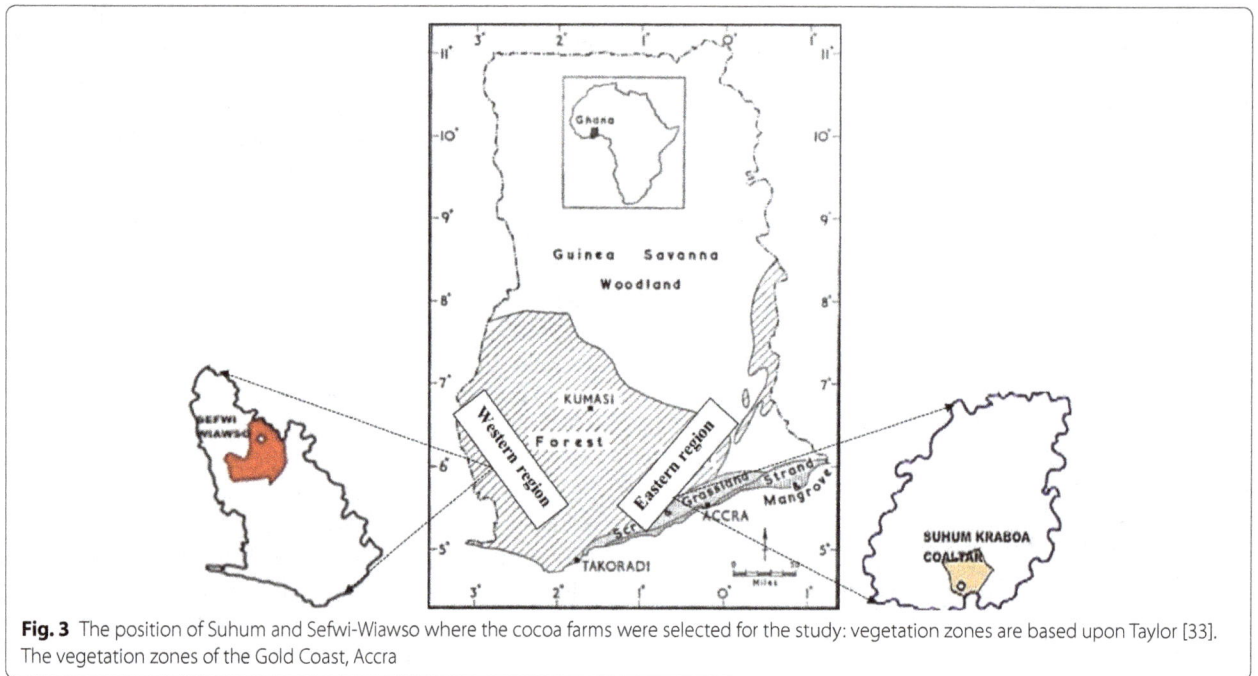

Fig. 3 The position of Suhum and Sefwi-Wiawso where the cocoa farms were selected for the study: vegetation zones are based upon Taylor [33]. The vegetation zones of the Gold Coast, Accra

(shaded, unshaded). Selected farms had cocoa stand ages of 10, 14, 25 and 28 years in the Eastern region (E) and 7, 13, 17 and 27 years in the Western region (W).

At each farm, plot sizes of 30 × 90 m were demarcated for sampling. Two 30-m transects dividing the plot into three of 30 × 30 m (~0.23 acre or 0.09 ha) subplots were demarcated to give three pseudo-replications of each farm. The common shade tree species identified on the cocoa farms included *Terminalia ivorensis*, *Terminalia superba*, *Entandrophragma cylindricum*, *Entandrophragma angolense*, *Newbouldia laevis*, *Persea americana*, *Celtis mildbraedii*, *Cola nitida*, *Carica papaya*, *Palmae* sp., *Spondia smombin*, *Ficus exasperate*, *Citrus sinensis* (L.) Osbeck, *Acacia mangium*, and other forest tree species. Avocado (*Persea americana*) was the dominant shade tree in cocoa farms found in the Western region whilst *Newbouldia laevis* was the dominant shade tree in the Eastern region's cocoa farms.

All trees were counted, and their diameters at breast height (DBH) measured, sorted and grouped into three diameter class sizes (upper, middle, and lower) relative to the DBH range of cocoa trees on the farms; 16 cocoa trees, comprising two cocoa trees per farm were randomly selected such that the diameter of one tree lay within the upper class and the other in the lower class for destructive sampling. The felled trees were each separated into trunks, branches and foliage (leaves, fruits); these parts were cut to smaller pieces, weighed in batches and then summed to give total component weight. Fresh leaf samples of the dominant shade trees found in each

region were also taken. Based on the measured DBH and the biomass per tree of the 16 cocoa trees that were destructively sampled across all the study sites, an allometric relation was developed using regression techniques to estimate standing cocoa tree biomass. The general equation from FAO [34], recommended by UNFCCC [35], was used to estimate the above-ground biomass of the shade tree species.

$$AgB = \exp\left[-2.134 + 2.530\ln(DBH)\right] \qquad (1)$$

where AgB denotes above-ground biomass, kg tree^{-1}, and DBH = diameter at breast height, cm.

Soil moisture and bulk density

Soil samples at 0–20, 20–40 and 40–60 cm depths were taken from a total of 16 plots comprising 2 micro-plots of (50 × 50 cm) that were established at random within the eight cocoa farms. Two core soil samples per depth were taken randomly at each micro plot using an auger after removing visible litter from the soil surface. Soil bulk density gives an indication of the level of soil compaction [36]. Soil bulk density and moisture contents at each sampling depth were determined on the undisturbed core samples, as outlined in Blake and Hartge [37].

Texture

Another set of soil samples from the same micro-plots was air-dried for 72 h, and ground to pass through a 2-mm mesh sieve to yield the fine earth fraction for chemical analysis. The soil particle size distribution was determined by laser granulometry, using a Coulter LS230 particle size analyser connected to a Windows-based computer [38–40].

Carbon concentration

Weights between 0.9–1.1, and 8.0–12.0 mg were taken respectively, from plant and soil samples the determination of C concentrations. The organic C concentration in the samples was determined using the Europa Roboprep connected to a VG 622 Mass Spectrophotometer.

Carbon stocks

There are various C pools, or compartments, within cocoa ecosystems. These include the soil C pool, the litter C pool and the woody biomass C pool in trees. The quantity of C stored in each pool is reported as the C stock, and the sum of the C stocks from the different pools constitutes the total ecosystems C stocks. On each farm, the total biomass-C stock was estimated as the sum of the C stocks in cocoa tree components (root, stem, branch, and leaf litter), floor litter and shade trees (if any) as expressed in Eq. (2). The cocoa tree component-C stocks were calculated as the product of the mean C concentration and the biomass per hectare [41]. The mean C concentration of leaves of shade trees was used as the average C concentration of the whole shade tree in estimating the C stock of the shade trees.

$$
\begin{aligned}
TotalC_{biomass} = \{ & [(\%C_{root} \times root_{biomass}) \\
& + (\%C_{stem} \times stem_{biomass}) \\
& + (\%C_{branch} \times branch_{biomass}) \\
& + (\%C_{leaf} \times leaf_{biomass})]_{cocoa} \\
& + (\%C_{litter} \times litter_{biomass}) \\
& + (\%C_{shade} \times Shade_{biomass}) \}
\end{aligned}
\qquad (2)
$$

Soil C stocks were also calculated using the formula:

$$
SOC = \sum \rho_i \times d_i \times \%C_i \qquad (3)
$$

where SOC denotes soil organic carbon stock (Mg/ha); ρ = soil bulk density (g/cm^3); i = 0–20, 20–40, and 40–60 cm sampling depth; d = depth over which the sample was taken (cm); and $\%C$ = soil carbon concentration (%). The total cocoa ecosystem carbon stock for each farm/system was then estimated as the sum of Eqs. (2) and (3).

Data analyses

The data were tested for normality using q–q plot with Anderson–Darling P values in MINITAB v16. Where the tested component C was found to be non-normal,

the appropriate transformation was determined with the help of Box-Cox transformation and optimal or rounded lambda that suggested one of the following transformational method as appropriate: square root, reciprocal square root, natural logarithm or inverse transformation method, according to the skewness of the data [42]. Specifically, litter, stumps and total ecosystem C data were normal (P > 0.05) without transformation; biomass and soil C were inversely transformed; cocoa tree C was transformed with square root and shade tree C was normalized using natural logarithms. The transformed data were analysed by the Linear MIXED Model of IBM SPSS statistics 20th edition to determine significant differences between Eastern and Western regions and between shaded and unshaded systems as well as the interactions on carbon stocks controlling for the ages (covariate) of cocoa farms. The means were then estimated by restricted maximum likelihood (REML) and back-transformed to maintain the original form of the measurement. Correlation analyses by Pearson's rank matrix were also carried out to determine any relationships among some of the ecosystem variables.

Authors' contributions

All authors (AMM, JSR, DM and AV) took part in the development of the main idea and the writing of the manuscript. The field data and laboratory analysis were done by AMM who also performed the majority of the statistical analysis. The data and analytical results were discussed among all four authors. All authors read and approved the final manuscript.

Author details

[1] CSIR-Savanna Agricultural Research Institute, Nyankpala, PO Box 52, Tamale, Ghana. [2] School of Archaeology, Geography and Environmental Science, University of Reading, Reading, UK.

Acknowledgements

This paper is an output from a Ph.D. study in the University of Reading, UK, with funds from the Ghana Education Trust Fund, and the authors are grateful for the permission granted to publish this work. The support by cocoa farmers in Ghana to use their fields for data collection is also gratefully acknowledged.

Competing interests

The authors declare that they have no competing interests.

References

1. Anim-Kwapong G, Frimpong E. Vulnerability of agriculture to climate change-impact of climate change on cocoa production. Final Report submitted to the Netherlands climate change studies assistance programme: NCAP; 2005. 2.
2. Stern NH. Stern review: the economics of climate change, vol. 30. London: HM treasury London; 2006.
3. IPCC. Climate change 2007 the physical science basis. Agenda. 2007;6:07.
4. IPCC. Climate change: the scientific basis. Cambridge: Cambridge University Press; 2001.
5. Marland G, Boden TA, Andres RJ, Brenkert A, Johnston C. Global, regional, and national fossil fuel CO$_2$ emissions. Trends: a compendium of data on global change; 2003. p. 34–43.

6. Seeberg-Elverfeldt C, Schwarze S, Zeller M. Payments for environmental services: incentives through carbon sequestration compensation for cocoa-based agroforestry systems in Central Sulawesi. Grauer; 2008.

7. Wade ASI, Asase A, Hadley P, Mason J, Ofori-Frimpong K, Preece D, Spring N, Norris K. Management strategies for maximizing carbon storage and tree species diversity in cocoa-growing landscapes. Agric Ecosyst Environ. 2010;138(3–4):324–34.

8. Gockowski J, Sonwa D. Cocoa intensification scenarios and their predicted impact on CO_2 emissions, biodiversity conservation, and rural livelihoods in the Guinea Rain Forest of West Africa. Environ Manage. 2011;48(2):307–21.

9. Dixon R. Agroforestry systems: sources of sinks of greenhouse gases? Agrofor Syst. 1995;31(2):99–116.

10. UNFCCC. Calculation of the number of sample plots for measurements with A/R CDM project activities version 02; 2009.

11. Oke D, Olatiilu A. Carbon storage in agroecosystems: a case study of the cocoa based agroforestry in Ogbese forest reserve, Ekiti State,Nigeria. J Environ Prot. 2011;2(8):1069–75.

12. Oke D, Odebiyi K. Traditional cocoa-based agroforestry and forest species conservation in Ondo State, Nigeria. Agric Ecosyst Environ. 2007;122(3):305–11.

13. Chagas T, O'Sullivan R, Bracer C, Streck C. Consolidating national REDD + accounting and subnational activities in Ghana. UNDP, Global Environment Facility, NORAD, and Gordon and Betty Moore Foundation; 2010. p. 36.

14. Smith JE, Heath LS. Identifying influences on model uncertainty: an application using a forest carbon budget model. Environ Manage. 2001;27(2):253–67.

15. Anglaaere LCN. Improving the sustainability of cocoa farms in Ghana through utilization of native forest trees in agroforestry systems, Ph.D. Thesis, University of Wales, Bangor; 2005. p. 340.

16. Kraenzel M, Castillo A, Moore T, Potvin C. Carbon storage of harvest-age teak (*Tectona grandis*) plantations, Panama. For Ecol Manage. 2003;173(1–3):213–25.

17. Losi CJ, Siccama TG, Condit R, Morales JE. Analysis of alternative methods for estimating carbon stock in young tropical plantations. For Ecol Manage. 2003;184(1–3):355–68.

18. Snowdon P, Raison J, Keith H, Ritson P, Grierson P, Adams M, Montagu K, Bi H-Q, Burrows W, Eamus D. Protocol for sampling tree and stand biomass. Canberra: Australian Greenhouse Office; 2002. p. 76.

19. IPCC. Good practice guidelines for National Greenhouse gas inventories. Hayama: Institute for Global Environmental Strategies (IGES); 2006.

20. Noordwijk MV, Rahayu S, Hairiah K, Wulan Y, Farida A, Verbist B. Carbon stock assessment for a forest-to-coffee conversion landscape in Sumber-Jaya (Lampung, Indonesia): from allometric equations to land use change analysis; 2002. p. 1–12.

21. Smiley G, Kroschel J. Temporal change in carbon stocks of cocoa–gliricidia agroforests in Central Sulawesi, Indonesia. Agrofor Syst. 2008;73(3):219–31.

22. Cifuentes-Jara M. Aboveground biomass and ecosystem carbon pools in tropical secondary forests growing in six life zones of Costa Rica, Ph.D. Thesis, Oregon State University; 2008. p. 195.

23. Dawoe E. Conversion of natural forest to cocoa agroforest in lowland humid Ghana: impact on plant biomass production, organic carbon and nutrient dynamics, Ph.D. Thesis, Kwame Nkrumah University of Science and Technology; 2009. p. 279.

24. Wood GAR, Lass R. Cocoa. Hoboken: Wiley-Blackwell; 2008.

25. de Oliveira Leite J, Valle RR. Nutrient cycling in the cacao ecosystem: rain and throughfall as nutrient sources for the soil and the cacao tree. Agric Ecosyst Environ. 1990;32(1):143–54.

26. Isaac M, Gordon A, Thevathasan N, Oppong S, Quashie-Sam J. Temporal changes in soil carbon and nitrogen in west African multistrata agroforestry systems: a chronosequence of pools and fluxes. Agrofor Syst. 2005;65(1):23–31.

27. Norgrove L, Hauser S. Carbon stocks in shaded *Theobroma cacao* farms and adjacent secondary forests of similar age in Cameroon. Trop Ecol. 2013;54(1):15–22.

28. Kürsten E, Burschel P. CO_2-mitigation by agroforestry. Water Air Soil Pollut. 1993;70(1):533–44.

29. Polzot CL. Carbon storage in coffee agroecosystems of southern Costa Rica: Potential applications for the clean development mechanism, M.Sc. Thesis, York University, Toronto; 2004. p. 162.

30. Isaac ME, Timmer VR, Quashie-Sam SJ. Shade tree effects in an 8-year-old cocoa agroforestry system: biomass and nutrient diagnosis of *Theobroma cacao* by vector analysis. Nutr Cycl Agroecosyst. 2007;78(2):155–65.

31. Jones J, Graham W, Wallach D, Bostick W, Koo J. Estimating soil carbon levels using an ensemble Kalman filter. Trans Am Soc Agric Eng. 2004;47(1):331–42.

32. Post WM, Kwon KC. Soil carbon sequestration and land-use change: processes and potential. Glob Change Biol. 2000;6(3):317–27.

33. Taylor CJ. The vegetation zones of the Gold Coast. Forestry Department Bulletin 4, Accra. 1952; pp. 48–51.

34. FAO. Estimating biomass and biomass change of tropical forests: a primer, in FAO Forestry Paper 134. Rome: FAO; 1997. p. 55.

35. UNFCCC. Monitoring methodologies for selected small-scale afforestation and reforestation project activities under the Clean Development Mechanism. Bonn; 2006.

36. Hillel D. Environmental soil physics: fundamentals, applications, and environmental considerations. New York: Academic Press; 1998. p. 19–201.

37. Blake G, Hartage K. Bulk density. Methods of soil analysis: part 1—physical and mineralogical methods, 1986(methodsofsoilan1). p. 363–75.

38. Buurman P, Pape T, Muggler C. Laser grain-size determination in soil genetic studies 1. Practical problems. Soil Sci. 1997;162(3):211–8.

39. Muggler C, Pape T, Buurman P. Laser grain-size determination in soil genetic studies 2. Clay content, clay formation, and aggregation in some brazilian oxisols. Soil Sci. 1997;162(3):219–28.

40. Arriaga FJ, Lowery B, Mays MD. A fast method for determining soil particle size distribution using a laser instrument. Soil Sci. 2006;171(9):663–74.

41. Subedi BP, Pandey SS, Pandey A, Rana EB, Bhattarai S, Banskota TR, Charmakar S, Tamrakar R. Forest carbon stock measurement: guidelines for measuring carbon stocks in community-managed forests; 2010. p. 69.

42. Hamilton LC. Modern data analysis: a first course in applied statistics. California: Brooks/Cole Pacific Grove; 1990.

10

Modelling forest carbon stock changes as affected by harvest and natural disturbances. I. Comparison with countries' estimates for forest management

Roberto Pilli[1*], Giacomo Grassi[1], Werner A. Kurz[2], Raúl Abad Viñas[1] and Nuria Hue Guerrero[1]

Abstract

Background: According to the post-2012 rules under the Kyoto protocol, developed countries that are signatories to the protocol have to estimate and report the greenhouse gas (GHG) emissions and removals from forest management (FM), with the option to exclude the emissions associated to natural disturbances, following the Intergovernmental Panel on Climate Change (IPCC) guidelines. To increase confidence in GHG estimates, the IPCC recommends performing verification activities, i.e. comparing country data with independent estimates. However, countries currently conduct relatively few verification efforts. The aim of this study is to implement a consistent methodological approach using the Carbon Budget Model (CBM) to estimate the net CO_2 emissions from FM in 26 European Union (EU) countries for the period 2000–2012, including the impacts of natural disturbances. We validated our results against a totally independent case study and then we compared the CBM results with the data reported by countries in their 2014 Greenhouse Gas Inventories (GHGIs) submitted to the United Nations Framework Convention on Climate Change (UNFCCC).

Results: The match between the CBM results and the GHGIs was good in nine countries (i.e. the average of our results is within ±25 % compared to the GHGI and the correlation between CBM and GHGI is significant at $P < 0.05$) and partially good in ten countries. When the comparison was not satisfactory, in most cases we were able to identify possible reasons for these discrepancies, including: (1) a different representation of the interannual variability, e.g. where the GHGIs used the stock-change approach; (2) different assumptions for non-biomass pools, and for CO_2 emissions from fires and harvest residues. In few cases, further analysis will be needed to identify any possible inappropriate data used by the CBM or problems in the GHGI. Finally, the frequent updates to data and methods used by countries to prepare GHGI makes the implementation of a consistent modeling methodology challenging.

Conclusions: This study indicates opportunities to use the CBM as tool to assist countries in estimating forest carbon dynamics, including the impact of natural disturbances, and to verify the country GHGIs at the EU level, consistent with the IPCC guidelines. A systematic comparison of the CBM with the GHGIs will certainly require additional efforts—including close cooperation between modelers and country experts. This approach should be seen as a necessary step in the process of continuous improvement of GHGIs, because it may help in identifying possible errors and ultimately in building confidence in the estimates reported by the countries.

Keywords: Net CO_2 emissions, Greenhouse gas inventories, European countries, Carbon Budget Model, Forest management, Harvest, Natural disturbances

*Correspondence: roberto.pilli@jrc.ec.europa.eu
[1] European Commission, Joint Research Centre, Institute for Environment and Sustainability, Via E. Fermi 2749, 21027 Ispra, VA, Italy
Full list of author information is available at the end of the article

Background

The United Nations Framework Convention on Climate Change (UNFCCC) and its Kyoto protocol (KP) recognize the role of forests in mitigating climate change. Emissions and removals from forests are included in the greenhouse gas inventories (GHGIs) submitted annually by developed countries to the UNFCCC, and typically represent by far the most important component of the "Land use, Land-use Change and Forestry" (LULUCF) sector. Inventories should follow the methodological guidance prepared by the Intergovernmental Panel on Climate Change (IPCC).

The forests in the European Union (EU, including 28 countries) cover about 165 Mha, they increased by about 4 % since 1990 and about 83 % of this area is available for wood supply [1]. According to the EU GHGI, between 1990 and 2012 the average annual sink of EU forests was about 435 Tg CO_{2eq}. $year^{-1}$, or about 9 % of the EU total emissions in the same period [2].

For the first commitment period of the KP (CP1, 2008–2012) the accounting of emissions and removals was mandatory for afforestation/reforestation and deforestation (AR and D, i.e. forest land-use changes since 1990) and voluntary for forest management (FM, i.e. forest existing before 1990). For the second commitment period of the KP (CP2, 2013–2020), significant revisions of accounting rules were agreed [3], as reflected in the latest IPCC guidance [4]. The major changes for the forest sector are: (1) the accounting of FM is now mandatory; (2) the FM accounting shall include the carbon (C) stock changes in the harvested wood products (HWP) pool; and (3) emissions and subsequent removals from natural disturbances may be excluded from the accounting. These changes represent new challenges for countries when developing their GHGIs.

Since the GHGIs represent the basis for assessing the effectiveness of any national climate policy, building confidence in their accuracy is of key importance for advancing the international efforts to mitigate climate change. While the GHGIs are subject to an UNFCCC expert review process, which aims to assess the adherence of GHGIs to IPCC guidance in terms of general reporting principles,[1] this expert review does not include an independent verification of the reported estimates. The verification activities should be performed by each country, as part of the process of improving the GHGI and build confidence in its reliability [5]. However, at the EU level, few countries report efforts or results of verification for the LULUCF sector [2]. In most cases, a real verification is very difficult due to the lack of truly independent and comparable data. For example, since GHGIs cover only emissions and removals from managed lands, an inherent mismatch

exists for LULUCF between GHGIs and estimates based on process studies or atmospheric methods [5]. As alternative, a largely independent comparison may be conducted between GHGIs and large-scale models (e.g. [6, 7]) that use data from National Forest Inventories (NFIs). While not a fully-independent verification, such comparisons may be very useful in building confidence in GHGI estimates and trends, improving scientific knowledge and identifying potential problems. The major challenges for this approach are to implement a model capable to reflect the latest IPCC guidance (e.g., including the HWP and natural disturbances, [4]) and to use adequate input data from the countries.

The general aim of this study is to implement a consistent methodological approach using an internationally well established forest carbon budget model to simulate for the period 2000–2012 the impacts of harvest and salvage logging, natural disturbances and land-use changes on forest CO_2 emissions and removals in all EU countries for which adequate information was available (26 countries out of 28). To this aim, the Carbon Budget Model (CBM) developed by the Canadian Forest Service [8] was used, as part of a broader effort for a comprehensive modelling framework for the forest sector [9]. The model was applied and validated at regional and national scales in Canada [10, 11] and Russia [12]. Furthermore, the CBM was successfully adapted to specific forest management conditions in Europe (e.g. uneven-aged forests, [13]), validated at regional level [14] and applied in one country case to estimate the C balance for FM [13] and AR [15].

Specific objectives of this paper are: (1) to validate the CBM against totally independent data available at the country level (for one case study) and to provide a detailed description of four representative country cases; (2) to compare FM estimates from CBM with each country's GHGI in terms of trends and levels of net CO_2 emissions for each forest C pools (living biomass, dead organic matter (DOM) and mineral soil); (3) to analyze how the main drivers affecting the living biomass (harvest and natural disturbances, including major storms and fires) affect the estimates obtained with the CBM and the GHGIs.

A companion paper [16] provides an analysis of the CBM results at the aggregate EU level, including net CO_2 emissions in the HWP pool and the impacts of forest-land use changes.

Results
Model validation

To validate our model's results with independent data sources (i.e., not used as input data by CBM), we first compared the mean annual increment and the average volume estimated by CBM (based on the equations applied by the model during the run and the values of

---

[1] i.e. Transparency, accuracy, completeness, consistency, comparability.

merchantable C stock provided for each species) with the additional, independent data reported in the Lithuanian GHGI (Fig. 1). A further comparison is made with the dead tree stems volume estimated by CBM and the values reported by NIR, based on a specific analysis until 2001 and on NFI permanent sample plots from 2002 to 2012.

The model's results can be further compared with other information for Lithuania, not fully independent of the input data used by CBM, because derived by the same data sources (i.e., NFI). Figure 2 reports the age class evolution estimated by CBM between 1996 and 2012, compared with the original age class distribution reported by NFI 2004–2008 (attributed to 2006).

In Fig. 3 (lower panel), the net CO_2 emissions estimated by CBM (further distinguished between living biomass, DOM and soil pools) are compared with the net emissions reported by the country's GHGI (in 2014) for the land use category forest land remaining forest land (FL–FL) (Lithuania, [18]). For Lithuania, our simulation starts in 1996 when, due to the effect of insect disturbances (see the Additional file: 1 for further details), we estimated a C source, consistent with the data reported by the country, and with the mean annual volume increment reported in Fig. 1. From 1997 to 2001, the model estimates an increasing C sink, mainly due to a reduction of the regular harvest, because of the salvage of logging residues. From 2002, the C sink decreases due to the increasing harvest demand (reported in the upper panel of Fig. 3) and, after 2007, the sink again increases following the decreasing amount of harvest. Further inter-annual variations are due to the effect of storms (in 2005 and 2007, according to the information by NIR), while the effect of fires is negligible. The interannual variability in net CO_2 emissions reported in the GHGI is considerably larger than estimated in the CBM. From 2007, the forest C sink reported by the country strongly increases, from -1.9 Mt CO_2 in 2007 to -8.0 Mt CO_2 in 2008 (i.e., $+300$ %), even if the total harvest demand decreases only slightly, from about 8.1 to 8.0 million m^3 (i.e., -1.2 %). This reduction was not observed in CBM results, which report only a slightly increase in the C sink between 2007 and 2012, which is consistent with the decreasing harvest rates.

CBM results vs country GHGIs

Net CO_2 emission estimates for the period 2000–2012 as estimated using the CBM and as reported by 26 EU countries in their 2014 GHGI show a wide range of patterns (Fig. 4). Data are for the area subject to FM^2 and are reported from an atmospheric perspective, where negative values represent a sink (CO_2 removals) and positive values a source (CO_2 emissions). Results focus on CO_2 and exclude organic soils. Non-CO_2 emissions (CH_4, N_2O) from forests may be important only for specific countries, in case of drained organic soils (not included in this paper) and in case of fires, for which we report results in terms of CO_2-eq for Portugal in the Additional file: 1.

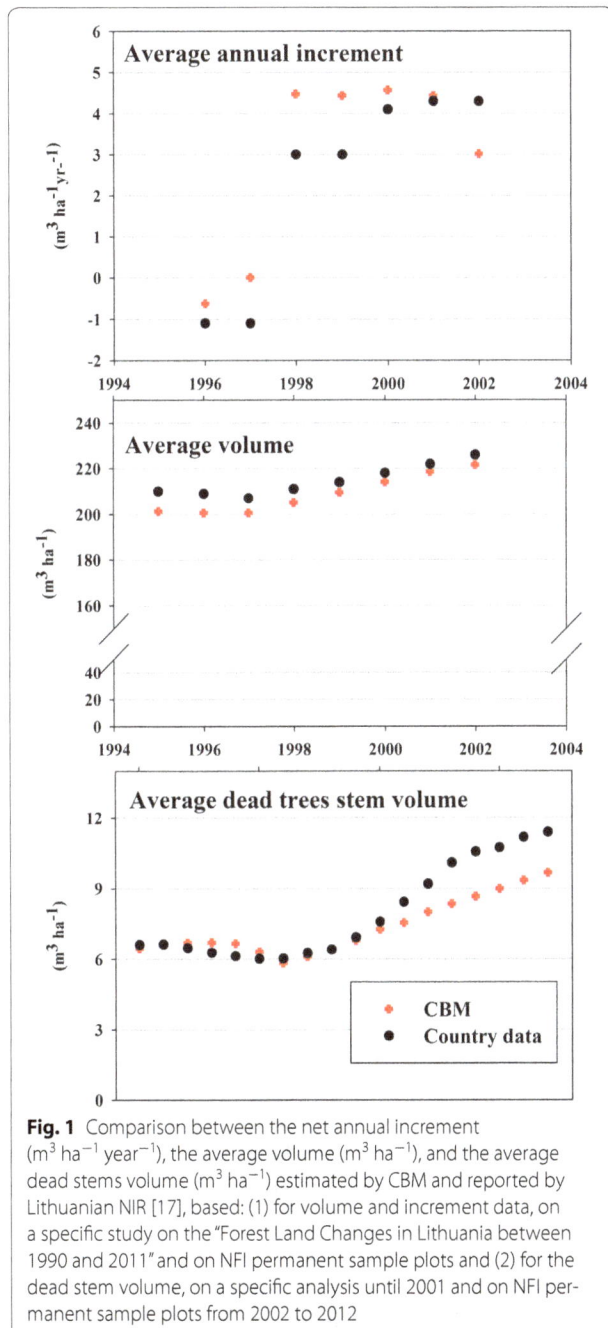

Fig. 1 Comparison between the net annual increment (m^3 ha^{-1} $year^{-1}$), the average volume (m^3 ha^{-1}), and the average dead stems volume (m^3 ha^{-1}) estimated by CBM and reported by Lithuanian NIR [17], based: (1) for volume and increment data, on a specific study on the "Forest Land Changes in Lithuania between 1990 and 2011" and on NFI permanent sample plots and (2) for the dead stem volume, on a specific analysis until 2001 and on NFI permanent sample plots from 2002 to 2012

---

[2] When available, FM country data from the KP-CRF tables was used for 2008-2012 (i.e., if FM had been elected during the first KP commitment period); alternatively, country data were taken from the Convention CRF tables using 'forest land remaining forest land' (FL remaining FL) as a proxy for FM.

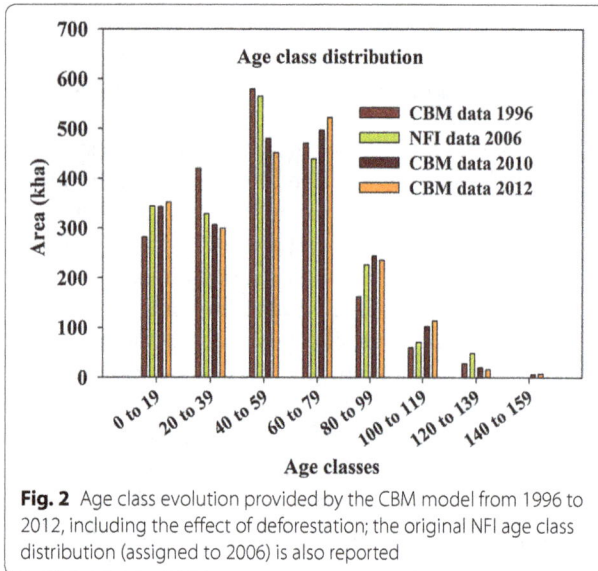

Fig. 2 Age class evolution provided by the CBM model from 1996 to 2012, including the effect of deforestation; the original NFI age class distribution (assigned to 2006) is also reported

The aggregated results at the EU level and including the harvested wood products (HWP) pool, afforestation/reforestation and deforestation, will be reported in a companion paper [16].

The results obtained from the GHGI and the CBM for these 26 countries can be assessed in terms of *level* and *trend*. For the *level*, we consider the match between CBM and each GHGI as "good" if the average net emission of CBM for the period 2000-2012 (Fig. 4) is within ± 25 %[3] compared to the GHGI. For the *trend*, Fig. 5 shows the correlation between CBM and each GHGI. In this case, we consider the match between CBM and each GHGI as "good" if the correlation is significant at P < 0.05.

Based on the match between CBM and GHGIs, in terms of *level* and *trend*, and on data reported in Figs. 4, 5, four different groups of countries may be distinguished:

A. Countries where CBM estimates and country data show a good match both in the *trend* and the *level*. This group includes nine countries: Croatia, Finland, Italy, Latvia, Lithuania, Portugal, Romania, Slovakia and Slovenia.

B. Countries where there is a good match in the *trend* but not in the *level*. This group includes five countries: Austria, Czech Republic, Estonia, Greece and Luxembourg,

C. Countries where there is a good match in the *level* but not in the *trend*. This group includes five countries: France, Germany, Spain, Sweden and United Kingdom.

D. Countries where the match is not good for the *level* and for the *trend*. This group includes seven countries: Belgium, Bulgaria, Denmark, Hungary, Ireland, Netherlands and Poland.

Figure 6 illustrates in more detail the results from four country cases (Discussed in "Country case studies" section.), each representative of the four groups above: Portugal (A), Austria (B), Germany (C) and Poland (D).

Discussion
Model evaluation
We implemented a consistent methodological approach to 26 EU countries, using the Carbon Budget Model to estimate the net CO_2 emissions for the period 2000–2012. To evaluate the capacity of the CBM to reproduce country data, our results can be compared with different data sources available at the country level, such as the age-class distribution reported by the NFI and the net CO_2 emissions reported by the country's GHGI. As expected, the comparison between the model results and the country GHGIs showed good agreements in level and trend for some countries and partially good for other countries. When the comparison was not satisfactory, in most cases we can identify possible reasons for these discrepancies. In many cases, however, these data are not fully independent from the NFI input data used by CBM. Where additional information is provided by independent studies (i.e., different datasets, not used by CBM), an independent validation of the model's output is possible. This is the case of Lithuania, where additional information on the living biomass increment, biomass volume and on the dead tree stem volume is available [17]. We select these parameters because increment is one of the main drivers affecting biomass growth estimated by the CBM, initial volume is the main parameter affecting biomass C stock at the beginning of the simulation and dead tree stem volume is the second major C pool with C stock changes over time, for the majority of the European countries (this is often due to the effect of natural disturbances). For Lithuania we verified that our estimates are consistent with these independent data sources. Of course, as highlighted by Vanclacy and Skovsgaard [19], the effective evaluation of a forest growth model is a complex and ongoing process, that could include additional independent validations performed at the regional level [14], sensitivity analysis of the main input data, and further comparison of our estimates with other data sources, including the country-specific GHGI data (see also other comparisons reported in the Additional file:1 for additional case studies). For Lithuania, the country's GHGI reports some peaks between 2000 and 2008,

---

[3] This value is in the lower part of the range of uncertainties typically reported by EU countries for FM emissions/removals (25-50 %, EU NIR 2014).

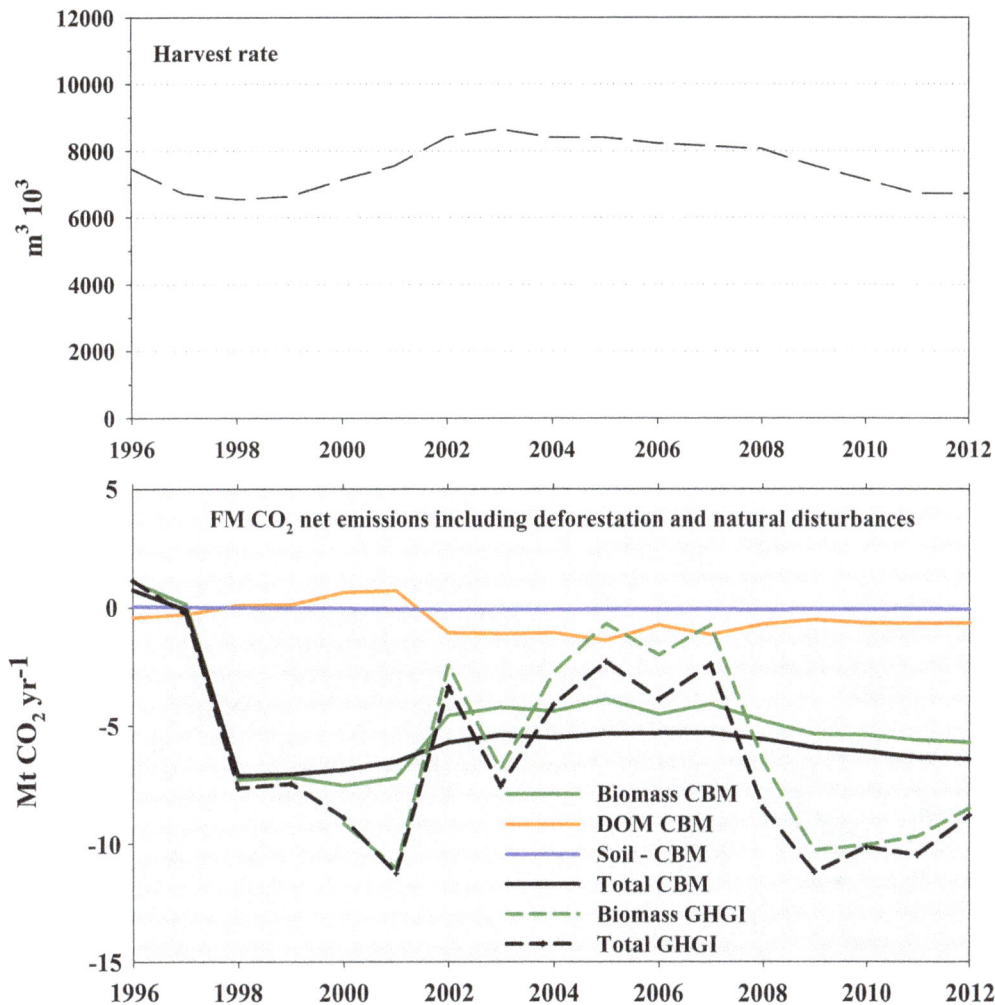

Fig. 3 The *upper panel* reports the harvest rate (m^3 10^3) applied to Lithuania by our study; the *lower panel* reports the net CO_2 emissions estimated by CBM (Mt CO_2 year^{-1}), further distinguished between living biomass, DOM (dead wood + litter) and soil pools and a comparison with the net CO_2 emissions reported by the country for the land use category FL–FL (Lithuania, [18]), assumed as a proxy of the FM area, for those counties where FM had not been elected during the first KP commitment period)

not highlighted by our model (see Fig. 3, lower panel). Apart from different assumptions on the area affected by storms and on the salvage of logging residues (we considered three main disturbance events, described in details in the Additional file: 1), these differences may be even due to the interannual statistical variability associated to the stock-change approach, that can exacerbate the real variability of the C stock changes [17]. Despite this different representation of the interannual variability, the overall match between the CBM results and the Lithuania's GHGI is good, i.e. the average of our results

is within ±25 % compared to the GHGI and the correlation between CBM and GHGI is significant at P < 0.05.

Country case studies

Based on comparisons of both level and trends in CO_2 emission estimates obtained from the CBM and the country GHGIs we partitioned the 26 countries into four groups, and we discuss one representative country for each group.

For Portugal, such as for other eight countries (Group A), the CBM estimates and country data show a good

(See figure on next page.)
Fig. 4 Comparison between the net CO_2 emissions from FM reported by the countries for the period 2000–2012 (in the 2014 GHGIs, [18]) and the CBM estimates. Data are reported from an atmospheric perspective, where negative values represent a sink (CO_2 removals) and positive values a source (CO_2 emissions)

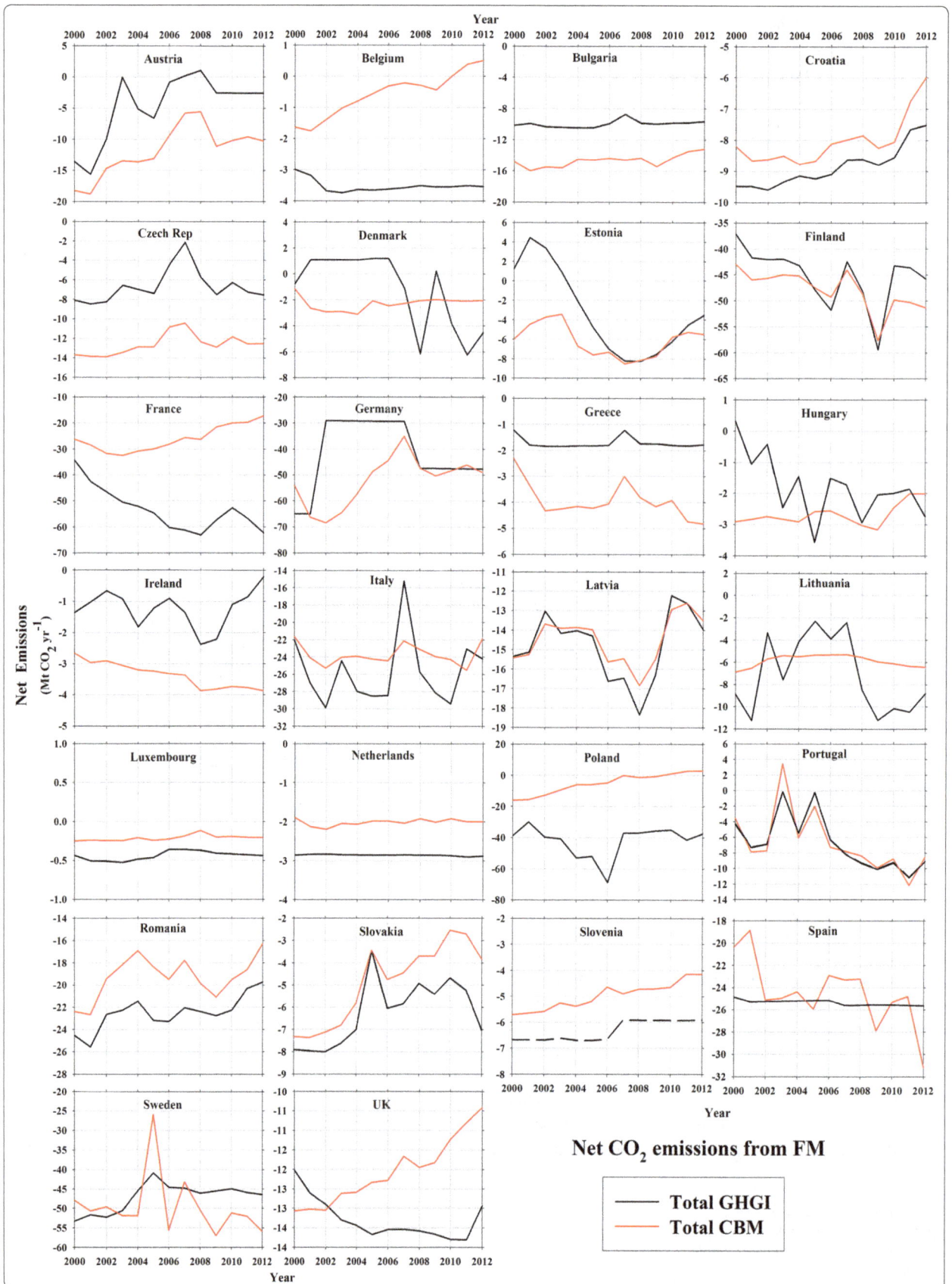

Net CO$_2$ emissions from FM

Total GHGI
Total CBM

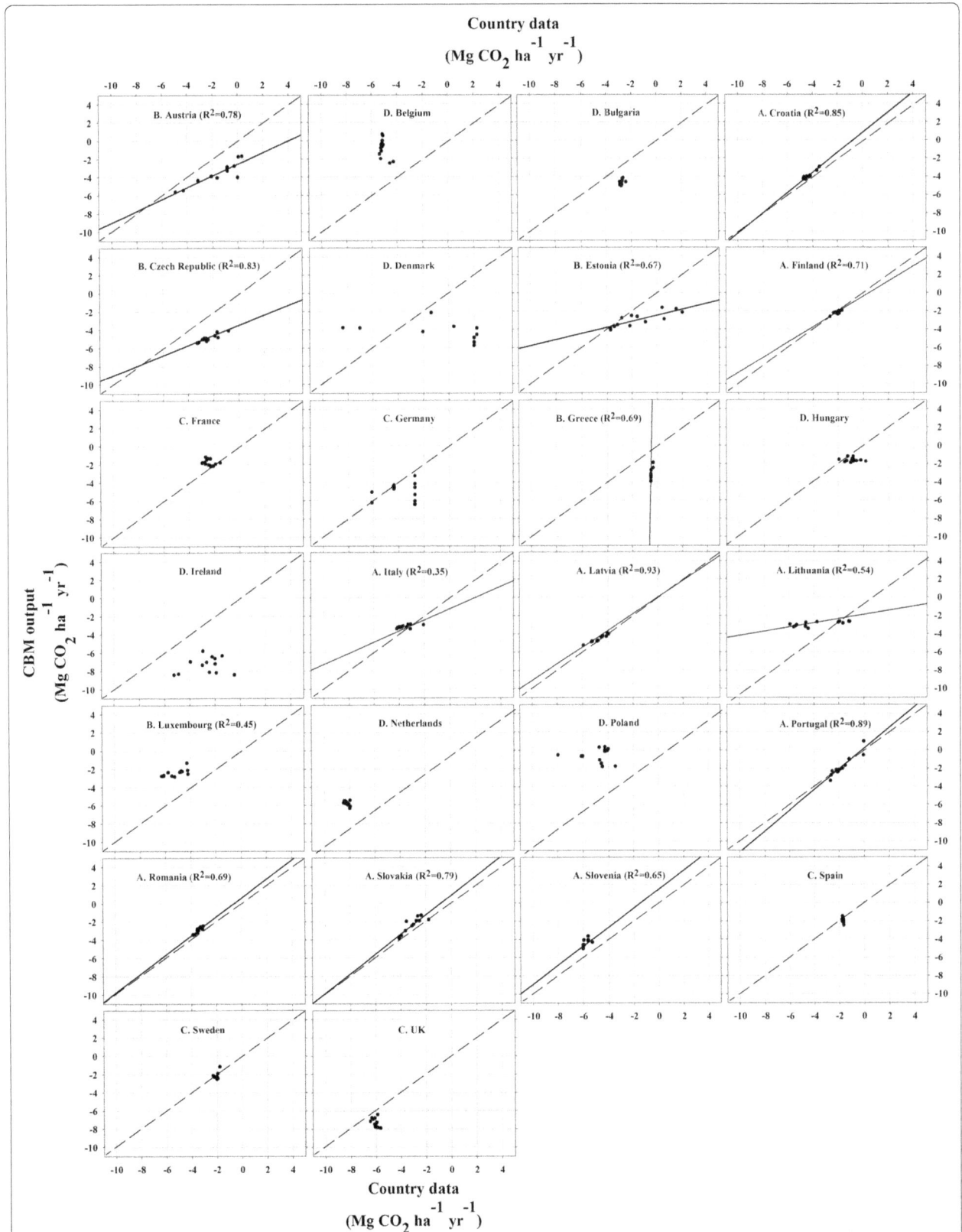

Fig. 5 Comparison between the net CO_2 emissions from FM (living biomass, DOM and mineral soil) as estimated by the CBM and reported in the countries' GHGIs. Each *point* represents one year for the period 2000–2012, as shown in Fig. 4. The *dashed line* is the 1:1 line. The *solid line* is the regression line, shown where the correlation between the CBM and GHGIs was statistically significant (P < 0.05)

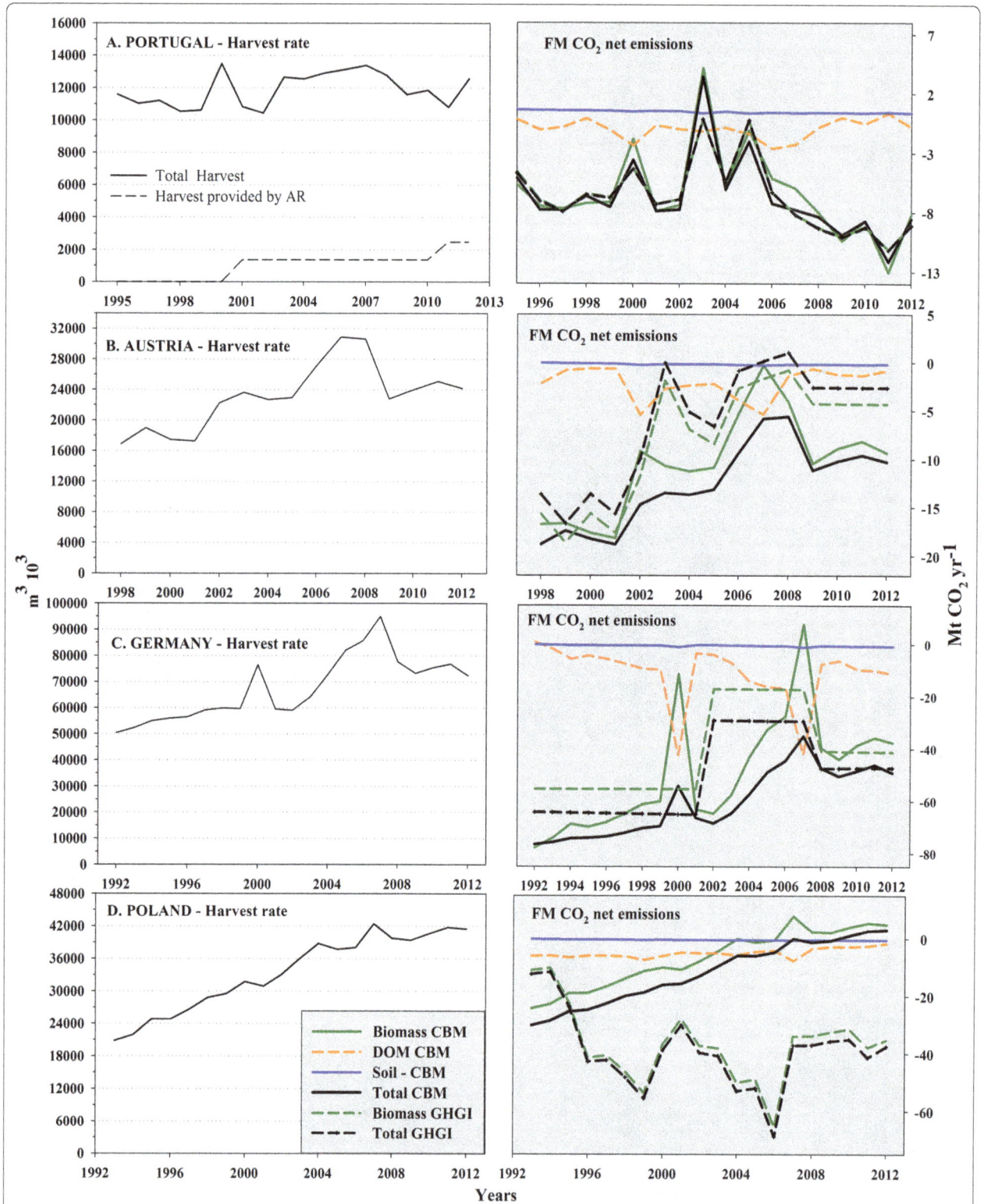

Fig. 6 Harvest rate (on the *left panels*, in m^3 10^3) and the main output provided by CBM for four representative case studies (Austria, Germany, Poland and Portugal). For each country we report the net CO$_2$ emissions estimated by CBM (Mt CO$_2$ year^{-1}, *right panel*), further distinguished between living biomass, DOM (dead wood + litter) and soil pools and a comparison with the biomass and the total net emissions reported by each country. When available, FM country data from the KP-CRF tables was used for 2008–2012 (i.e., if FM had been elected during the first KP commitment period); alternatively, country data were taken from the Convention CRF tables using 'forest land remaining forest land' (FL remaining FL) [18]. For Portugal, the amount of harvest provided by afforestation (AR) is also reported (panel **a**, *left* panel)

match both in the *trend* and the *level*. The C balance of this country is strongly affected by inter-annual variations in harvest demand and direct fire emissions (Additional file: 1 for further details). The total C sink estimated by CBM is slightly lower than the reported values but it has the same trend and the differences decrease with time (in 2011 we reported the same values). These differences may be due to the relative amount of harvest provided by Eucalyptus plantations accounted as AR (from less than 15 % of the total amount of harvest in 2002 to about 25 % in 2011, as highlighted in the harvest's panel of Fig. 6, panels A). As expected, DOM and living biomass pools showed an opposing pattern: when fires kill trees and decrease the biomass C stock, we observe an increase in DOM C pools (i.e., the transfer of C to dead wood and litter add more C than is lost from these pools during the fire).

For Austria, such as for other four countries included in Group B, the CBM estimates show a good match in the *trend* but not in the *level*. In these cases, the different level may be caused by a number of reasons (different conversion factors, different input data, etc.). In the case of Austria, the CBM simulation represents the impact of various natural disturbances. The biomass C balance estimated by CBM (Fig. 6, panels B) follows the same trend that is reported by the country until 2006 and it is strongly affected by the inter-annual variations due to the impact of storms and insect attacks. Indeed, we highlighted a significant statistical correlation ($r = 0.77$) between the total C sink reported by the country and the amount of volume damaged by bark beetle between 1998 and 2007 (see Additional file: 1 for further details). In 2003 and 2005 however, the total C sink reported by the country is considerably lower than our estimates. This may be due to different assumptions about the effect of natural disturbances in specific years. Overall, the biomass C sink estimated by CBM is consistent with the reported trend (Fig. 6). As expected, the DOM C sink has an opposite trend compared with the living biomass. Storms and insect attacks moved C from the living biomass to the dead wood pool and, subsequently salvage logging moved C to the products pool. Yet this impact was not reported by the country's data (which report a stable C source from DOM pools, equal on average to + 1.8 Mt CO_2 year^{-1} between 1998 and 2012). This may also explain the differences between our estimates on the total C sink and the values estimated by country: for example, in the CBM in 2007 a strong reduction of the living biomass pools due to a storm is compensated by a corresponding increase in the DOM pools. After 2008, due to different assumptions about the average amount of harvest and about the effect of natural disturbances (country's data report a constant amount of harvest equal

to about 25 million m^3 from 2009 to 2012) our estimates are not comparable with the country because we used different harvest rates.

Germany (Group C, including five countries), represents an example where there is a good agreement in the *level* but not in the *trend*. We use it to illustrate the difference between the stock-change approach used in the GHGI and the gain-loss method used in the CBM. This methodological difference has a strong impact on the inter-annual variability of estimates as affected by harvest and natural disturbances. Overall, the total C sink estimated by CBM follows the same trend provided by the country (Fig. 6, plot C), even if the correlation is not significant at $P < 0.05$ (i.e., the threshold considered by our study). Due to the stock-change approach, the national sink estimates report three annual values, each applied to the inventory period over which observed stock changes have been annualized [20]. Compared to the reported values, our estimates show a larger inter-annual variability (in particular for the living biomass and DOM pools) due to the storms that occurred in December 1999 (assumed as 2000) and 2007. As expected, the CBM reports opposite trends in the biomass and DOM pools due to the transfer of C from living biomass to the dead wood pool. From 2008 to 2012, our estimates are fully consistent with the data reported by Germany. Further details on natural disturbances and the evolution of the age-class distribution are reported in the Additional file: 1.

For Poland, such as for other 6 countries included in Group D, the estimates differ significantly for both the *trend* and the *level*, for reasons that will require further analysis. For this country, the CBM estimates a decreasing C sink, consistent with a strong increase of the total amount of harvest reported by FAO statistics (see the left panel of Fig. 6, panel D). In contrast, Poland reports an increasing sink with increasing harvest rate. According to our estimates, the DOM pool (not reported by the country) is a C sink, because of the amount of residues left after harvest (i.e., moved from living biomass to DOM). In addition storms in 1999 and 2007 also moved C from living biomass to the dead wood pool (see Additional file: 1 for further details).

CBM results *vs.* GHGIs: impact of carbon pools coverage, harvest and natural disturbances

A first, potentially relevant factor, to be considered when comparing the CBM results with the GHGIs, is the inclusion of C pools. The CBM includes all forest C pools (living biomass, DOM and mineral soils) for all countries, but DOM and soil pools are not reported in some GHGIs. While all 26 countries report living biomass, seven do not report DOM and 14 do not report mineral soils [2]. The mineral soil is in most cases neither a large sink or source (in the CBM, and in the GHGIs). In contrast, the CBM estimates of net CO_2 emissions for DOM pools can be large when natural disturbances occur. Nevertheless, differences in the reported C pools help to explain the observed differences between the CBM and the GHGIs in only a few cases (e.g., Czech Republic, Romania, Slovakia and Slovenia). It is therefore necessary to extend the analysis to the impact of the main drivers of net CO_2 emissions, i.e. harvest and natural disturbances as indicated in both the CBM and GHGIs results.

The net CO_2 emissions from living biomass are generally correlated with the three main drivers: harvest rate, area affected by fires and area affected by storms in both the estimates from the CBM and the GHGI (Fig. 7).

The correlations shown in Fig. 7, demonstrate that for 21 out of 26 countries there is, as expected, a clear negative correlation (generally with r < −0.5) for both CBM and the countries' GHGIs, i.e. more harvest decreases the biomass sink (see for example, Croatia, Finland, Latvia, Lithuania, Portugal, Romania and Slovenia).

Within this group, in three cases (Germany, Estonia, and Slovakia) the correlation between biomass net emissions and the area affected by disturbances is negative for CBM and, surprisingly, is positive for the countries. For Estonia and Slovakia the differences may be due to different assumptions on the effect of storms on the living biomass or DOM pools (i.e., the amount of biomass moved to DOM or removed with salvage logging). For Germany, the main reason appears to be the stock-change approach applied to consecutive NFIs [20]: this approach does not capture the inter-annual variations within a measurement period caused by natural disturbances.

In other cases, despite both the CBM and the GHGIs showing a similar (negative) correlation between the biomass net CO_2 emissions and both harvest and natural disturbances, overall the match between modelled trends and the GHGI is not good (see Fig. 4). For Austria, the main difference lies in different assumptions about the mineral soil pool (which is a source in the GHGI) and partly about the DOM, since the match between the CBM and the GHGI is good for the living biomass. For France, the discrepancy between the CBM and the GHGI requires further investigation, especially with regard to possible differences about harvest assumptions and increment. For Greece, the total FM sink estimated by the CBM is higher than the values reported by the country, but different assumptions on the effect of fires (for example on the amount of biomass burned and the distribution of fires between the FM area and the unmanaged forest area) could explain some these differences. The FM sink reported by Hungary is considerably lower than our estimate and it shows a higher inter-annual variability,

for reasons that are not yet understood. For Ireland, the total sink reported by the GHGI has the opposite trend (i.e., a decreasing C sink) compared with our estimates. Ireland did not elect FM under the CP1 therefore the values reported for this country were derived from the FL remaining FL land use category and a certain amount of harvest is certainly provided by afforestation [21]; this may explain the differences observed. The sink reported by Luxembourg is considerably higher than our estimates and does not seem compatible with the harvest rate applied by CBM. The FM sink estimated by CBM for Spain is overall quite similar to the country GHGI; the main difference is that CBM shows inter-annual variability due to fires and harvest rates, while the stock-change approach implemented by Spain's GHGI masks this variability [22]. Emissions from forest fires estimated by CBM are generally lower than the CO_2 emissions reported by Spain. This is probably due to different assumptions on the amount of biomass and DOM burned. For Sweden, the differences detected on the trend may be due to the effect of storms (above all in 1999 and 2005) and an overestimate on the biomass C stock by CBM. A special case is the lack of any correlation for Italy, where the main driver of the inter-annual variability in biomass net CO_2 emissions is clearly fire (r < −0.80), as also highlighted by [13].

For 5 out of 26 countries (Belgium, Denmark, the Netherlands, Poland and UK), the correlation between biomass net CO_2 emissions and harvest rate is negative (r < 0) for CBM and, surprisingly, is positive (r > 0) for the country GHGI. In principle, this discrepancy may be explained by three reasons. First, the harvest rate applied by our study is different from the harvest reported by the country in its GHGI; even if we always tried to be consistent with the harvest reported by countries, some differences may exist due to inconsistency between different data sources (e.g. see [23]). Second, other factors (e.g. natural disturbances or rapid changes in net increment not included in our study) are a more important driver of biomass net CO_2 emissions compared to harvest; although this case does not seems very likely, it cannot be totally ruled out. Third, the estimation method used by the country in its GHGI masks the effect of harvest on the biomass carbon stock change.

For both the Netherlands and UK, a good match in both the trend and the level existed between CBM and the 2013 GHGIs, suggesting that some recent changes (in input data and/or method) were implemented for the 2014 GHGI. For Denmark, although the known most relevant storms (1999/2000 and 2005) were considered by CBM, the overall correlation between CBM and Denmark GHGI is poor (see again Fig. 4). This could potentially be explained by the method used by Denmark, where a

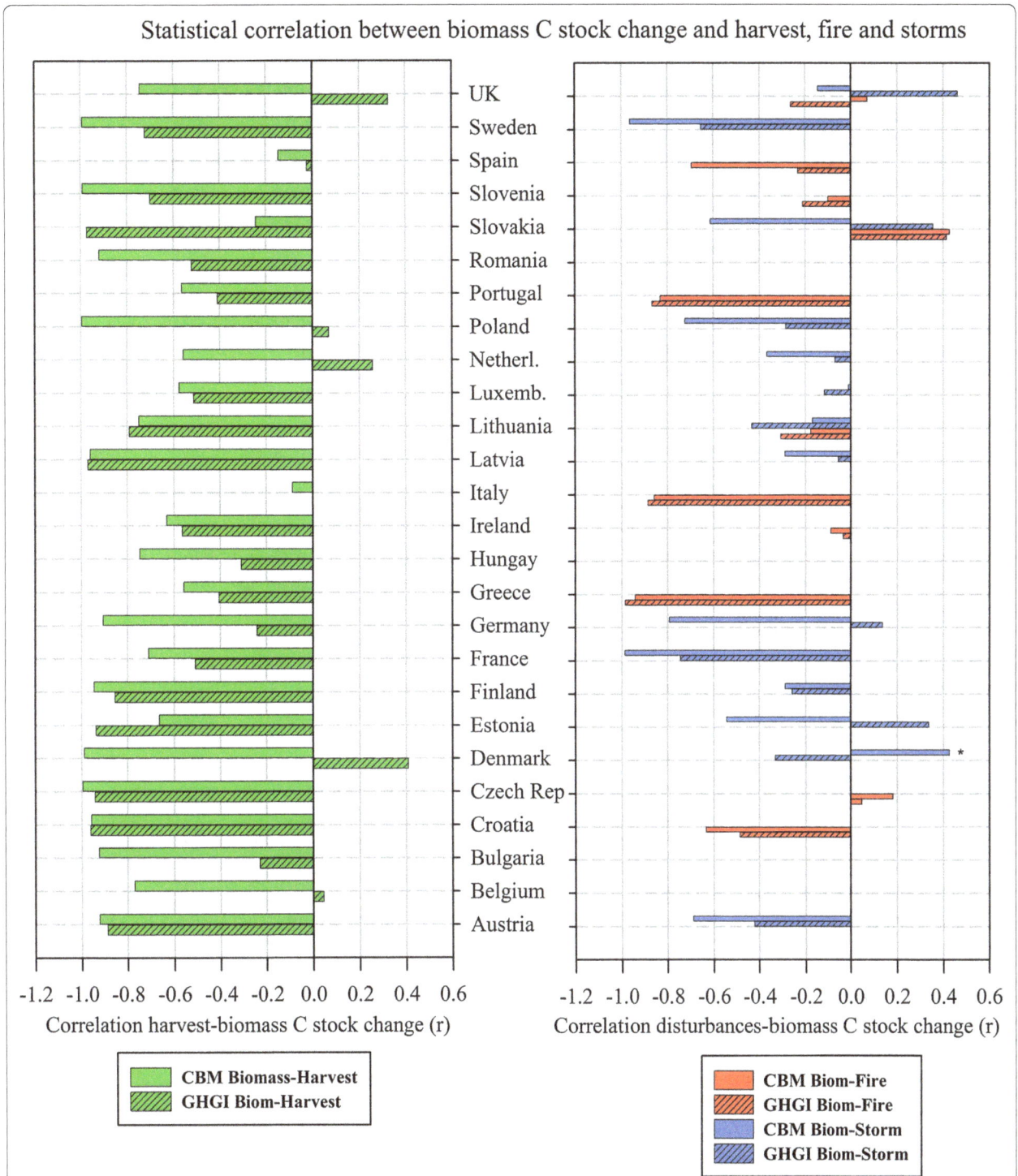

Fig. 7 Correlation (r) between the biomass net CO_2 emissions (estimated by CBM and reported in the countries' GHGIs) and the total amount of harvest (m^3 year^{-1}; *left panel*) and the area affected by fires and storms or insect attacks (ha year^{-1}, *right panel*). In some cases, the correlation is not statistically significant (*), due to the number of available observations

stock-change approach is implemented every year based on the information collected annually from the NFI [24]. It is possible that the interannual statistical variability of

data associated to this approach overrides all the other factors considered and exacerbates the real interannual variability of C stock changes. For Belgium and Poland further analysis is needed, to explain the observed differences between our results and the country's estimates. These may be due to the lack of data or to some incorrect assumption on the input data (i.e., the harvest).

Summary of the main differences between CBM and countries' estimates

Since the CBM and the GHGIs typically share most of the basic input data (e.g., forest area, timber volume and net increment, taken from the NFIs), we briefly discuss the level of independence of input data. Forest area will be strongly correlated between the CBM and the GHGIs, because we used whenever possible the area used by the GHGI for FM (or, for countries that did not elect FM, for FL remaining FL).

The methods used to estimate the emissions/removals per unit of area—typically the major source of uncertainty of GHGIs—differ. Eleven out of 26 countries use the *stock-change approach* in their GHGIs [5], implemented either every year (using any available new data) or at the end of each NFI cycle [25]. In these cases, the degree of independence between the CBM and GHGIs is very high because the GHGIs typically do not use net increment and harvest values (i.e. the most important drivers for the sink estimated by the CBM). Furthermore, even in the 15 countries that use the *gain-loss approach* [5]—the approach also used by the CBM—the steps needed to obtain CO_2 emission/removals are complex and introduce uncertainty, e.g. converting net increment minus disturbance losses (harvest, storm, fire) into the sink estimate. For example, the most recent data from NFI typically used in GHGIs (e.g. on net increment) are not always publicly available, and in several cases require interpretations and/or assumptions. Equations used by CBM to convert volume into C are totally independent from GHGIs. Harvest rates for the 26 countries used by the CBM are based on FAO statistics (which often require interpretation and/or adjustments, see [23], but the GHGIs may use either FAO or other national-level statistics. In summary, in most cases the methods to estimate emission/removals should be seen as largely independent between the CBM and GHGIs.

Few studies compared model results with European countries' GHGIs. The main comparison may be done with [7], where two models (EFISCEN and G4 M) were applied in 24 EU countries for the period 2000 to 2008, with the discussion focused on six countries. In comparison to that study, our analyses cover a longer period (2000–2012) and 26 countries and we include the DOM and mineral soils pool dynamics and the explicit simulation of the impact of natural disturbances. Beyond these differences—which in several cases allowed CBM to obtain a better match with GHGIs—most of the conclusions from Goen et al. [7] are valid also for our study,

e.g. (1) in several cases (i.e., for Germany), the estimation method used in the GHGIs (stock-change vs gain-loss) explains most of the differences observed, and (2) in the remaining cases, the differences seem to have country-specific reasons, like the amount of harvest used and the way harvest losses are treated.

In addition to the above, another essential aspect is the recalculations performed annually by the countries as part of the continuous process of improving their GHGIs. Our study therefore represents a "picture" in a rather dynamic process as future changes to GHGI may affect our conclusions. The frequency of the recalculations in the LULUCF sector is high: according to EU countries' GHGIs submitted between 2010 and 2014 (including the time series 1990–2008 and 1990–2012, respectively), and focusing only on FL–FL, on average every year 5–6 countries out of 26 revised emissions of the previous GHGI by 10–25 % (in terms of absolute level of emissions), and another 5–6 countries revise emissions by more than 25 %. This means that every year more than a third of the countries analyzed in this study show substantial recalculations compared to their previous GHGIs, with the biggest changes usually for the more recent years. These recalculations are due to a number of reasons (e.g. new input data, addition of pools or gases, correction of previous errors, change in methods, etc.), linked to country internal processes or to recommendations provided by the UNFCCC expert review teams. The magnitude of these recalculations is consistent with the information available on uncertainties from countries' GHGIs, which for FM in most cases fall in the range of 25–50 % [2].

Overall, given the frequency and the magnitude of the changes in GHGIs—and the associated uncertainties—for a modeler it is challenging to capture all the latest data and methods used (including possible errors) in 26 different GHGIs; an improved process to share updated information by country on an ongoing basis would certainly help. Nevertheless, the large amount of work completed by implementing the CBM in 26 countries allowed us obtain satisfactory results in most of the countries analyzed, and to understand the reasons for differences in many of the remaining cases.

Conclusions

This study implemented a consistent methodology to estimate the GHG balance in the managed forests of 26 EU countries using the CBM to estimate the historical (2000 to 2012) net CO_2 emissions from forest management (Sensu Kyoto, i.e. forest existing before 1990) as affected by harvest and natural disturbances (storms, fires and insects). In terms of number of countries, C pools and type of disturbances simulated, to our knowledge this is the most comprehensive study of its kind to date.

The comparison of CBM results with the data reported by the countries in their GHGIs shows a good match (both in the trend and in the level) in nine cases, a partially good match (either for the trend or the level) in an additional ten cases, and an un-satisfactory match in the remaining seven cases. A successful independent country-level validation of the CBM has also been performed.

Our study confirms that, in the short period (and excluding possible effects of climate change), the main factors driving the forest C sink of Europe's managed forests are the harvest rates and natural disturbances (storms for most countries). When these factors are considered in a consistent way, i.e. the gain-loss method is used in both the CBM and the GHGIs, the trends of net CO_2 emissions are very similar. Where the comparison between the CBM and the GHGIs was not fully satisfactory (for the trend and/or for the level), in most cases we provided possible explanations for the discrepancies observed, including: (1) representation of the interannual variability due to harvest and natural disturbances: while it is well simulated by the CBM, it may be masked if the country uses the stock-change approach for the GHGIs; (2) a different treatment of non-biomass pools (not reported by several countries, or reported using different assumptions compared to the CBM), or of CO_2 emissions from fires, natural mortality or other parameters (e.g. harvest residues). Beyond these explanations, some cases—e.g. where the GHGI counter intuitively reports an increasing biomass sink associated with a trend of increasing harvest rates—clearly deserve further analysis, to identify the possible cause of the discrepancy. In general, the results of the comparisons were good in those countries where the input data for the model were based on accessible recent statistics. Finally, when analyzing the discrepancies between the CBM results and the GHGIs, it should be noted that the frequent update cycle and recalculations of GHGIs can only be reflected in the model results if national statistics on harvest and disturbance rates are readily available for the model analyses.

Overall, this study documents a promising foundation for the use of the CBM both as tool to help countries in estimating the forest C dynamics (e.g., including natural disturbances) and as a potential tool to support the verification of GHGIs at the EU level using a consistent methodological approach for all countries. A systematic comparison of the CBM with the GHGIs will certainly require additional efforts—that will require close cooperation between modelers and country experts—and caution should be applied when interpreting these first results. Nevertheless, this application of consistent methods makes a useful contribution to the continuous improvement of GHGIs, because it may help in identifying possible errors, in increasing scientific understanding and ultimately in building confidence in the estimates of emissions and removals reported by the countries by increasing consistency, transparency and completeness of the estimates.

Methods

The Carbon Budget Model (CBM-CFS3) and the main input data

The CBM is an inventory-based, yield-data driven model that simulates the stand- and landscape-level C dynamics of above- and below-ground biomass, dead organic matter (DOM: litter and dead wood) and mineral soil [8]. The model, developed by the Canadian Forest Service (the model description is available to the following URL: http://www.nrcan.gc.ca/forests/climate-change/carbon-accounting/13107), was recently applied to the Italian forests, in order to test the CBM for different European silvicultural systems, proposing a novel approach to include uneven-aged forest structures [13].

Because this work applies the same general assumptions used in the Italian case study, we provide only a short description of the model, highlighting the specific methodological assumptions related to the present study. Further details of the model can be found in [8], and its applications to European countries are found in [13–15].

The spatial framework applied by the CBM conceptually follows reporting method 1 ([4]) in which the spatial units are defined by their geographic boundaries and all forest stands are geographically referenced to a spatial unit (SPU). We considered 26 administrative units (i.e., European countries, as reported by Table 1) and 35 climatic units (CLUs, as defined by [26]) for a total of 910 SPUs. The CLU's mean annual temperatures, range from −7.5 to +17.5. Each SPU was linked to a CLU through the information provided by Corine Land Cover.

The total managed forest area of the 26 EU countries represented here covers about 138 Mha (i.e., about 82 % of the EU forest area). Two EU countries excluded from the analysis are Cyprus (no NFI data available) and Malta (very small forest area, mainly covered by shrub lands).

Within a SPU, each forest stand is characterized by age, area and seven classifiers that provide administrative and ecological information, the link to the appropriate yield curves, and parameters defining the silvicultural system such as forest composition and management type (MT), and the main use of the harvest provided by each SPU, as fuelwood or industrial roundwood. For each country, these parameters were mainly derived from NFIs. According to country-specific information, MTs may include even-aged high forests, uneven-aged high forests, coppices and specific silvicultural systems such as clear-cuts (with different rotation lengths for each forest type, FT), thinnings, shelterwood systems, and partial cuttings.

Table 1 Summary of the main parameters applied by the CBM model for each country

Country	Original NFI year	Time step 0 (years)	CBM FM area (Mha)[a]	Harvest rate (av. 2000–2012, Mm³)	County specific biomass equations
Austria	2008	1998	3.2	22.9	X
Belgium	1999	1999	0.7	4.3	
Bulgaria	2000	2000	3.2	5.3	
Croatia	2006[b]	1996	2.0	4.6	
Czech Republic	2000	2000	2.6	17.0	X
Denmark	2004	1994	0.5	2.3	
Estonia	2000	2000	2.1	7.9	
Finland	1999	1999	21.7	55.0	
France	2008	1998	14.6	54.9	
Germany	2002	1992	10.6	74.7	X
Greece	1992[b]	1992	1.2	1.6	
Hungary	2008	1998	1.6	6.2	X
Ireland	2005	1995	0.5	2.8	
Italy	2005	1995	7.4	10.2	X
Latvia	2009	1999	3.2	15.8	X
Lithuania	2006	1996	2.0	7.7	
Luxembourg	1999	1999	0.1	0.3	
Netherlands	1997	1997	0.3	1.2	
Poland	1993	1993	8.9	37.8	
Portugal	2005	1995	3.6	12.2	X
Romania	1985	1985	6.6	17.2	X
Slovakia	2000	2000	1.9	9.0	
Slovenia	2000	2000	1.1	3.3	
Spain	2002	1992	12.6	16.8	
Sweden	2006	1996	22.6	79.5	
United Kingd.	1997	1997	2.5	9.8	
EU			137.9	480.7	8 countries

The table reports the NFI original reference year; the year since the model was applied; the FM area used by CBM at time step 0; the average harvest rate used; the countries where specific equations to convert the merchantable volume into aboveground biomass were selected. Two countries were not modeled: Cyprus (no NFI data available) and Malta (very small forest area, mainly covered by shrub lands)

[a] FM area used by CBM at time step 0. According to KP rules, FM is the area of forest in 1990, decreased by any subsequent deforestation. The FM area is taken from the official submissions made by countries to UNFCCC/Kyoto Protocol [18, 29], giving priority to data from KP-CRF tables when available (i.e., if FM had been elected during the first KP commitment period), or alternatively taking data from the Convention CRF tables (using 'forest land remaining forest land' in 1990 as a proxy for FM). To obtain FM area at time step 0, the D area reported by all countries under the Kyoto Protocol was used. Please note that CBM runs did not include forests reported as "not productive" (e.g., 0.4 Mha in Austria, 0.02 Mha in Bulgaria, 5 Mha in Sweden) and overseas territories (8.2 Mha in France)

[b] Analysis based on data from Forest Management Plans

Species-specific, stand-level equations [27] convert merchantable volume production into aboveground biomass, partitioned into merchantable stem wood, other (tops, branches, sub-merchantable size trees) and foliage components [8]. Where additional information provided by NFIs or by literature was available (see last column in Table 1), country-specific equations were selected to convert the merchantable volume into aboveground biomass [13]. If no data were available, we used the same equations selected for other countries and similar forest types (FTs, defined according to the main species). Below-ground biomass is calculated using the equations provided by [28] and the annual dead wood and foliage input

is estimated as a pool-specific turnover rate (percentage) applied to the standing biomass stock.

Forest inventories typically contain no or only insufficient data on stocks in DOM and soil C pools. The model therefore uses an initialization process to estimate the size of all DOM pools at the start of the simulation and then, following IPCC guidance, links DOM dynamics to biomass dynamics. Inputs from biomass to DOM pools result from biomass litterfall and turnover as well as natural and human-caused disturbances. The DOM parameters were first calibrated in the Italian cases study (see [13], Appendix E for further details), then validated on a specific study at regional level [14] and, if necessary, further modified for specific countries, such as Finland and Sweden.

We use two sets of yield tables (YT) in these analyses [13]. Historical YTs derived from the standing volumes per age class reported by the NFI represent the impacts of growth and partial disturbances during stand development. Current YTs derived from the current annual increment reported in country NFIs represent the stand-level volume accumulation in the absence of natural disturbances and management practices.

To implement the CBM to uneven-aged FTs (when this forest structure was observed in a country), all the uneven-aged forest area was allocated to a reference age class, with the average volume equal to the volume reported by the NFI for these stands. Starting from this age class, a decreasing percentage increment was applied to the subsequent (older) age classes. We assumed that, after a certain number of years, equal to species-specific cutting cycles defined at country level, each uneven-aged stand was disturbed and moved back to the initial reference age class [13]. This approach was tested through a number of simulations in which we varied different parameters. Overall, we simulated (1) a faster (but decreasing) re-growth phase during the first period following the partial cut and (2) a decreasing growth phase during the following years.

Since this study aimed to be as comparable as possible with countries' information reported to the UNFCCC and its KP, the model was applied individually to each country and we modeled 'forest management' (FM) as the forests existing in 1990 minus any deforestation (D) since 1990. Forest area in 1990 and deforestation rates were obtained, respectively, from the 2014 GHGIs submitted by each country to the UNFCCC and to the KP [29]. The start year of the simulations (time step 0) varied between countries. FM area was reduced, during the model run, due to D between 1990 and time step 0. The D area within each country was distributed proportionally to the area of each FT. Table 1 shows the country-specific FM area at the start of model runs.

In order to provide a comparable dataset for all the EU countries, covering the period 2000–2012, when the NFI reference year was after the year 2000 (see Table 1), the original NFI age-class distribution (for even-aged forests) was rolled back by 10 years (see [13] for further details).

Harvest rate

To provide a consistent estimate of the harvest demand for all 26 EU countries, historical data on harvest were obtained from FAO statistics [30]. For some countries, the original FAOSTAT data were slightly modified to ensure consistency with other information provided by countries under the KP. The country-specific modifications applied to the original FAOSTAT data (in most cases due to different treatment of the bark fraction) are described in [23].

FAOSTAT data (modified where necessary) were further distinguished at the country level, between four compartments: Industrial Roundwood (IRW, i.e., the portion of roundwood used for the production of wood commodities) and Fuelwood (FW, i.e., wood for energy use) and between coniferous and non-coniferous (i.e., for our analysis, broadleaved) species groups [30]. For each compartment, we defined in CBM: (1) the FTs (i.e., broadleaved species for IRW and FW broadleaved species, and coniferous species for IRW and FW coniferous species), (2) the MTs (for example coppices for FW from broadleaved species) and (3) the silvicultural practices (for example thinnings for FW from coniferous species) providing the total amount of wood expected each year (the harvest target).

We assumed that the harvest rate was entirely satisfied by the FM area, considering that the possible amount of harvest provided by lands afforested or reforested (AR) since 1990 was generally negligible [15], with the exception of Portugal (see the Additional file: 1 for details).

Natural disturbances

For each country, the historical effects of storms and ice (15 countries), fires (11 countries) and insect attacks (i.e., bark beetles attacks, for 2 countries) were analysed (see Table 2 for details). We assumed that that natural disturbances occurred on the FM area, excluding possible disturbances on the afforested area.

The effect of storms was evaluated using the data reported by the FORESTORMS database [31] provided by the European Forest Institute and by specific additional information available at the country level. Depending on the available information, the effect of each event was modelled according to (1) the amount of forest biomass damaged by storm and eventually salvage logged and/or (2) the amount of area affected by the disturbance event. In the first case, we mainly modified the 'disturbance matrix' that describes the proportion of C transferred between pools and to the forest product sector or released to the atmosphere [8], in order to be consistent with the disturbance impact reported by the FORESTORMS database. In the second case, we verified that the amount of forest area affected by the disturbance event was consistent with the area reported by this database. In some cases, such as for Sweden, both these criteria were verified.

More specific information on the methodological assumptions applied to represent storms and insect attacks are reported in the Additional file: 1 for some representative case study. Since the information available on these disturbances may vary considerably by country,

Table 2 Overview of countries with natural disturbance events simulated by the CBM (*F* fire, *S* storms and ice sleets, *I* insect attacks), with information on input data used for storms (country data, National Inventory Reports, NIR or the FORESTORMS database [31] and the average annual burned area

Country	Natural disturb.	Storms, ice and insect disturbances		Fires
		Source	Vol. affected[a] (Mm³ year⁻¹)	Area burned[b] (kha year⁻¹)
Austria	S + I	Vol. based on country data	4.1	–
Belgium	–			–
Bulgaria	–			–
Croatia	F			2.3
Czech Rep.	F			0.5
Denmark	S	Vol. based on country data	0.5	–
Estonia	S	Area and vol. based on NIR	0.7	–
Finland	S	Vol. based on FORESTORMS	0.6	–
France	S	Area and vol. based on FORESTORMS	18.3	–
Germany	S	Vol. based on FORESTORMS	6.2	–
Greece	F			6.0
Hungary	–			–
Ireland	F			0.4
Italy	F			35.0
Latvia	S	Vol. based on FORESTORMS	0.7	–
Lithuania	S + F + I	Vol. based on the NIR + FORESTORMS	0.2	0.3
Luxembourg	S	Vol. based on FORESTORMS	<0.1	–
Netherlands	S	Vol. based on FORESTORMS	<0.1	–
Poland	S	Vol. based on FORESTORMS	0.4	–
Portugal	F			49.1
Romania	–			–
Slovakia	S + F	Vol. based on FORESTORMS + country data	0.8	0.6
Slovenia	S + F	Vol. based on country data	<0.1	0.1
Spain	F			35.3
Sweden	S	Vol. based on FORESTORMS + country data	7.1	–
United K.	S + F	Vol. based on FORESTORMS	<0.1	3.5
	22 countries		39.6*	134.0

[a] Average volume affected by storms, ice and insects between 2000–2012, as reported by the input data used by CBM. The interannual variations of these disturbances can vary considerably among countries (i.e., in many cases disturbances are concentrated in few big events). In some cases, further damages were considered before 2000

[b] Average area affected by fires between 2000–2012, mainly based on the data reported by National Inventory Reports*

our assumptions were adapted to the conditions in each country.

Fire disturbances were modelled according to the amount of area affected by fire, as reported by national statistics, proportionally distributed between different FTs or according to further information provided by literature (mainly, the National Inventory Reports) The disturbance matrix associated with fires was modified according to specific country-level information, to account for salvage of logging residues, commonly applied in some Mediterranean countries (i.e., Portugal). More specific information on the methodological assumptions applied to these disturbances is reported in the Additional file:1 for Portugal. As in the case of

storms, our model assumptions were adapted to the specific country's conditions. When relevant (e.g., for Latvia), we also included the burning of harvest residues after a clearcut.

Model validation

For Lithuania, the information provided by CBM, based on Lithuania's NFI used as input data for the model, can be also compared and validated against some independent data, derived by specific studies[4] on living and dead tree volumes in forest land, reported by Lithuania's NIR

---

[4] "Study 1, "Forest Land Changes in Lithuania during 1990–2001" ([17], page. 349).

[17]. Further details on the methodological assumptions are reported in the Additional file: 1.

Abbreviations

AR: afforestation and reforestation; C: carbon; CBM: Carbon Budget Model; CP1: first commitment period; CP2: second commitment period; D: deforestation; DOM: dead organic matter; EU: European Union; FL: forest land; FM: forest management; FRA: forest resources assessment; FT: forest type; FW: fuelwood; GHG: greenhouse gas; GHGI: greenhouse gas inventory; HWP: harvested wood product; IPCC: Intergovernmental Panel on Climate Change; IRW: industrial roundwood; KP: Kyoto protocol; LULUCF: land use, land-use change and forestry; MT: management type; NFIs: National Forest Inventories; NIR: National Inventory Report; SPU: spatial unit; UNFCCC: United Nations Framework Convention on Climate Change; YT: yield table.

Authors' contributions

RP carried out the data analysis, in collaboration with NG. GG and WAK helped in the design of the study and the interpretation of results, and together with RP wrote the manuscript, in collaboration with RAV. All authors read and approved the final manuscript.

Author details

[1] European Commission, Joint Research Centre, Institute for Environment and Sustainability, Via E. Fermi 2749, 21027 Ispra, VA, Italy. [2] Natural Resources Canada, Canadian Forest Service, Victoria, BC V8Z 1M5, Canada.

Acknowledgements

This paper was prepared in the context of the Contract no. 31502, Administrative Arrangement 070307/2009/539525/AA/C5 between JRC and DG CLIMA. Further information was collected in the context of the AA 071201/2011/611111/CLIMA.A2. The analysis performed for each country was generally based on data public available and on additional information collected at country level, in collaboration with many colleagues and experts for each country. We especially thank Stephen Kull, Scott Morken and the Carbon Accounting Team for their indispensable technical support during this study and our colleagues, Giulia Fiorese, Viorel Blujdea and Tibor Priwitzer, who provided useful comments and suggestions.

We also thank two anonymous reviewers, who provided useful comments and suggestions to improve the manuscript.

The views expressed are purely those of the authors and may not in any circumstances be regarded as stating an official position of the European Commission or of Natural Resources Canada.

Competing interests

The authors declare that they have no competing interests.

References

1. FOREST EUROPE, UNECE, FAO. State of Europe's Forests 2015. Status and trends in sustainable forest management in Europe, 2015. URL (Access Mar 2016): http://www.foresteurope.org/docs/fullsoef2015.pdf.
2. EU NIR. Annual European Community greenhouse gas inventory 1990–2012 and inventory report 2014. Submission to the UNFCCC Secretariat. European Environment Agency, Technical report No 09/2014. URL (Access Feb 2016): https://www.google.it/search?q=Annual+European+Community+greenhouse+gas+inventory+1990%E2%80%932012+and+inventory+report+2014&ie=utf-8&oe=utf-8&gws_rd=cr&ei=fwi7VoSoJIXAOoCTroAF.
3. UNFCCC, Decision 2/CMP.7 on Land use Land use Change and Forestry, 2011. URL (Access Feb 2016): http://unfccc.int/resource/docs/2011/cmp7/eng/10a01.pdf.
4. IPCC (Intergovernmental Panel on Climate Change). Revised supplementary methods and good practice guidance arising from the Kyoto protocol. Hiraishi T, Krug T, Tanabe K, Srivastava N, Baasansuren J, Fukuda M, Troxler TG, editors. Switzerland: IPCC; 2014.
5. IPCC, Intergovernmental Panel on Climate Change. IPCC Guidelines for National Greenhouse Gas Inventories. In: Eggleston S, Buendia L, Miwa K, Ngara T, Tanabe K, editors. Agriculture, forestry and other land use, vol. 4. Japan: Hayama; 2006.
6. Böttcher H, Verkerk PJ, Mykola G, Havlik P, Grassi G. Projection of the future EU forest CO_2 sink as affected by recent bioenergy policies using two advanced forest management models. GCB Bioenergy. 2012;4(6):773–83.
7. Groen T, Verkerk PJ, Böttcher H, Grassi G, Cienciala E, Black K, Fortin M, Köthke M, Lehtonen A, Nabuurs G-J, Petrova L, Blujdea V. What causes differences between national estimates of forest management carbon emissions and removals compared to estimates of large-scale models? Environ Sci Policy. 2013;33:222–32.
8. Kurz WA, Dymond CC, White TM, Stinson G, Shaw CH, Rampley G, Smyth C, Simpson BN, Neilson E, Trofymow JA, Metsaranta J, Apps MJ. CBM-CFS3: a model of carbon-dynamics in forestry and land-use change implementing IPCC standards. Ecol Model. 2009;220:480–504.
9. Mubareka S, Jonsson R, Rinaldi F, Fiorese G, San Miguel J, Sallnas O, Baranzelli C, Pilli R, Lavalle C, Kitous A. An integrated modelling framework for the forest-Based bioeconomy. IEEE Earthzine. 2014;7(2):908802.
10. Kurz WA, Apps MJ. A 70-year retrospective analysis of carbon fluxes in the Canadian forest sector. Ecol Appl. 1999;9:526–47.
11. Stinson G, Kurz WA, Smyth CE, Neilson ET, Dymond CC, Metsaranta JM, Boisvenue C, Rampley GJ, Li Q, White TM, Blain D. An inventory-based analysis of Canada's managed forest carbon dynamics, 1990–2008. Glob Chang Biol. 2011;17:2227–44.
12. Zamolodchikov DG, Grabovsky VI, Korovin GN, Kurz WA. Assessment and projection of carbon budget in forests of Vologda Region using the Canadian model CBM-CFS (in Russian, with summary in English). Lesovedenie. 2008;6:3–14.
13. Pilli R, Grassi G, Kurz WA, Smyth CE, Bluydea V. Application of the CBM-CFS3 model to estimate Italy's forest carbon budget, 1995 to 2020. Ecol. Modell. 2013;266:144–71.
14. Pilli R, Grassi G, Cescatti A. Historical analysis and modeling of the forest carbon dynamics using the Carbon Budget Model: an example for the Trento Province (NE, Italy). Forest @. 2014;11:20–35.
15. Pilli R, Grassi G, Moris JV, Kurz WA. Assessing the carbon sink of afforestation with the Carbon Budget Model at the country level: an example for Italy. Forest. 2014;8:410–21.
16. Pilli R, Grassi G, Kurz WA, Moris JV, Viñas RA. Modelling forest carbon stock changes as affected by harvest and natural disturbances. II. EU-level analysis including land use changes, Carbon Balance and Management. 2016. **(submitted).**
17. Lithuania. Lithuania's National Inventory Report 2014. URL (last Access Mar 2015). http://unfccc.int/national_reports/annex_i_ghg_inventories/national_inventories_submissions/items/8108.php.
18. KP CRF tables, 2014. URL (last Access Mar 2015): http://unfccc.int/national_reports/annex_i_ghg_inventories/national_inventories_submissions/items/8108.php.
19. Vanclay JK, Skovsgaard JP. Evaluating forest growth models. Ecol Model. 1997;98:1–12.
20. Germany. National Inventory Report for the German Greenhouse Gas Inventory 1990–2012. Federal Environment Agency, 2014. URL (last Access Feb 2016): http://unfccc.int/national_reports/annex_i_ghg_inventories/national_inventories_submissions/items/8108.php.
21. Ireland. Ireland National Inventory Report 2014. EPA Environmental Protection Agency. URL (Access Mar 2016): http://unfccc.int/national_reports/annex_i_ghg_inventories/national_inventories_submissions/items/8108.php.
22. Spain 2014. Inventario de emisiones de gases de efecto invernadero de Espana años 1990-2012. Ministerio de Agricultura, Alimentacio y Medio Ambiete. URL (Access Mar 2016): http://unfccc.int/national_reports/annex_i_ghg_inventories/national_inventories_submissions/items/8108.php.
23. Pilli R, Fiorese G, Grassi G. EU Mitigation Potential of harvested wood products. Carbon Balance Manage. 2015;10:6.

24. Denmark. Denmark's National Inventory Report, 2014. Aarhus University, 2014. URL (last Access Feb 2016): http://unfccc.int/national_reports/ annex_i_ghg_inventories/national_inventories_submissions/items/8108. php.

25. Blujdea V, Raul AV, Federici S, Grassi G. The EU greenhouse gas inventory for LULUCF sector: I. Overview and comparative analysis of methods used by EU member states. Carbon Manag 2016;6(5–6):247–59.

26. Pilli R. Calibrating CORINE land cover 2000 on forest inventories and climatic data: an example for Italy. Int J Appl Earth Obs. 2012;9:59–71.

27. Boudewyn P, Song X, Magnussen S, Gillis MD. Model-based, volume-to-biomass conversion for forested and vegetated land in Canada. Canadian Forest Service, Victoria, Canada, 2007 (Inf. Rep. BC-X-411). URL (Access Mar 2015): http://cfs.nrcan.gc.ca/publications/?id=27434.

28. Li Z, Kurz WA, Apps MJ, Beukema SJ. Belowground biomass dynamics in the Carbon Budget Model of the Canadian Forest Sector: recent improvements and implications for the estimation of NPP and NEP. Can J For Res. 2003;33:126–36.

29. UNFCCC CRF tables. UNFCCC Common reporting format tables, 2014. URL (last Access Feb 2016): http://unfccc.int/national_reports/annex_i_ ghg_inventories/national_inventories_submissions/items/8108.php.

30. FAOSTAT. FAOSTAT data, 2013. URL (last access March 2015): http:// faostat3.fao.org/home/index.html#DOWNLOAD.

31. Gardiner B, Blennow K, Carnus JM, Fleischer P, Ingemarson F, Landmann G, Lindner M, Marzano M, Nicoll B, Orazio C, Peyron JL, Reviron MP, Schelhaas MJ, Schuck A, Spielmann M, Usbeck T. Destructive storms in European forests: past and forthcoming impacts. Final report to European Commission—DG Environment, 2010. URL (Access Mar 2015): http:// www.efiatlantic.efi.int/portal/databases/forestorms/.

11

Comparison of national level biomass maps for conterminous US: understanding pattern and causes of differences

N. Neeti[1*] and R. Kennedy[2]

Abstract

Background: As Earth observation satellite data proliferate, so too do maps derived from them. Even when two co-located maps are produced with low overall error, the spatial distribution of error may not be the same. Increasingly, methods will be needed to understand differences among purportedly similar products. For this study, we have used the four aboveground biomass (AGB) maps for conterminous US generated under NASA's Carbon Monitoring System. We have developed systematic approach to (1) assess both the absolute accuracy of individual maps and assess the spatial patterns of agreement among maps, and (2) investigate potential causes of the spatial structure of agreement among maps to gain insight into reliability of methodological choices in map making.

Results: The comparison of the four biomass maps with FIA based total biomass estimates at national scale suggest that all the maps have higher biomass estimate compared to FIA. When the four maps were compared among each other, the result shows that the maps S and K have more similar spatial structure whereas the maps K and W have more similar absolute values. Although the maps K and W were generated using completely different methodological workflow, they agree remarkably. All the maps did well in the dominant forest type with maximum agreement between them. The comparison of difference between maps S and K with regional maps suggests that these maps were able to capture the disturbance and not so much regrowth pattern.

Conclusions: The study provides a comprehensive systematic approach to compare and evaluate different real data products using examples of four AGB maps. Although ostensibly the four maps map the same variable, they have different spatial distribution at different scale. Except the 2003 map, one can use other maps at the coarser spatial resolution. Finally, the disparate information available through different maps indicates a need for a temporal framework for consistent monitoring of carbon stock at national scale.

Keywords: CMS, AGB, FIA, Carbon

Background

Because forests provide important ecosystem services and play a role in the global carbon cycle, forest characteristics such as biomass, tree height, and percent forest cover have been mapped using a variety of remote sensing techniques [1–7]. As data increase in availability, and as mapping techniques proliferate, many similar products are becoming available for the same region—often at different spatial scales and derived using different techniques. In an ideal world, multiple mapped estimates of the same biophysical variable should largely agree in the same location and time. Presumably, all credible maps are verified against a more reliable source to estimate overall error, but the spatial distribution of errors may not be the same for different maps. Thus, when two maps of the same quantity are compared at a given location, they may disagree. From an end-user perspective, this is a problem: different maps may have quite different implications for carbon accounting, for example, but users are given no guidance about which map to choose. From a scientific perspective, this may be an opportunity:

*Correspondence: neeti@teriuniversity.ac.in
[1] Department of Natural Resources, TERI University, New Delhi, India
Full list of author information is available at the end of the article

patterns of disagreement among maps may provide insight into how datasets and techniques perform under varying conditions.

Reconciling maps will become an increasingly important activity within NASA's Carbon Monitoring System (NASA-CMS) [8]. NASA-CMS is a broad initiative to apply NASA's synoptic view of Earth systems to the monitoring of carbon. NASA-CMS activities run the gamut from local-scale mapping of carbon state [9, 10] to global scale mapping of carbon flux [11]. Often, projects at different scales produce similar products, or products that could be compared against other projects through simple manipulation (e.g., multiple estimates of biomass over time could be compared against estimates of flux). As a mix of both regional scale and global scale projects, NASA-CMS will be increasingly faced with disparate estimates of carbon states and fluxes at different scales.

Recognizing that different estimates of carbon-related maps co-occur, several researchers have recently reported on comparisons among different biomass maps [12–14]. In some cases, comparisons among maps were made with fine-resolution reference data (e.g., [13, 14]), but such data are not always available. In other cases, maps were compared at the same spatial resolution (e.g., [12]) but this precludes comparisons generated at different resolution. In fact, there are often situations where we need to identify a map which is most close to the truth in the absence of fine-resolution reference data while understanding various methodological and input grain size differences between various available maps. Indeed, the notion of multiple maps may imply that one is better, or that one more closely matches truth, but such a uniform truth is rarely feasible, and therefore we need spatial distribution of uncertainty. In practice, the assessment of spatially-distributed map data can be considered an effort to paint a picture of spatial uncertainty which is carried out by describing relative error among maps or measurements using a suite of complementary quantitative measures [15]. However, there exist several challenges in assessing a map and comparing different maps. First, the unit of analysis in a raster-based map is a square pixel that is arbitrary when we compare to any phenomenon in the landscape. Second, raster based maps are often generated at different scales, so comparison requires cell by cell alignment while ensuring the information in individual maps remain same. Moreover, most of the techniques developed to compare real variable maps can be used only for pairwise comparison, limiting the ability of simultaneous comparison of multiple maps to assess the spatial pattern of similarity and dissimilarity.

As the NASA-CMS matures, mapped estimates of both carbon biomass and flux will be available at different scales for the same location, requiring quantitative comparisons among many maps. Indeed, NASA-CMS has already produced one forest biomass map for the conterminous United States, but already several versions existed developed by other groups. For users, guidance must be given about the relative merits of these maps under different conditions, and under what conditions a given map may be avoided. For developers and scientists, however, comparisons should be structured to leverage the differences in methods to provide insight into possible improvements or best practices. Thus, using these several forest biomass maps as a test case, we report on strategies and methods both to describe differences among maps for users, and to evaluate possible sources of disagreement.

We therefore had two broad objectives:

(1) Descriptive: Assess both the absolute accuracy of individual maps and assess the spatial patterns of agreement among maps.
(2) Evaluative: Investigate potential causes of the spatial structure of agreement among maps to gain insight into reliability of methodological choices in map making.

To achieve these objectives, we introduce a comprehensive and systematic approach to compare multiple maps generated at different scales (extent and spatial resolution). This comprehensive approach not only evaluates accuracy of individual maps while describing the similarity and dissimilarity between the spatial structure of various maps, but also systematically explores effects of scale and various causes of change.

Methods

Our systematic approach consists of several steps (Fig. 1). The foundational step is a thorough assessment of key map-making steps. This provides a context for the descriptive phase, and helps guide hypotheses to be tested in the evaluative phase.

Next, the descriptive phase seeks to quantify how well individual maps agree with reference data and with each other. Two key strategies are important. First, in addition to evaluating pairwise differences, as is done in other comparative studies, we advocate simultaneous comparison across all maps to understand underlying spatial patterns of agreement and potentially isolate outlier maps. Second, we aggregate our measures of agreement and disagreement to ecologically coherent mapping regions, recognizing that some maps may perform better in particular ecological contexts and that users may only be interested in this regional scale.

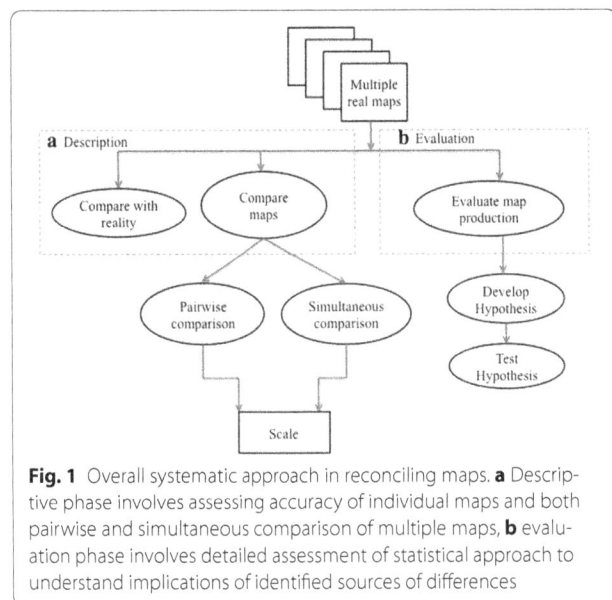

Fig. 1 Overall systematic approach in reconciling maps. **a** Descriptive phase involves assessing accuracy of individual maps and both pairwise and simultaneous comparison of multiple maps, **b** evaluation phase involves detailed assessment of statistical approach to understand implications of identified sources of differences

Finally, the evaluative phase seeks to test specific sources of error in the map making process. Here, we must draw inference from comparisons without the benefit of manipulation. Thus, rather than simply developing a range of comparative metrics, this phase places comparisons into specific test of methodological contrast among maps. The key is to hypothesize how differences in methodology would lead to specific differences among pairs of maps, and then test those hypotheses. It is the combination of metric and map combination that allows greater inference: the observation of an effect in a single map is likely uninformative, but hypothesis-driven comparison of that effect between maps may provide greater insight.

Assessment of map-making steps

We examined the four national-scale maps of forest biomass that were available at the beginning of our study (Fig. 2). Papers detailing map production are given by Saatchi et al. [16], Kellndorfer et al. [17], Blackard et al. [18], and Wilson et al. [19]. For parsimony, we hereafter refer to these maps by their first letter, i.e. S, K, B, and W, respectively. All maps are generally considered usable, either through self-reported accuracies or through community use. Map W is reported to have strong agreement with plot based estimates of biomass (agreement coefficient ~0.99) and to have a strong goodness of fit (within 90 % confidence interval). The average absolute error for the map B ranges between 40 and 60 metric tons per hectare, except for the higher-biomass areas of the Pacific Northwest (163 metric tons per hectare). The accuracies for map S and K are not known, but they, like all of the

maps, are already being used as input in different studies [10, 20, 21].

Like many spatial mapping exercises in remote sensing, the general approach for producing AGB maps is to extrapolate high quality training data from a small sample of locations to a large, contiguous space. Typically, values measured at the training samples are linked with data values from geospatial datasets at the locations where they intersect. Statistical models built at those locations are then used to extrapolate to the rest of the map. Maps can differ in how they handle training data, which geospatial datasets are used, and what statistical models are built.

We identified six methodological sources of disagreement in how the four biomass maps were generated (Table 1). For each, we developed expectations for how that difference may manifest itself in the resultant maps. The first source is the forest mask used to define the total area for which biomass is estimated. Conservative forest masks would lead to lower estimates of total biomass, but a more liberal forest mask would include a higher proportion of marginal forest and thus likely reduce average biomass density. The second source of potential disagreement between the maps is spatial resolution. Map resolution interacts with the fundamental spatial structure of the landscape: If the former is finer than the latter, the map can adequately capture the range of variability of the landscape. A third contrast among maps is the remote sensing source data used for extrapolation. Maps using optical data may be less able to capture high biomass than maps using radar data, as optical data are known to saturate at lower biomass compared to radar (longer wavelength such as L- and P-band) [22–25]. The sensitivity of L- band SAR data for biomass estimation increases if used along with forest height generated using InSAR [25]. The fourth source of difference is the statistical technique used for extrapolation (parametric vs. non-parametric). Non-parametric techniques could be expected to perform better at extreme values (low and higher biomass region) than parametric techniques, especially if the extreme values deviate greatly from a normal distribution. The fifth source of difference is year of mapping. Maps produced in different years would be expected to differ both because of intervening disturbance and growth, and because the pool of training data could be different.

The sixth area of potential disagreement rests not with the methodologies to extrapolate, but with the training data themselves. In the case of forest AGB in the conterminous US, the training data come from the US Forest Service's Forest Inventory and Analysis (FIA) program. At each of thousands of field plots, field crews measure details of trees using a regular sampling and mensuration

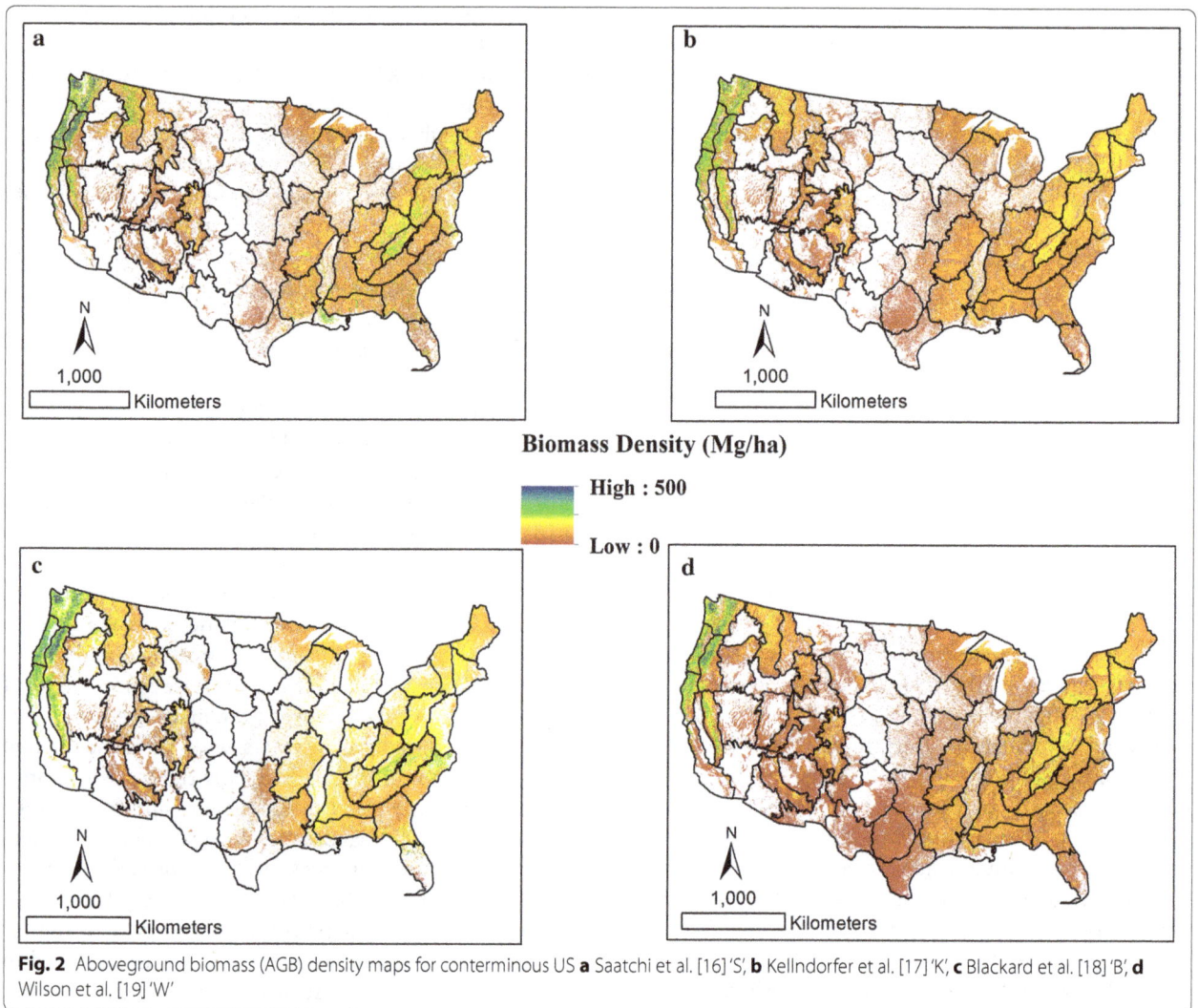

Fig. 2 Aboveground biomass (AGB) density maps for conterminous US **a** Saatchi et al. [16] 'S', **b** Kellndorfer et al. [17] 'K', **c** Blackard et al. [18] 'B', **d** Wilson et al. [19] 'W'

Table 1 Sources of disagreement among the four national AGB maps

AGB maps	Forest mask	Grain size (m)	Spatial predictors	Extrapolation technique (parametric/non-parametric)	Year of generation of map	Allometric equation
Saatchi et al. [16] (S)	NLCD 2006	93	F(MODIS, Landsat, L-band PALSAR, GLAS, topography)	Maximum Entropy (Parametric)	2005	Component ratio method
Kellndorfer et al. [17] (K)	NLCD 2001	30	F(C-band InSAR, Landsat, NLCD canopy density, structure, SRTM)	Regression tree (Random forest, Non-parametric)	2000	Regional method
Blackard et al. [18] (B)	In house	250	F(MODIS, climate variables variables, topography)	Regression tree (Cubist, Non-parametric)	2003	Regional method
Wilson et al. [19] (W)	No Forest mask	250	F(MODIS, climate variables, topography, level III ecoregions)	Phenological gradient nearest neighborhood (Semi-parametric/Semi-nonparametric)	2009	Component ratio method

protocol. These raw tree data can then be converted to estimates of plot-level biomass using allometric equations that relate tree characteristics to known biomass measured destructively at a sample of locations (and often as an entirely separate effort). Allometric equations varied across the four maps studied here, with some using regional-specific equations and others a more nationally-consistent component ratio method (CRM) [26]. The CRM approach has been shown to lead to lower biomass estimates then the regional approach [26], and thus maps based on the CRM would thus be expected to show an overall bias toward lower values.

Descriptive comparisons

Descriptive comparisons provide users guidance about which maps are more accurate, and where on the landscape (at both national and regional scales) the maps agree and disagree when analyses carried out at pixel level.

Comparison with ground reference data

Because the FIA program is tasked with providing defensible estimates of forest resources at the national scale, estimates from FIA plot data are the de facto standard against which any forest resource maps must be compared. Although FIA plot data are used as inputs in various points of the mapping process for all four national maps tested here, this does not guarantee that summarized estimates of biomass will agree, since maps extrapolate the FIA information differently.

Comparison with FIA data requires appropriate use of those data. The goal of the FIA sampling design is to provide good estimates at aggregated administrative levels: The smallest unit is typically the US county or parish, but state-level comparisons are more common and robust, especially when the analyses is carried out at the MODIS scale [27]. Additionally, normal users do not have access to actual plot locations, and thus data are typically available and readily usable only at county or larger scales.

Thus, for our comparisons we compared plot and map data at the aggregation level of the state. Map data aggregation was a simple matter of summing biomass density estimates to the state level, utilizing each map's own forest mask for the aggregation footprint. For plot data, we acquired state level FIA AGB data from the FIA 2013 database [28] at the appropriate time step for the FIA collection strategy of the state. For forests in the eastern US, we used the time frame of 2003–2007, as that was the most consistent time frame without double counting any trees across different states. For forests in the western US, which are on a 10 year repeat cycle, we used data from 2000 onwards for the west side as that was the most consistent data available through the website across

all the states at the time of analyses. The estimates have already converted the plot-level tree measurements to the plot scale, and then used sample-design considerations to scale the plot-level data to the state level. Three considerations of the FIA data are relevant for later comparisons. First, the FIA uses the CRM approach to calculate biomass from tree measurements. Second, the inventory cycles occur at 5 and 10 year cadences in the east and west, respectively, meaning that an estimate at a given time will be a different mix of older and more recent plots in the east and the west. Finally, the extrapolation scaling factors are based on forestland masks developed by the US Forest Service (and also used in the B and W maps).

Once biomass had been aggregated, we produced three descriptive products. First, we summed forest biomass among all maps and all plots at the conterminous US scale. Second, we used simple regression of aggregated map biomass at the state level against FIA plot biomass (n = 48 states in conterminous US). Finally, we calculated the Wilmott's index of agreement (d):

$$d = 1 - \frac{\sum_{i=1}^{n}(X_i - Y_i)^2}{(|X_i - \underline{X}| + |Y_i - \underline{X}|)^2}X$$

where X_i was the biomass estimate for state i estimated from FIA plot data and Y_i was the same estimated from the maps [29]. The X variable is considered to be the truth variable, which is appropriate here because FIA plot data are the national standard for forest monitoring. Wilmott's index of agreement is symmetric, bounded and does not over penalize for disagreement. High agreement is indicated by values close to 1.0. There exist many other indices for such pair wise comparison such as mean square error (MSE) and its root (RMSE) to more specialized metrics [e.g., agreement coefficient (AC)] to comparing distribution. The indices such as MSE, RMSE have been critiqued for being unbounded and asymmetric [30] while the AC is found to highly sensitive to outlier. The comparison of distributions could simply yield the uninteresting finding of differences with little insight on the reason behind the differences. Spatial structure of agreement and disagreement is critically useful for understanding whether disagreement is potentially related to specific issues.

Comparison of multiple maps

After descriptive comparison against a trusted data source (FIA based estimates), the next step was to show where maps agreed and disagreed. We used principal components analysis (PCA) for simultaneous comparison among all maps. PCA is a spectral decomposition technique commonly used in remote sensing to remove the redundancy from multi-spectral images, but it can also be used to identify the dominant pattern common among various maps [31]. The six orientation modes (O, P, Q, R, S, T) commonly used for PCA differ in their definition of statistical variables and observations [32]. In this case, statistical observations are samples in space and statistical variables are various continuous maps, therefore, R-mode PCA is used for the analysis.

Implementation of the PCA took place in the R statistical package [33], and required some basic preparation of the datasets. First, because the maps reported biomass in different unit systems and different map projections, we aligned all biomass map values to the same system at the pixel scale, and then aggregated (taking mean of x by x window size followed by nearest neighborhood resampling, x = 8 for map S, x = 2) the finer-scale maps (S and K) to the 240 m grain size of the coarser maps to make them comparable to the other two maps. The coarser resolution maps (B and W) were resampled from 250 m to 240 m resolution by using nearest neighborhood approach. Second, we clipped all maps to the forest mask area common to all forest masks, as "no-data" would not be informative in the PCA analysis. Finally, we stacked all four maps into a single, four-layer image. Once these preparatory steps had been taken, we ran the PCA analysis on both the entire conterminous US data (at the 240 m pixel scale) and at the scale of each of the 66 mapping regions [34]. The regional scale analysis is useful because the PCA statistical space is defined by the range of variation in the whole dataset, and thus the broad US-wide comparison may obscure patterns that would be relevant to users at the regional scale.

Results of the PCA were interpreted in two ways. First, spatial patterns in the first and second axis images show where on the landscape the maps agree and disagree. Because PCA axis 1 identifies the vector through the multivariate space that explains the most variation, it can be interpreted as the overall pattern of biomass agreement across all maps. PCA axis 2 is orthogonal to the first axis, and thus indicates the dominant sources of disagreement among the maps. Second, the correlation coefficients of each map contributing to those two axes can give insight into which maps are agreeing or disagreeing with the other maps. A map with high correlation coefficient on the first axis is one whose spatial patterns agree with the other maps; a map with high correlation coefficient on the second axis is one whose spatial patterns disagree with the other maps.

Scale of agreement

If maps generally agree with truth data when aggregated to the mapping region scale, but show patterns of disagreement at the pixel scale, then a natural question from users is whether intermediate scales of resolution may make maps more comparable. To evaluate this question, we conducted pair-wise map comparison to analyze the impact of spatial resolution on map agreement. As with the PCA analysis ("comparison of multiple maps" section), we first ensured that all maps were clipped to the same footprint and were at the same starting pixel resolution (acknowledging that the 30 m and 90 m products were already degraded for the first comparison), and then we aggregated from the starting 240 m resolution to pixel resolutions of 480, 720 and 960 m. For each of those resolutions, we evaluated paired map agreement with the Wilmott's index of agreement at mapping region scale, as all the pixels in a mapping region are expected to have similar ecological condition. Critically, because no map could be considered the truth map, we conducted this analysis on all pairwise comparisons. High median d-scores across all mapping regions suggest consistent strong agreement, while high range of d-scores suggests spatial patterns of variability across mapping regions.

Evaluative comparisons

Evaluative comparisons were designed to test whether patterns of agreement and disagreement could provide insight into sources of error in map production. Because each map was produced using a different suite of methods (Table 1), each map needed to be compared individually to the other maps. To test the six different possible sources of differences, we used a range of different pairwise map comparisons, both on the original data and on maps of differences between pairs of maps. Additionally, for certain tests we brought in ancillary data on forest type, topographic position, and regional biomass time series. We first describe the basic data manipulations, and then describe how these were used to evaluate the six different sources of map difference.

Data manipulations

Cumulative biomass   The simplest data manipulation was simple summing of biomass across all pixels in a given map to estimate total forest biomass at the national scale. Total biomass at the national scale had already been calculated for the descriptive phase (see section on "cumulative biomass" section).

Mapping region scale biomass density   Within each mapping region, we calculated the cumulative frequency distribution of biomass density values of a map for all pixels in each map's forest mask. From these, we identified the 10th-, 50th- and 90th percentile values for the mapping region. We also calculated mean biomass density by mapping region.

Difference maps   For all pairwise combinations of the four national-scale maps, we developed maps of biomass difference at the pixel scale. As with the PCA analysis, the maps were first clipped to a common forest mask, and if the pixel sizes of the two maps were different, the higher resolution map was resampled and aggregated to the lower resolution map cell size (see details on aggregation above). Differencing was achieved using simple image algebra on a cell by cell basis. Spatial patterns in the difference maps were then related to spatial patterns of ancillary data.

Ancillary data We sought to understand whether spatial patterns in map differences were related to spatial patterns in other geospatial datasets. For these, we obtained the spatial predictor data layer and manipulated the map as necessary (using mean/mode aggregation and/or nearest neighborhood resampling) to ensure cell alignment among maps. We obtained through the ORNL website [35] a forest type land cover map with cells of 250 m resolution, for the nominal mapping year 2001. We utilized all forest cover types for analysis. Separately, we obtained a forest age map with cell size of 1 km for the nominal mapping year 2006 [36] and grouped pixels into young (<40 years old), immature (40–80 years old), mature (80–140 years), and old (>140 years). Finally, for topographic variables, we obtained SRTM elevation at 30 m resolution and derived slope. We grouped pixels according to three elevation categories (<500 m, 500–1500 m, and >1500 m), and, where appropriate, two slope categories: relatively benign (0 to <30 degree slope) and extreme (30–50 degree slope). Within mapping regions of interest, we summarized the difference maps according to the zones delineated by the groups in those ancillary data sources.

To assess impact of year of map generation, we required ancillary biomass map data where technique of production was held constant across time. At present no such map exists at the national scale, and thus we used a regional-scale map product [37]. That product generated yearly biomass estimates from 1990 to 2010 for the high-biomass states of Washington, Oregon, and California. Because the core of that approach was a change detection approach at the Landsat scale, those maps explicitly represent the effects of growth and disturbance on biomass. We refer to these maps as KennXXXX, where XXXX represented the year of map. We differenced KennXXXX maps for the years corresponding to each of the map pairs in the national efforts, and compared distributions of the differences in the consistently-produced maps to the differences in distributions in the national maps. Note that we not using this map as a truth dataset, but simply as a means of holding constant the means of production across time to isolate the impacts of growth and disturbance.

Evaluative tests

The data manipulations were then used in evaluative tests of each of the six identified sources of map error. We emphasize that evaluative tests utilize metrics that by themselves could be uninformative—they key is to compare the metrics among map pairs, driven by hypotheses about how map production could lead to differences.

Forest mask Each map used a slightly different mask to identify potential forested pixels. If the forest mask were the only factor affecting map results, we would expect that maps with a conservative forest mask would show a lower total biomass when aggregated to the national scale.

To test, we simply summed biomass nationally for all four maps and evaluated relative to the forest mask area of each map.

Spatial resolution Spatial resolution determines the scale of pattern which can be captured by a given map. If spatial patterns of forest biomass vary meaningfully at a scale finer than that of a given map, that map will be unable to capture the high and low biomass values, and thus would be expected to have a compressed range of biomass relative to the actual landscape [24]. When comparing among maps of different resolution, the impact of resolution will matter if the scale of meaningful variation is intermediate between resolutions of the maps.

To test, we compared 10th and 90th percentile values by mapping region ("mapping region scale biomass density" section) for all pairs of maps. If pixel resolution were a driving factor, we would expect to see compressed ranges in both the W and B maps relative to the S and K maps.

Source sensor data Because the signal retrieved from a passive optical sensor typically saturates at biomass levels lower than that of an active sensor such as radar (e.g., L-Band SAR, C-band InSAR) [26, 27], we might expect that maps involving radar would be able to track biomass better in high-biomass mapping regions even though there may be difference between various radar data depending on frequency and polarization. To test, we compared the 90th percentile values by mapping region for all pairs of maps. If sensor source were a driving factor, we would expect the W and B maps to show a compressed upper end relative to the S and K maps.

Similarly, because older forests have more structurally complex canopies, we might expect an active-sensor approach to better capture biomass in older forests, or in generally higher-biomass forest types [38]. To test, we compared difference maps grouped by forest type and age for mapping regions where high biomass and older forests were more prevalent. If active sensors performed better in these systems, we might expect the S and K maps to show higher biomass (net positive difference) vs the B and W maps in older forests and in higher biomass type forests.

Finally, because of the side-looking nature of radar, we might expect the radar-derived maps to perform more poorly in topographically complex areas [39]. To test, we compared difference maps grouped by elevation and slope categories (noted in section on the ancillary data in the method above). If error introduced by side-looking radar were an issue, we would expect the S and K maps to show greater difference with the optical maps in either high elevation or high slope areas.

Extrapolation technique   When linking the observed covariate (here, biomass) to predictor data (here, geospatial data), parametric regression techniques explicitly attempt to minimize variability in the covariate, often resulting compressed prediction ranges when extrapolation is later performed [40] low values are overestimated and high values are underestimated. Non-parametric approaches may not suffer from this extrapolation issue. To test, we examined the 10th and 90th percentiles of biomass distributions. If extrapolation technique were an issue, we would expect the W and S maps to show ranges compressed relative to the K and B maps.

Nominal year of map   National maps were produced in different years, and thus some differences between any pair of maps would be caused by intervening growth and disturbance in the period between the two maps. Thus, the observed difference between maps convolves both map production differences and real differences in the landscape. To test, we compared difference maps to differences in the KennXXXX map pairs for the same period. Because the KennXXXX maps explicitly capture disturbance and growth processes using a consistent technique, they serve as a proxy for the expected differences between maps for any given pair of years. If two maps agree but differ only in disturbance and growth, we would expect the distributions of biomass difference to be similar to that in the proxy maps. Substantial departures from that expected distribution would indicate that year of map production was not the only cause of difference in the maps.

Allometric equations   Because the CRM method used nationally has been shown to produce lower estimates of biomass than regionally-specific allometric equations, we would expect mean and median values of biomass to be lower in those maps using CRM approaches. To test, we compared median and mean values by mapping region. If allometric equations were a critical difference, we would expect the S and W maps to be consistently lower than the K and B maps.

Results and discussion
Descriptive comparisons
Comparison with ground reference data
All four maps estimate slightly higher national AGB than does the summed FIA plot data based estimates for conterminous US (Table 2). Of the four, map S's estimate is most similar to FIA and map B's most dissimilar.

When regressing biomass on a state-by-state basis (Fig. 3), map S shows the closest alignment with the 1:1 line (slope: 1.02), but with greater scatter than both maps K and W (r^2 = 0.98, 0.99). Map B remains substantially different at the state scale, particularly for medium to high biomass states. Wilmott's d suggests that map W is most similar to the FIA estimates at state scale, followed closely by map S.

Comparison across multiple maps
The national-scale PC1 image shows patterns of biomass that are generally in agreement among all four maps (Fig. 4a), with correlations strong between PC1 and all four source maps (Fig. 4c). As expected, forest biomass is highest in west-coast forests, lower in parts of the mountainous eastern interior forests and lowest in the interior west and north central states. Patterns of biomass disagreement (PC 2) are more spatially heterogeneous (Fig. 4b), and appear to be dominated by map B (Fig. 4d). The difference of map B is in the PCA analysis is consistent with its difference against the other maps in the state level comparison with FIA plot data (Fig. 3). Relative to PC comparisons at the national scale, mapping-region scale PC analyses can show more nuanced or even opposing patterns. Mapping region 1 (Washington state) and mapping region 52 (Wisconsin and Michigan state) provide useful examples. In mapping region 1 (Fig. 5), patterns of biomass agreement (PC 1, Fig. 5a, c) are as expected, with high biomass forests in the wetter western portions of the state and lower in the drier eastern portions of the state. Consistent with the national scale analysis, Map B dominates the disagreement vector (PC 2, Fig. 5b, d). However, in mapping region 52, Map W and map B dominate the disagreement vector (PC 2,

Table 2 Total forest area and biomass for conterminous US for the four AGB maps based estimates and FIA based estimates

AGB map	Forest area (Million Ha)	Total biomass (Gt)
K	414.8	27.4
B	260.47	29.6
S	350.23	26.7
W	486.14	27.2
FIA	–	25.4

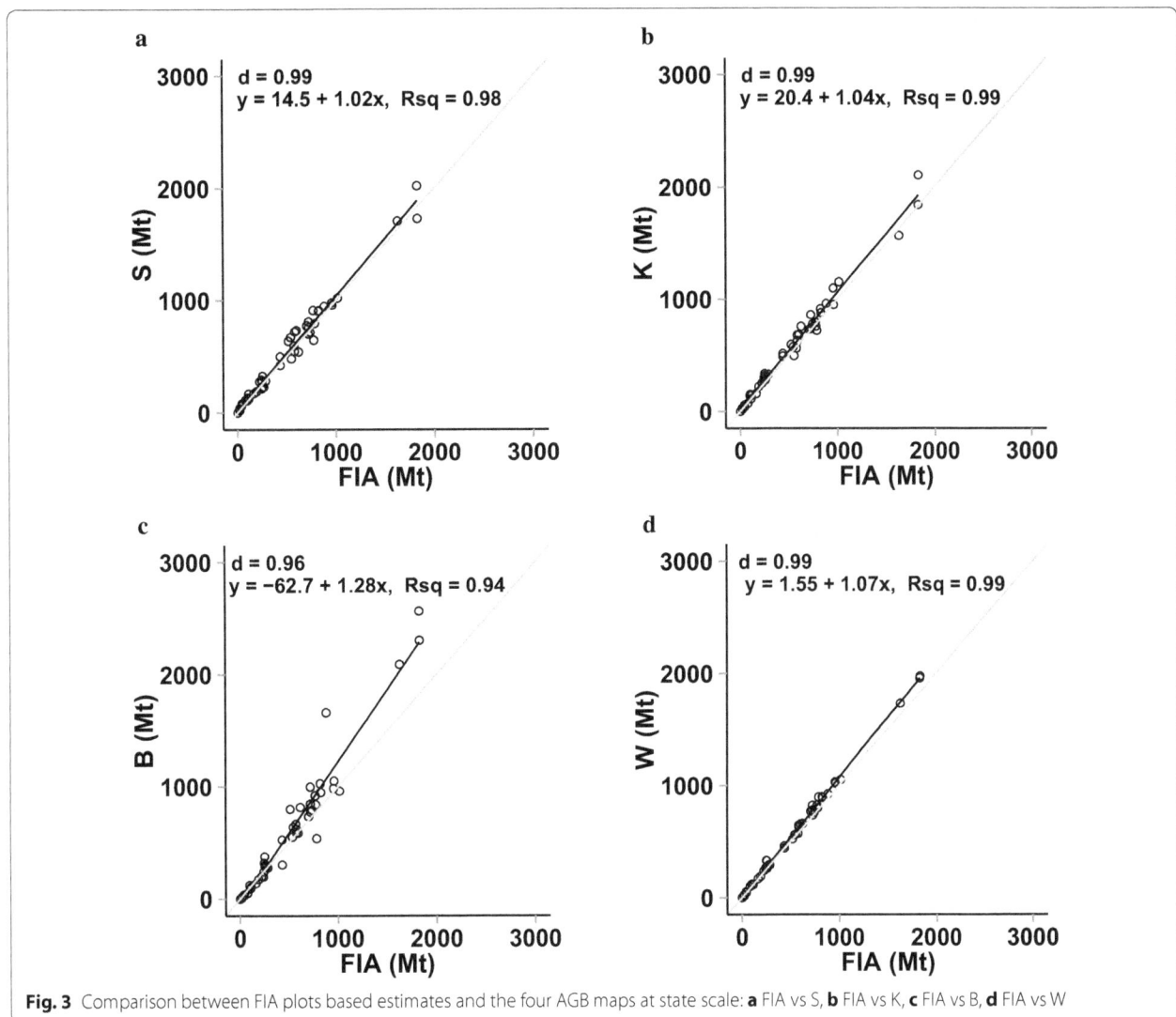

Fig. 3 Comparison between FIA plots based estimates and the four AGB maps at state scale: **a** FIA vs S, **b** FIA vs K, **c** FIA vs B, **d** FIA vs W

Fig. 6b) even though the patterns of biomass agreement are largely as expected (Fig. 6a, c). The map W and map B has completely opposing patterns in the mapping region, and the spatial pattern is not at all related to the other two maps (Fig. 6b, d). Thus, the guidance for which map is less reliable may vary with mapping region, and points to the need for local users to develop local-scale comparisons.

Scale of agreement
For all maps but Map B, the Wilmott's measure of agreement between pair-wise maps suggests that agreement increases at coarser spatial resolution, potentially leveling off at the top end of the range of spatial resolution tested here (Fig. 7a). There is high consistency across all pair-wise comparisons at multiple spatial resolution (range of d values less than 0.4) except for the comparison with the

map B. For Map B, similarity did not change with increasing resolution, and the range of disagreement was high (range of d values near 0.8). However, similar to mapping region results for PCA, the agreement between maps varies with regions (Fig. 7b). The comparisons of map S with the other three maps at 240 m resolution suggest that it has maximum agreement with the map K across all the mapping regions. The agreement of map S with B decreases in the eastern mapping regions (Fig. 7b).

Evaluative comparisons
Given that these maps are being compared post hoc, no experimental manipulations of method can be designed to unambiguously identify causes of change. However, by designing evaluative tests to compare specific maps against each other under specific hypotheses of change,

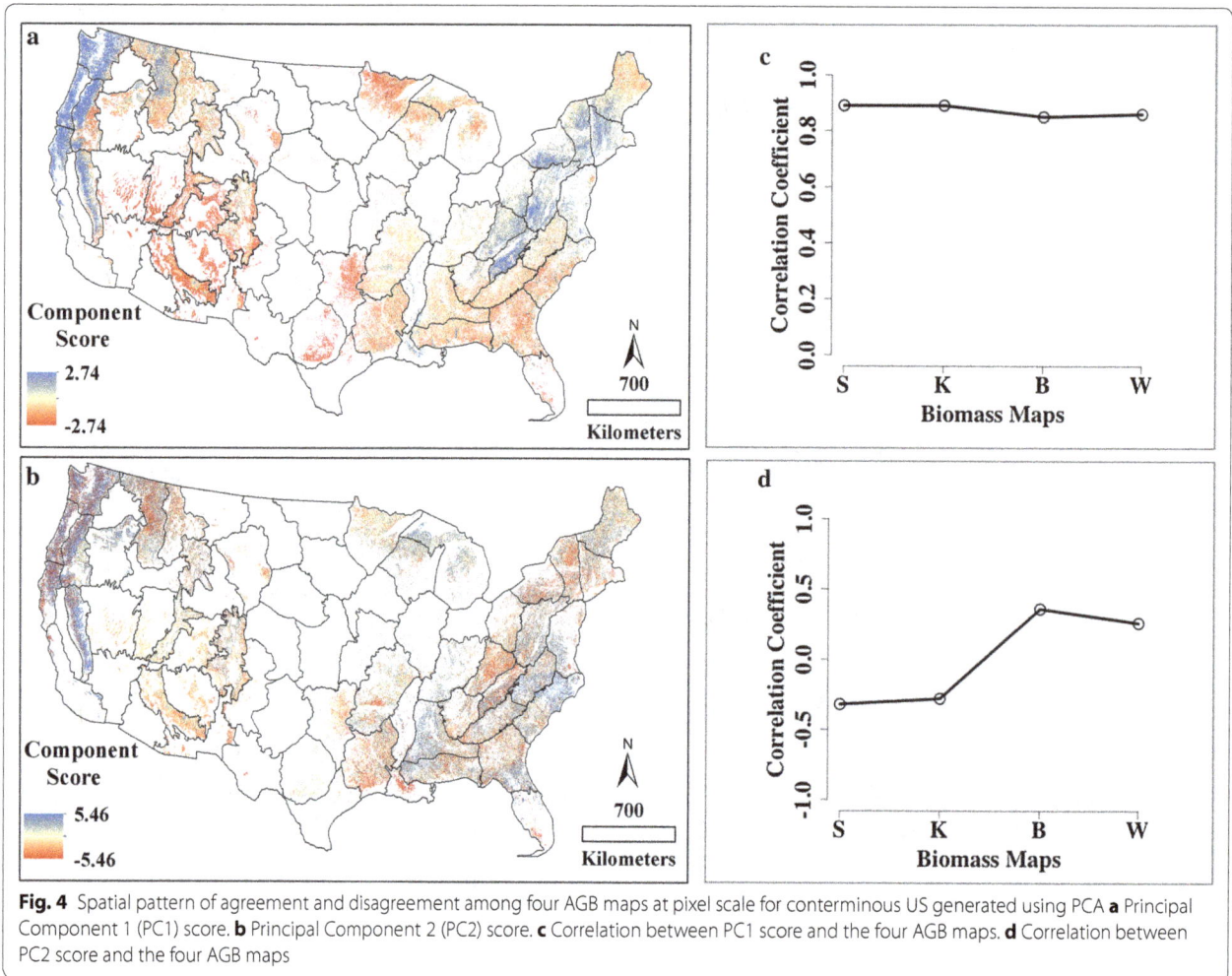

Fig. 4 Spatial pattern of agreement and disagreement among four AGB maps at pixel scale for conterminous US generated using PCA **a** Principal Component 1 (PC1) score. **b** Principal Component 2 (PC2) score. **c** Correlation between PC1 score and the four AGB maps. **d** Correlation between PC2 score and the four AGB maps

we can paint a richer picture of possible sources of difference. Thus, we first describe the results of the metric comparisons here, and then follow with an assessment of how those comparisons should be interpreted on a pairwise basis to test potential methodological sources of error.

Cumulative biomass
The same cumulative biomass results conducted for the descriptive phase apply to the evaluative phase (Table 1; Fig. 3). Notably map S and map W agree well with FIA, and maps W and B have higher estimates. This contrast is likely related to the commonality of allometric equation (CRM approach) between the FIA and S and W maps.

Mapping regions scale biomass density
Pairwise comparisons between the four maps at low, medium and high biomass density quantiles at mapping region scale parse out differences among maps more closely. In general, maps K and W track the 1:1 line at

low, median, and high biomass, suggesting agreement across the biomass distribution and across all mapping regions (Fig. 8d, h, k). Map S agrees with Maps K and W at the median of the biomass distribution (Fig. 8d), but is generally higher than Maps K and W at both high and low ends of the range (Fig. 8g, j). Map B is higher than all other maps in nearly all situations, and particularly appears to have relatively higher biomass at the low end of the biomass range (Fig. 8g–i).

Difference maps and relation to ancillary data
We evaluated difference maps in relation to ancillary data for all 66 mapping regions. At the US scale, the difference between the map S and K suggests that the map S has higher biomass estimates compared to the map K at higher biomass regions, and lower biomass at lower biomass regions (Fig. 9a). Again, we use mapping region 1 to illustrate core principles and findings. In mapping region 1, Map S has higher biomass estimates than does Map K in the wetter, western part of

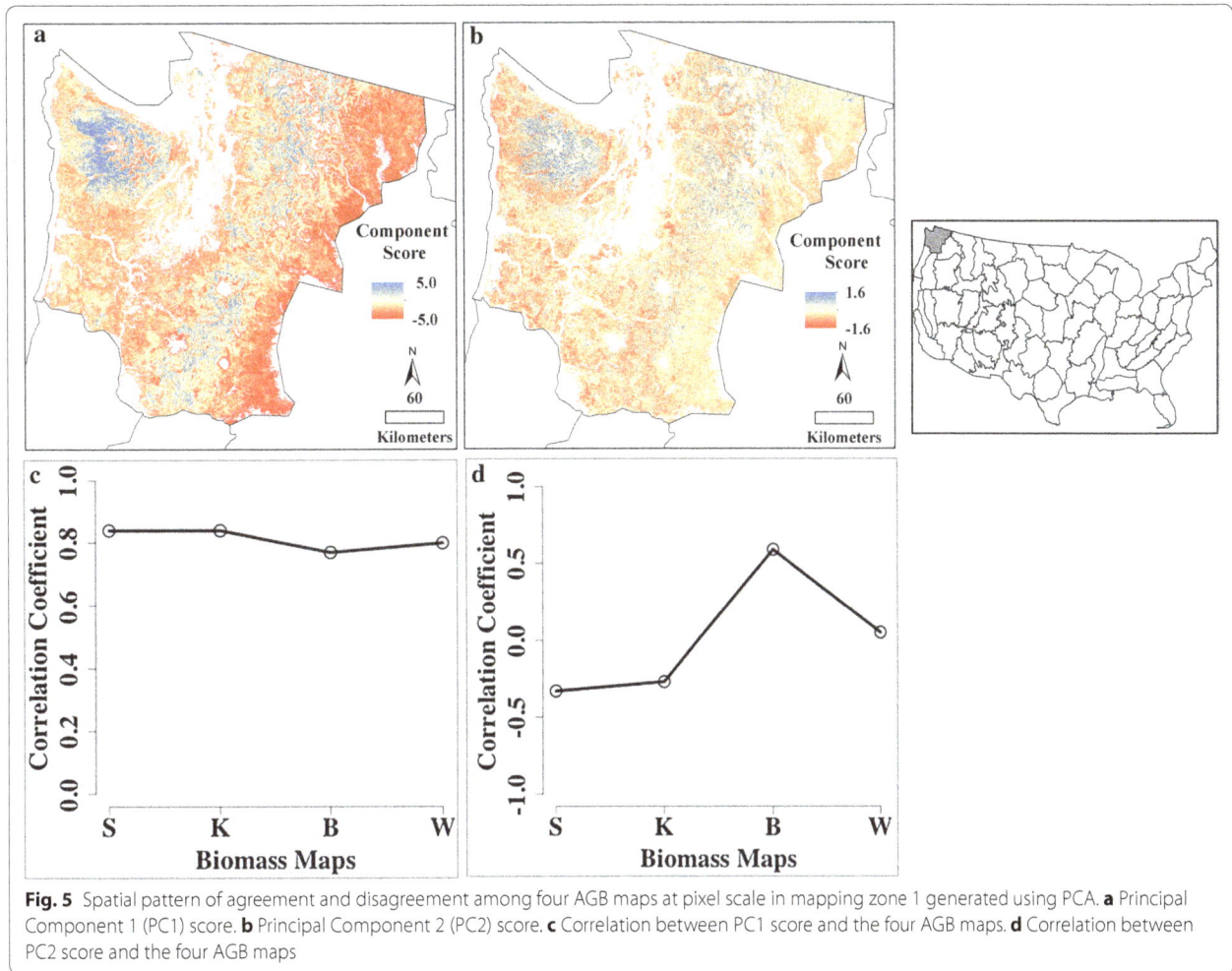

Fig. 5 Spatial pattern of agreement and disagreement among four AGB maps at pixel scale in mapping zone 1 generated using PCA. **a** Principal Component 1 (PC1) score. **b** Principal Component 2 (PC2) score. **c** Correlation between PC1 score and the four AGB maps. **d** Correlation between PC2 score and the four AGB maps

the region, but that pattern is generally flipped when compared to Maps B or W (Fig. 9 b–d). Forest characteristics (Fig. 10a, b) and topography (Fig. 10c, d) were then related on a pixel by pixel level to the difference map, and difference median and range summarized by forest type (Fig. 11a), forest age (Fig. 11b), elevation (Fig. 11c), and slope (Fig. 11d). As before, Map B was an outlier in most comparisons in this mapping region. In most of the comparison with map B, the median biomass difference is farther from 0, the range of difference is larger with higher biomass estimates across the mapping region.

Testing sources of disagreement

Metrics of comparison described above are the core from which tests can be developed. Again, the goal of the evaluative comparisons is to use specific map-pair comparisons by metric to infer whether certain methodological choices are contributing to map disagreement.

Forest mask

If forest mask area were a fundamental reason for disagreement, we would expect maps with conservative forest area to show less biomass. But comparing total forest and total biomass (Section on cumulative biomass above, Table 2), there is little evidence that forest mask was an important factor driving the difference among maps. Indeed, the map with the highest biomass estimates (Map B) had the most conservative forest mask.

Spatial resolution

For tests of the impact of spatial resolution, we compared the high-resolution maps (S and K) with the coarser-resolution maps (W and B). If spatial resolution were an issue we would expect coarse resolution maps to show compressed data ranges at both high and low biomass values relative to the higher resolution maps. Map S appears to show this effect, as it shows greater range in both the 90th and 10th percentiles compared to the W map (Fig. 8j, g). However, map K did not show this effect

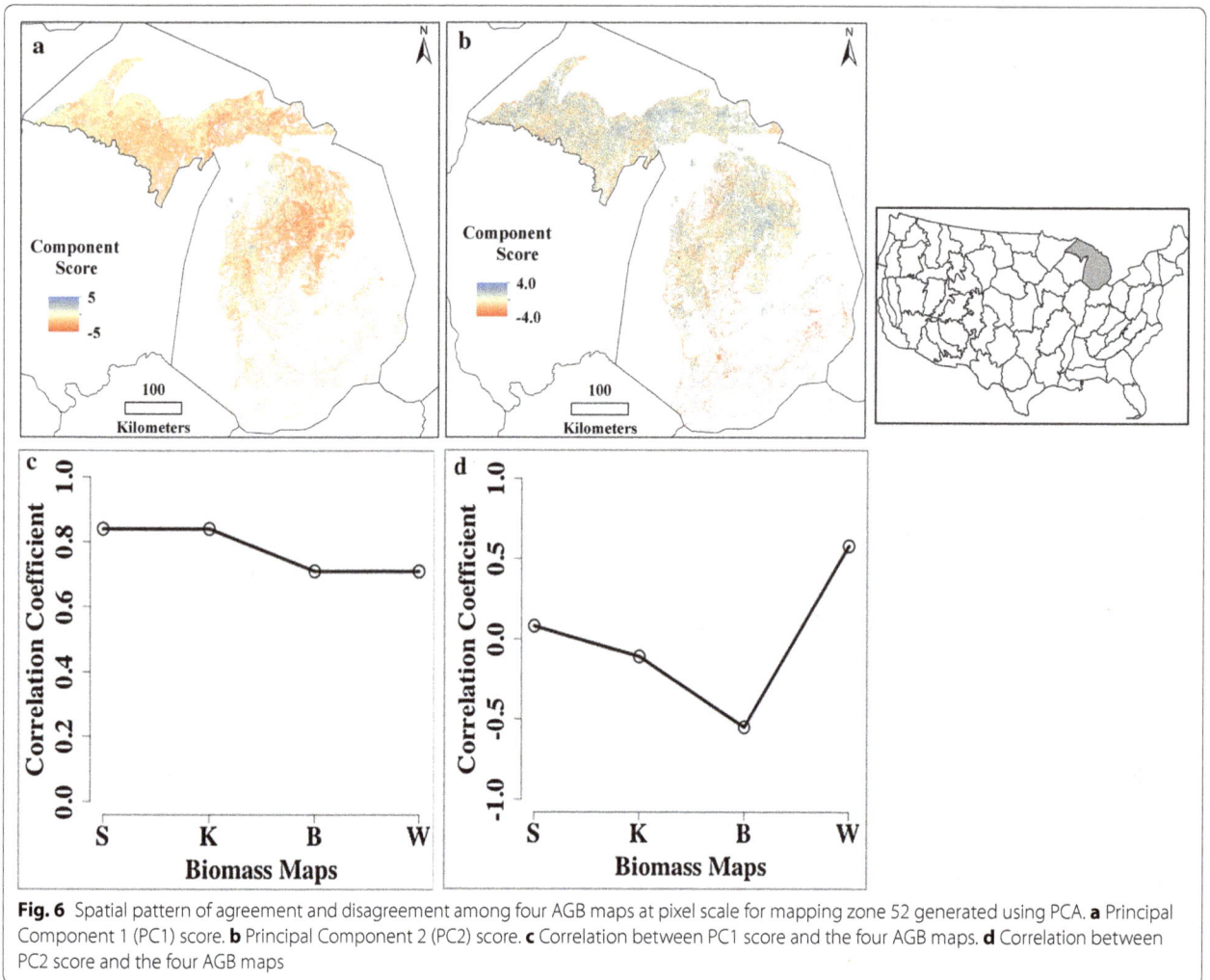

Fig. 6 Spatial pattern of agreement and disagreement among four AGB maps at pixel scale for mapping zone 52 generated using PCA. **a** Principal Component 1 (PC1) score. **b** Principal Component 2 (PC2) score. **c** Correlation between PC1 score and the four AGB maps. **d** Correlation between PC2 score and the four AGB maps

in comparison to map W (Fig. 8h, k), suggesting that resolution alone does not explain the differences in the maps.

Source sensor data

If active sensors are able to map higher biomass without saturating, we would expect maps W and B to saturate at high biomass relative to both map S and K. Map S does appear to estimate higher biomass (90th percentile) than all other maps (Fig. 8j), but this effect does not occur with the other map created using an active sensor (Map K). However, this effect may be associated with the use of L-Band PALSAR in the map S. Map K uses C-band InSAR, which is less sensitive to the higher biomass compared to the higher wavelength [23]. Thus, our results are consistent with the notion that a specific type of active sensor may provide greater response to high biomass areas.

This pattern can be partially corroborated by focusing on the high-biomass mapping region 1 (Fig. 11a). There, map S shows higher biomass estimates compared to K (that uses C-band radar data for height measurement) for high biomass forests (Douglas fir or Hemlock/Sitka spruce; Fig. 11a). However, the contrast does not hold for high biomass age classes (>80 years old; Fig. 11b). Moreover, when compared to optical-based maps, the map K map does not consistently show higher biomass estimates in high biomass forests (Douglas fir or Hemlock/Sitka spruce; Fig. 11a) nor in high biomass age classes (>80 years old; Fig. 10b).

Moreover, we see no evidence for topographic impacts on radar-derived maps. If topographic conditions were an issue for side-looking radar, maps S and K might show dampened range in regions of higher topography. But instead, we find that the range of difference values between either the S or K maps and the other maps show

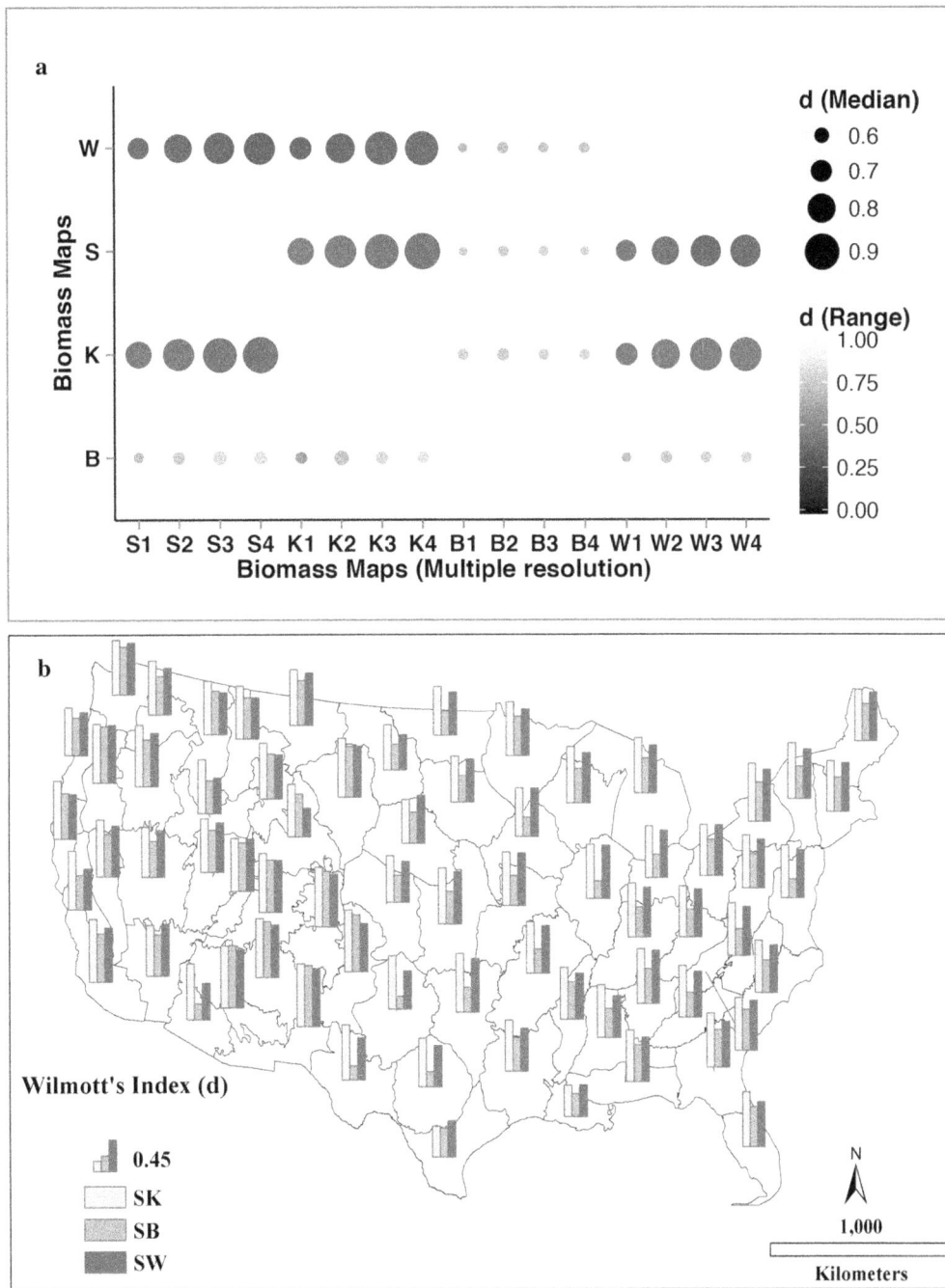

Fig. 7 Pairwise comparison between the four AGB maps **a** multiple resolution comparison at mapping zone scale where X1 = 240 m, X2 = 480, X3 = 720 m, X4 = 960 m; X = AGB map, **b** Comparison at 240 m at pixel level within each mapping zone

no patterns with topographic condition (Fig. 11d). We acknowledge that these maps also began as higher resolution sources, and thus errors of topography may be offset by improvements from resolution.

Extrapolation technique

We see no evidence that extrapolation technique influenced differences among the maps. Again, we focus on the high and low end of the biomass range, and ask

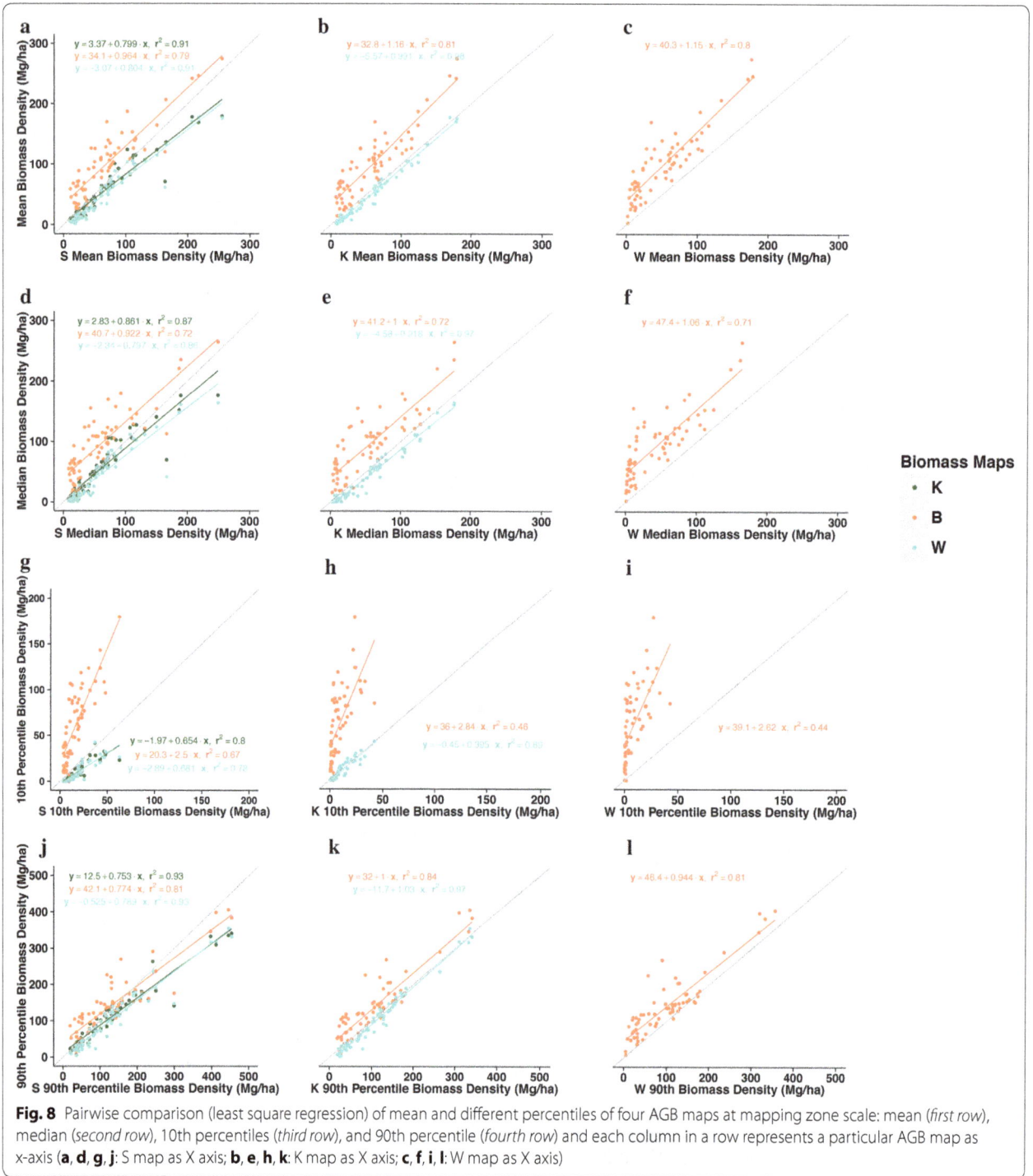

Fig. 8 Pairwise comparison (least square regression) of mean and different percentiles of four AGB maps at mapping zone scale: mean (*first row*), median (*second row*), 10th percentiles (*third row*), and 90th percentile (*fourth row*) and each column in a row represents a particular AGB map as x-axis (**a**, **d**, **g**, **j**: S map as X axis; **b**, **e**, **h**, **k**: K map as X axis; **c**, **f**, **i**, **l**: W map as X axis)

whether maps that use parametric or semi-parametric approaches (maps S and W) are compressed relative to counterparts that use nonparametric approachs (maps B and K). In this case, neither the S nor W map shows compressed ranges in the 10th or 90th percentiles of biomass distributions relative to the other maps (Fig. 8). This

argues that extrapolation technique may not play a large role in determining differences among maps.

Nominal year of map

Because maps differed in nominal year represented, we would expect possible differences due to disturbance and

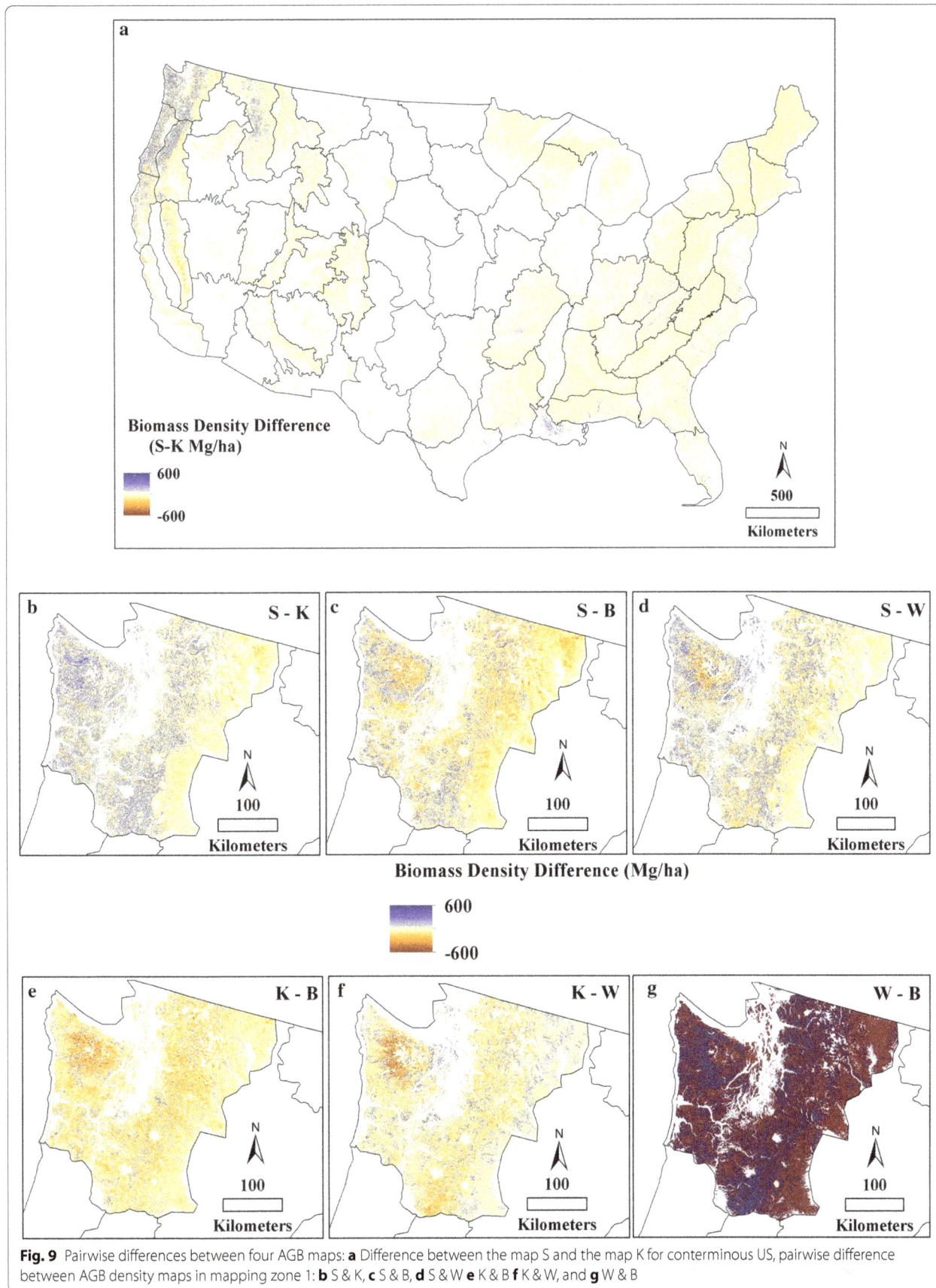

Fig. 9 Pairwise differences between four AGB maps: **a** Difference between the map S and the map K for conterminous US, pairwise difference between AGB density maps in mapping zone 1: **b** S & K, **c** S & B, **d** S & W **e** K & B **f** K & W, and **g** W & B

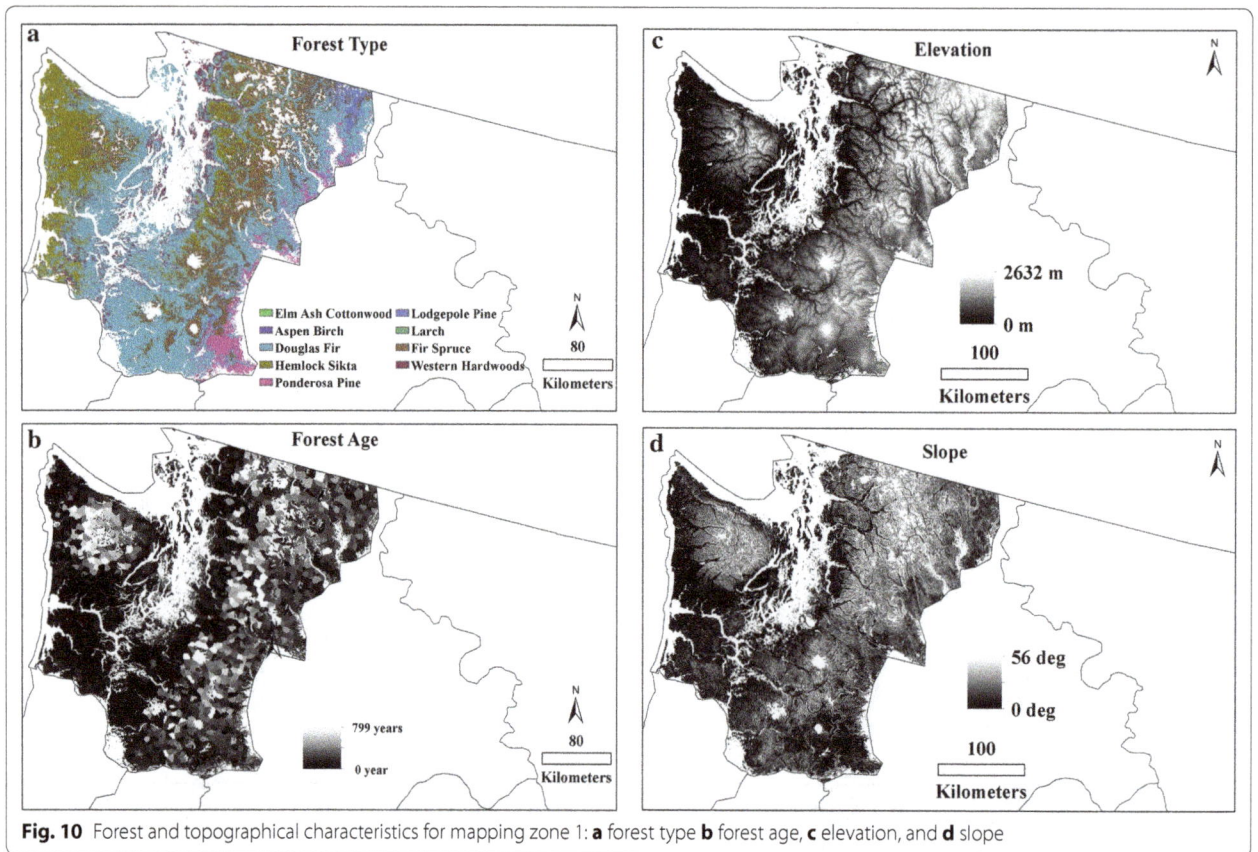

Fig. 10 Forest and topographical characteristics for mapping zone 1: **a** forest type **b** forest age, **c** elevation, and **d** slope

growth of forests between map years. We tested against a regional scale map whose production focused on disturbance and growth, and that had consistent methods across time (Fig. 12). Although weak, evidence suggests that between 40 and 45 % of the total difference between national maps was consistent with growth and disturbance in the regional-scale map (Table 3).

Allometric equations

Allometric equation impacts may have played a role in some of the differences among maps. Median and mean biomass estimates from Map B (using regional scale equations) were consistently higher than all other maps (Fig. 8a–f), and Map W's estimates (using CRM equations) were lower than Map K's. However, Map S did not fit the pattern, showing higher median values than both Map K and W.

Conclusions

There exist several AGB maps generated using different set of input data and methodology. Although these maps describe the same biophysical variable, they vary both in quantity and spatial pattern. We provide a systematic approach to describe and evaluate the differences between maps at multiple scales, while assessing the accuracy of individual mapping effort at an aggregated scale. This study emphasizes that comparison between maps need to be structured with specific hypothesis and tests. Moreover, we suggest that simple directives about which map to use are perhaps overly simplistic. Indeed, the answer for defining various tests depends on the need of the user and scale of comparison and therefore one needs to have set of methodology at different scale with different identified steps. The results from this systematic analysis on comparison of four national level AGB maps suggest that the absolute accuracy and spatial pattern of agreement vary with scale (both spatial resolution and spatial extent). Three (S, K and W) of the four maps largely agreed, and the two maps (K and W) generated with quite different methodological workflows agreed remarkably well. One map generated with high resolution, active sensor data appeared to capture a greater range of data. Finally, the map that did not agree at the pixel scale continued to disagree even with aggregation, indicating aggregation alone may not make maps similar.

When compared with FIA based AGB summed at national scale—which is also basis for national level carbon accounting—all four maps slightly overestimate

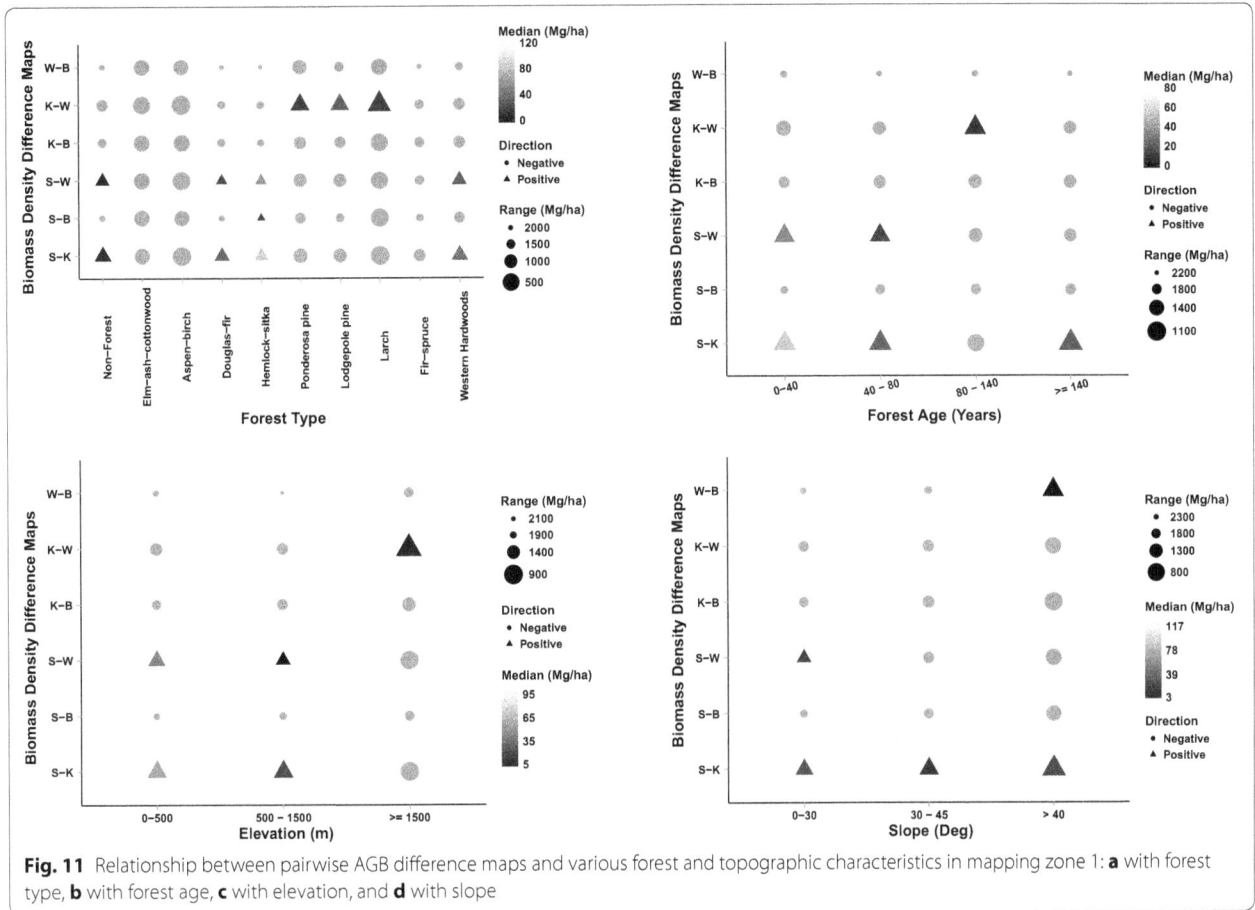

Fig. 11 Relationship between pairwise AGB difference maps and various forest and topographic characteristics in mapping zone 1: **a** with forest type, **b** with forest age, **c** with elevation, and **d** with slope

biomass. However, the total summed AGB for map S is most similar to FIA based estimates, and would therefore argue that the map S could be chosen among the four maps for national level analysis. However, the map W is most similar to FIA based estimates when aggregated to the state scale. The comparison among biomass maps at pixel level (240 m resolution) using R-mode PCA suggests that spatial structure of agreement varies at national and regional scale (spatial extent). Map B disagrees most with respect to the other maps when analyzed at national scale, but it is not always true when analyzed at the mapping region scale. Thus, one needs to look at both regional and broad scale differences before making decision about using one map over others for carbon accounting. The spatial structure of the maps S and K have maximum agreement at national scale PCA analysis, thus they agree well with each other as well as FIA based estimates.

The spatial structure of agreement also varies with the spatial resolution. Except for map B, the agreement among other maps increases as spatial resolution coarsens. Maximum agreement occurs in the northwest (higher biomass) mapping region of the country. The changes in agreement

with aggregation varied between maps, and were not same for the three maps (leaving map B aside).

The structure of the agreement and disagreement among the AGB maps achieved by testing the full suite of potential sources of differences provides evidence in support of causes of differences. For example, all the maps did best in the dominant forest type of a given region, but variability was found in non-dominant forest types. The comparison of maps with regional efforts provided information about how well these maps were able to capture regrowth and disturbance pattern. For example, the comparison of the difference between the maps K and S with regional map suggest that the locations where there is lower biomass in S compared to K is mainly due to disturbance. However, the difference between the two could not capture regrowth well. Thus, sequences of maps capture both actual change and the combined effects of each map's error, and thus argues against using sequences of these maps for spatial monitoring of biomass over time.

Our strategy for testing the full suite of potential error sources shows how remarkably consistent some maps were. Notably, the W and K maps differed in nearly every methodological approach, yet agreed closely at

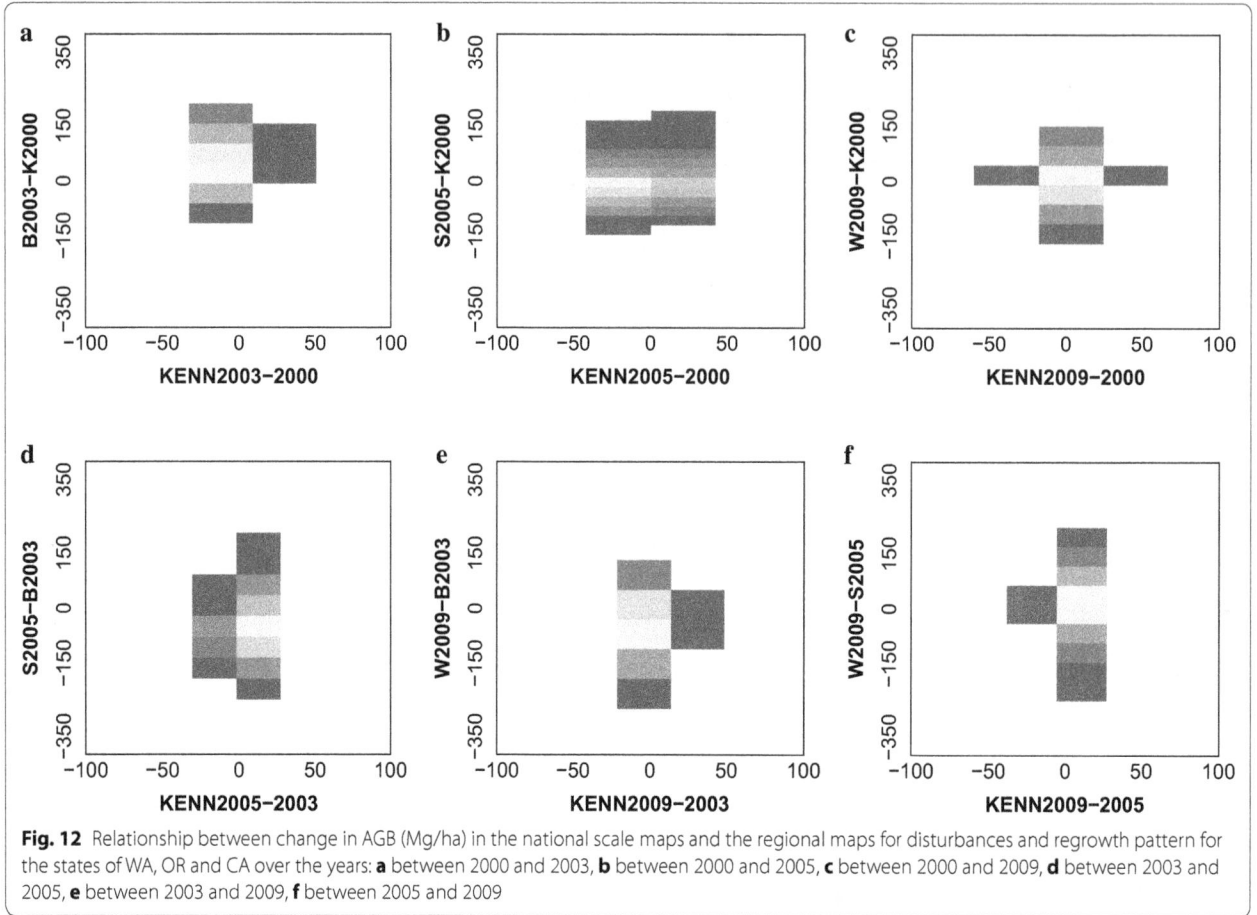

Fig. 12 Relationship between change in AGB (Mg/ha) in the national scale maps and the regional maps for disturbances and regrowth pattern for the states of WA, OR and CA over the years: **a** between 2000 and 2003, **b** between 2000 and 2005, **c** between 2000 and 2009, **d** between 2003 and 2005, **e** between 2003 and 2009, **f** between 2005 and 2009

Table 3 Total biomass estimates in million tonnes in three western states (OR, WA, and CA) over four years: Comparison between national map and regional map based estimates

States	2000 (Mt)	2003 (Mt)	2005 (Mt)	2009 (Mt)
WA	K: 1564.88	B: 2091.94	S: 1812.82	W: 1732.58
	Kenn: 1547.72	Kenn: 1537.92	Kenn: 1526.93	Kenn: 1509.06
OR	K: 1838.67	B: 2564.11	S: 2150.93	W: 1957.12
	Kenn: 1765.63	Kenn: 1734.37	Kenn: 1727.32	Kenn: 1708.72
CA	K: 2106.35	B: 2305.67	S: 1818.43	W: 1974.56
	Kenn: 1815.66	Kenn: 1808.30	Kenn: 1810.52	Kenn: 1798.77

the mapping region scale. The magnitude of W and K matches closely even though the spatial structure differs whereas the spatial structure of S and K agrees (simultaneous comparison using PCA) but magnitude differs (pairwise comparison) when compared at ecoregion scale.

While most of the maps generally agree at broad spatial scales, spatial patterns of disagreement at the local scale are notable. PCA analyses at both national and

mapping region scale clearly show that the disagreement (PC2) has spatial pattern, and the improvement of agreement with spatial aggregation (Fig. 11) corroborates the notion that pixel-level estimates vary considerably among maps. Thus, a user at the regional scale would be advised to evaluate local scale variation among all maps before choosing one to get insight on the sources of differences (error or physical change in the landscape). With the understanding of differences, one can use ensemble approach to have an accurate map of aboveground biomass map.

Thus, this study provides guidance how to approach comparison of multiple maps systematically by designing specific steps for various hypothesis for describing and evaluating the spatial pattern of differences between maps.

Abbreviations
NASA-CMS: National Aeronautics and System Administration Carbon Monitoring System; MSE: mean square error; RMSE: root mean square error; AGB: above ground biomass; FIA: Forest Inventory and Analysis; CRM: Component Ratio Method; PCA: Principal Components Analysis; ORNL: Oak Ridge National Laboratory.

Authors' contributions

The study was designed by NN and RK. NN performed the analysis and produced the figures. NN and RK wrote the manuscript and have equal contribution. Both authors read and approved the final manuscript.

Author details

[1] Department of Natural Resources, TERI University, New Delhi, India. [2] College of Earth, Ocean, Atmospheric Sciences, Oregon State University, Corvallis, OR, USA.

Acknowledgements

This work was part of NASA funded CMS project (Grant number NNX12AP76G, PI-Robert Kennedy) and a USDA NIFA funded project (Grant number 2011-67003-20458, PI-Robert Kennedy). We are thankful to Sassan Saatchi, Sangram Ganguly, Barry Tyler Wilson, Rachael Riemann and Josef Kellndorfer for providing their datasets and clarifying questions regarding the datasets.

Competing interests

The authors declare that they have no competing interests.

References

1. Saatchi SS, Houghton RA, Alvala RCDS, Soares JV, Yu Y. Distribution of aboveground live biomass in the Amazon basin. Glob Change Biol. 2007;13:4.
2. DeFries RS, Hansen MC, Townshend JRG, Janetos AC, Loveland TR. A new global 1-km dataset of percentage tree cover derived from remote sensing. Glob Change Biol. 2000;6:2.
3. Powell SL, Cohen WB, Healey SP, Kennedy RE, Moisen GG, Pierce KB, Ohmann JL. Quantification of live aboveground forest biomass dynamics with landsat time-series and field inventory data: a comparison of empirical modeling approaches. Remote Sens Environ. 2010;114:5.
4. Turner DP, Cohen WB, Kennedy RE. Alternative spatial resolutions and estimation of carbon flux over a managed forest landscape in western Oregon. Landscape Ecol. 2000;15:5.
5. Simard M, Zhang K, Rivera-Monroy VH, Ross MS, Ruiz PL, Castañeda-Moya E, Twilley RR, Rodriguez E. Mapping height and biomass of mangrove forests in Everglades National Park with SRTM elevation data. Photogramm Eng Remote Sens. 2006;72:3.
6. Ni X, Zhou Y, Cao C, Wang X, Shi Y, Park T, Choi S, Myneni RB. Mapping forest canopy height over continental China using multi-source remote sensing data. Remote Sensing. 2015;7:7.
7. Yang T, Wang C, Li GC, Luo SZ, Xi XH, Gao S, Zeng HC. Forest canopy height mapping over China using GLAS and MODIS data. Science China Earth Sciences. 2015;58:1.
8. http://carbon.nasa.gov. Accessed 1 Jan 2016.
9. Zhang G, Ganguly S, Nemani RR, White MA, Milesi C, Hashimoto H, Wang W, et al. Estimation of forest aboveground biomass in California using canopy height and leaf area index estimated from satellite data. Remote Sensing Environ. 2014;151:44.
10. Huang W, Swatantran A, Johnson K, Duncanson L, Tang H, Dunne JO, Hurtt G, Dubayah R. Local discrepancies in continental scale biomass maps: a case study over forested and non-forested landscapes in Maryland, USA. Carbon Balance Manag. 2015;10:1.
11. Liu J, Bowman K, Lee M, Henze D, Bousserez N, Brix H, et al. Carbon monitoring system flux estimation and attribution: impact of ACOS-GOSAT XCO2 sampling on the inference of terrestrial biospheric sources and sinks. Tellus B. 2014;. doi:10.3402/tellusb.v66.22486.
12. Mitchard ET, Saatchi SS, Baccini A, Asner GP, Goetz SJ, Harris NL, Brown S. Uncertainty in the spatial distribution of tropical forest biomass: a comparison of pan-tropical maps. Carbon Balance Manag. 2013;8(1):1.
13. Avitabile V, Herold M, Henry M, Schmullius C. Mapping biomass with remote sensing: a comparison of methods for the case study of Uganda. Carbon Balance Manag. 2011;6(1):1.
14. Huang W, Swatantran A, Johnson K, Duncanson L, Tang H, Dunne JO, Hurtt G, Dubayah R. Local discrepancies in continental scale biomass maps: a case study over forested and non-forested landscapes in Maryland, USA. Carbon Balance Manag. 2015;10(1):1.
15. Riemann R, Wilson BT, Lister A, Parks S. An effective assessment protocol for continuous geospatial datasets of forest characteristics using USFS forest inventory and analysis (FIA) data. Remote Sens Environ. 2010;114:10.
16. Saatchi, S, Yifan Y, Fore A, Neumann M, Chapman B, Nguyen M, et al. CMS US forest biomass. 2005. (Received data through personal communication).
17. Kellndorfer J, Walker W, LaPoint L, Bishop J, Cormier T, Fiske G, Kirsch K, Westfall J. NACP Aboveground Biomass and Carbon Baseline Data (NBCD 2000), USA. 2000. Data Set. http://daac.ornl.gov.
18. Blackard JA, Finco MV, Helmer EH, Holden GR, Hoppus ML, Jacobs DM, Lister AJ, Moisen GG, Nelson MD, Riemann R. Mapping US forest biomass using nationwide forest inventory data and moderate resolution information. Remote Sens Environ. 2008;112:4.
19. Wilson BT, Woodall CW, Griffith DM. Imputing forest carbon stock estimates from inventory plots to a nationally continuous coverage. Carbon Balance Manag. 2013;8:1.
20. Raciti SM, Hutyra LR, Newell JD. Mapping Carbon Storage in Urban Trees with Multi-Source Remote Sensing Data: Relationships between Biomass, Land Use, and Demographics in Boston Neighborhoods. Sci Total Environ. 2014;500:501.
21. Krankina ON, DellaSala DA, Leonard J, Yatskov M. High-biomass forests of the Pacific Northwest: who manages them and how much is protected? Environ Manag. 2014;54:1.
22. Steininger MK. Satellite estimation of tropical secondary forest aboveground biomass: data from Brazil and Bolivia. Int J Remote Sens. 2000;21:6–7.
23. Lu D. The potential and challenge of remote sensing-based biomass estimation. Int J Remote Sens. 2006;27:7.
24. Lu D, Chen Q, Wang G, Liu L, Li G, Moran E. A survey of remote sensing-based aboveground biomass estimation methods in forest ecosystems. Int J Digit Earth. 2016;9(1):63–105.
25. Saatchi S, Marlier M, Chazdon RL, Clark DB, Russell AE. Impact of spatial variability of tropical forest structure on radar estimation of aboveground biomass. Remote Sens Environ. 2011;115(11):2836–49.
26. Heath LS, Hanson MH, Smith JE, Smith WB, Miles PD. Investigation into Calculating Tree Biomass and Carbon in the FIADB Using a Biomass Expansion Factor Approach. In 2008 Forest Inventory and Analysis (FIA) Symposium October 21–23, 2008; Park City, UT. Edited by: McWilliams W, Moisen G, Czaplewski R. USDA Forest Service, Rocky Mountain Research Station. Proc. RMRS-P-56CD. US Department of Agriculture, Forest Service, Rocky Mountain Research Station. Fort Collins, CO, 2009.
27. Zhang X, Kondragunta S. Estimating Forest Biomass in the USA Using Generalized Allometric Models and MODIS Land Products. Geophys Res Lett. 2006;33:9.
28. http://www.fia.fs.fed.us/tools-data/. Accessed on Sep 01, 2012.
29. Willmott CJ, Ackleson SG, Davis RE, Feddema JJ, Klink KM, Legates, O'donnell J, Rowe CM. Statistics for the evaluation and comparison of models (1978–2012). J Geophys Res Oceans. 1985;90:C5.
30. Ji L, Gallo K. An agreement coefficient for image comparison. Photogramm Eng Remote Sens. 2006;72:7.
31. Neeti N, Eastman JR. Characterizing implications of two-dimensional space–time orientations for principal component analysis of geographic time series. Int J Remote Sens. 2015;36:1.
32. Cattell RB. The data box: its ordering of total resources in terms of possible relational systems. 67–128. RB Cattell. In: Handbook of Multivariate Experimental Psychology, 1966.
33. Team, R Core. R: A language and environment for statistical computing. Vienna, Austria. 2014. http://www.R-Project.Org.
34. Homer CG, Gallant, A. Partitioning the conterminous United States into mapping zones for Landsat TM land cover mapping. 2001, USGS Draft White Paper available at http://landcover.usgs.gov.
35. Ruefenacht B, Finco MV, Nelson MD, Czaplewski R, Helmer EH, Blackard JA, Holden GR, Lister AJ, Salajanu D, Weyermann D. Conterminous US and alaska forest type mapping using forest inventory and analysis data. Photogramm Eng Remote Sens. 2008;74:11.

36. Pan Y, Jing MC, Birdsey R, McCullough K, He L, Deng F. Age structure and disturbance legacy of North American forests. Biogeosciences. 2011;8:3.

37. Kennedy RE, Ohmann JL, Gregory MJ, Roberts HM, Yang Z, Cohen W, et al. In: Prep an empirical, integrated forest carbon monitoring system. http://geotrendr.ceoas.oregonstate.edu/data/

38. Fernandez-Ordonez Y, Jesus S-R, Leblon B. Forest Inventory using Optical and Radar Remote Sensing. In: Advances in Geoscience and Remote Sensing. InTech. 2009. doi:10.5772/8330

39. Bayer T, Winter R, Schreier G. Terrain influences in SAR backscatter and attempts to their correction. IEEE Trans Geosci Remote Sens. 1991;29:3.

40. Cohen WB, Maiersperger TK, Gower ST, Turner DP. An improved strategy for regression of biophysical variables and landsat ETM+ data. Remote Sens Environ. 2003;84:4.

12

Carbon uptake by mature Amazon forests has mitigated Amazon nations' carbon emissions

Oliver L. Phillips[*][†] ⓘ, Roel J. W. Brienen[†] and the RAINFOR collaboration

Abstract

Background: Several independent lines of evidence suggest that Amazon forests have provided a significant carbon sink service, and also that the Amazon carbon sink in intact, mature forests may now be threatened as a result of different processes. There has however been no work done to quantify non-land-use-change forest carbon fluxes on a national basis within Amazonia, or to place these national fluxes and their possible changes in the context of the major anthropogenic carbon fluxes in the region. Here we present a first attempt to interpret results from ground-based monitoring of mature forest carbon fluxes in a biogeographically, politically, and temporally differentiated way. Specifically, using results from a large long-term network of forest plots, we estimate the Amazon biomass carbon balance over the last three decades for the different regions and nine nations of Amazonia, and evaluate the magnitude and trajectory of these differentiated balances in relation to major national anthropogenic carbon emissions.

Results: The sink of carbon into mature forests has been remarkably geographically ubiquitous across Amazonia, being substantial and persistent in each of the five biogeographic regions within Amazonia. Between 1980 and 2010, it has more than mitigated the fossil fuel emissions of every single national economy, except that of Venezuela. For most nations (Bolivia, Colombia, Ecuador, French Guiana, Guyana, Peru, Suriname) the sink has probably additionally mitigated all anthropogenic carbon emissions due to Amazon deforestation and other land use change. While the sink has weakened in some regions since 2000, our analysis suggests that Amazon nations which are able to conserve large areas of natural and semi-natural landscape still contribute globally-significant carbon sequestration.

Conclusions: Mature forests across all of Amazonia have contributed significantly to mitigating climate change for decades. Yet Amazon nations have not directly benefited from providing this global scale ecosystem service. We suggest that better monitoring and reporting of the carbon fluxes within mature forests, and understanding the drivers of changes in their balance, must become national, as well as international, priorities.

Keywords: Amazonia, Carbon balance, Carbon sink, Sequestration, Land use change, Climate change, Tropical forests, Ecosystem service

Background

Biospheric processes of carbon exchange exert significant control on the evolution of the atmospheric carbon dioxide burden, and hence on the rate of global climate change itself. Over recent decades on average less than half of anthropogenic carbon dioxide emissions have accumulated in the atmosphere, with the balance apportioned to large sinks of the order of ca. 2.5 Pg C yr^{-1} each in the oceans and on land [e.g., 7, 13]. Nevertheless, the terrestrial sink and the terrestrial fluxes, apart from those due to fossil fuel emissions, remain poorly constrained and are often computed simply as the residual of the better quantified fluxes into the ocean and those due to direct anthropogenic processes [e.g., 7]. The terrestrial sink also exhibits substantial inter-annual variation, which is largely driven by variations in temperature and

*Correspondence: o.phillips@leeds.ac.uk
[†]Oliver L. Phillips and Roel J. W. Brienen have contributed equally to this work
School of Geography, University of Leeds, Leeds LS2 9JT, UK

moisture particularly in the tropics [e.g., 56, 57]. Both the large long-term terrestrial sink and its strong inter-annual variation indicate potentially critical roles for the planet's most productive terrestrial ecosystems to modify and respond to anthropogenic climate change.

As the world's largest tropical forest by extent the Amazon is a leading candidate for influencing the long-term terrestrial carbon balance and fluxes, their inter-annual fluctuations, and any trend in the terrestrial sink. Its remoteness challenges attempts to map and monitor its carbon function but several lines of measurement evidence illustrate the significance and climate sensitivity of its carbon fluxes. For example, eddy covariance measurements of canopy gas exchange suggest that the landscape-scale carbon balance of natural Amazon forests is seldom in balance on sub-annual timescales [e.g., 49], while aircraft measurements of atmospheric carbon dioxide concentrations and inverse modelling of the trajectory of air parcels reveal strong inter-annual differences and drought sensitivity at the basin-scale [24]. Satellite-based assessment of deforestation and fire confirm large but spatially and temporally very variable emissions from the loss of biomass [e.g., 5, 30, 53].

Permanent plots in which the lives of individual trees are tracked are a key technology for investigating the biomass fluxes and net balance of forests worldwide [e.g., 40]. On a per-unit-area basis, the net fluxes within mature forests are expected to be much smaller than these from deforestation, degradation, and regrowth processes, but such small changes in mature forests may nevertheless scale to large values when integrating over bigger regions. While efforts to track the behaviour of Amazon forests on the ground are sparser than in most temperate regions, the total on-the-ground monitoring effort has nevertheless increased several-fold since the early 1980's, to encompass more than 300 plots by the 2000's using standardized protocols. By the late 1990's this long-term network was already suggesting that mature Amazon forests were not in balance [43]. The expanding measurement base has continued to support the inference of a large, long-term carbon sink into forest biomass, also showing that while the sink results from productivity exceeding mortality, both the rate of growth and the rate of death have tended to increase [e.g., 35], and that the sink extends beyond Amazonia to other tropical forests [e.g., 36]. Most recently, Amazon tree growth rates have stalled, but tree mortality has continued to accelerate, so that the net balance of the two—the biomass carbon sink—has declined [10]. The reasons for this continued increase in mortality remain uncertain. It has been proposed that faster growth may lead to faster tree death [e.g., 11, 42], while evidence also suggests that recent intense droughts in parts of Amazonia are directly

responsible for killing enough trees to shut down the biomass sink for periods of a year or more [e.g., 51], via mechanisms such as carbon starvation or hydraulic failure [15, 50]. The ground data from the Amazon RAINFOR network are also consistent with atmospheric GHG profiles [24] in showing both the sensitivity of the carbon balance of intact Amazon forests to drought in 2005 and 2010, and the continued net sink of hundreds of millions of tons in non-drought years [20, 41].

In sum, observations indicate that the remaining old-growth forests in Amazonia have contributed a large net biomass sink from the atmosphere to the land, albeit one that appears to be in decline as a result of different processes. There has however remarkably been no effort to quantify such net fluxes on a regional or national basis within Amazonia, or to place them and their possible changes directly in the context of major anthropogenic carbon fluxes in the region. Addressing this major gap is important for at least three reasons. First, historically, if Amazonia has provided a large environmental service to the global climate, then the net carbon emissions of the Amazon nations—Brazil, Bolivia, Colombia, Ecuador, French Guyana, Guyana, Peru, Suriname, Venezuela—may be greatly over-estimated. Typically, national and international assessments simply omit the behaviour of intact forest ecosystems for example while Brazil's reporting to the UNFCCC includes gross deforestation for all land, carbon removal from the atmosphere is only estimated for managed lands. Second, the renewed emphasis on national reporting of all carbon fluxes following the Paris 2015 climate agreement means that it may well be advantageous for tropical forest nations to examine the behaviour of their old-growth forests extremely carefully. And third, while world leaders have set an ambition of limiting global temperature rise to 1.5 °C above pre-industrial levels, in practice this may only be accomplishable if the biosphere cooperates and provides large net sinks into natural and managed ecosystems worldwide.

Here, we aim to interpret the latest RAINFOR findings in a much more biogeographically, politically, and temporally differentiated way. Our specific objectives are to:

1. Provide a biogeographically differentiated (i.e., region-by-region) assessment of the Amazon forest carbon sink over the last three decades;
2. Provide a politically differentiated (i.e., country-by-country) assessment of the carbon sink over the last three decades.
3. Evaluate the magnitude and trajectory in relation to national anthropogenic carbon emissions (fossil fuels and deforestation) and in relation to estimated land-use related fluxes within Amazonia.

This is the first attempt to evaluate the results on natural forest dynamics from the RAINFOR network in the context of national estimates of fossil-fuel emissions and of land-use change disturbance. The data sources for each of these processes differ greatly. While large anthropogenic and natural disturbance processes are best detected and quantified via remote-sensing methods [e.g., 14, 16], in equatorial forests natural large disturbances and subsequent recovery do not appear to substantially impact large-scale long-term biomass dynamics [17, 26]. Detecting the small changes within mature forests instead typically requires direct tree-by-tree measurements to track the identity, growth, and death of individual trees. Based on such an approach, our analysis here seeks to provide an assessment of the net ("natural") fluxes as measured in plots to the climate change research and policy communities, by biogeographic and political unit. We thus reanalyse the most up-to-date pan-Amazon dataset of biomass dynamics [10], decade-by-decade, and at the level of biogeographical region and nation state, and compare these fluxes with independent estimates of carbon fluxes from land use change and fossil fuel combustion.

Methods: summary

Here we summarize our overall approach. Later, in the Additional file 'Detailed Materials and Methods', we describe the methodological process in more detail.

We use the plot-by-plot and census-by-census data which were recently analysed to derive overall, Amazon-wide fluxes and trends [10]. These data represent the efforts of more than 100 collaborators in the RAINFOR network (Amazon Forest Inventory Network), using 309 long-term plots in 71 distinct sites across mature Amazon forests. Spatially, we limit our analysis here to the hydrographic Amazon basin plus the contiguous moist forests of the Guiana Shield, so we exclude 11 extra-Amazonian plots in northwest South America presented in Brienen et al. [10]. Temporally, we analyse for three successive decades, the 1980's, 1990's, and 2000's.

We analyse the behaviour of these structurally mature, "old-growth" forest sites in three ways, reporting always our estimates of the net biomass carbon balance together with its estimated uncertainty derived from these plot measurements. We thus estimate the net sink firstly by time across the Amazon, then by biogeographical region across the Amazon, and finally by nation (and by time) across the Amazon. For all the time-differentiated analyses, for simplicity we break down the results by decadal units. For the biogeographically-differentiated analyses we follow a recent approach [19] that divided the lowland tropical forests of South America into five different regions based on biogeographic and biogeochemical evidence to take account of known major

ecosystem discontinuities within the region (see Additional file 1: Fig. S1). For our national analyses, we used the biogeographically-based estimates of mean and uncertainty of carbon balance in each region to estimate the area-weighted mean mature Amazon forest carbon balance per country, based on the area of forest represented in each biogeographical region in each nation.

For all these analyses we rely on best estimates of mature forest area as mapped for each country for the year 2000 in the Global Land Cover product [8]. These values were projected forward in time to 2011 and back in time to 1980, by deriving estimates of annualized change rates in Amazon forest area for each country from available sources (see "Methods" section). GLC 2000 land cover class area uncertainties are not available for South American countries, but to provide an alternative and very conservative lower bound to the sink estimates, we repeated all the above analyses using the 'intact forest landscape' (IFL) product [45], which, excluding all landscapes which may have direct human impacts, defines IFLs as unbroken expanses of natural ecosystems within areas of current forest extent, without signs of significant human activity, and having an area of at least 500 km^2 [45].

To compare with the fossil fuel emissions we use a global compilation of national data reported by CDIAC [9]. To estimate deforestation-related emissions, a number of alternative sources are available but no single source provides year-by-year estimates of deforestation-based carbon emissions for all Amazon countries throughout. We therefore developed a hybrid approach, described in "Methods" section, identifying preferred sources based primarily on satellite-based analyses with explicit methodologies [e.g., 46, 53] over nationally compiled statistics [e.g., 18], where possible accounting for estimated non-uniform density of carbon in forests across the Amazon. We also explored an alternative source [25] to assess whether the deforestation estimate we used was likely to be conservative or not, for the period and location for which a direct comparison of estates is possible (2001–2010 Amazon forests).

Finally, for other land-use changes—including fragmentation and edge effects, logging, fire, secondary re-growth and subsequent disturbance—information is much less systematically available through time and across nations, and measurement uncertainties are greater. Given the measurement difficulties and the uneven coverage of available estimates we do not attempt to derive time trends in these processes, and we make a number of necessarily simplifying assumptions (see "Methods" section). Where appropriate we add independent uncertainties in quadrature [e.g., 3], and use a conversion factor of 0.47 to derive the carbon content in tropical biomass [1].

Results

Across the Amazon basin there has been a significant, sustained, but declining net carbon sink into mature forest biomass (Fig. 1). Decade-by-decade, Amazon forests gained biomass at a similar rate during the 1980's and the 1990's, at about 500 Tg C per year, although the better sampling in the 1990's results in much greater confidence in the magnitude of the sink during the 1990's than the 1980's (see error bars in Fig. 1). The sink slowed by more than a third during the first decade of the twenty-first century, to ca. 300 Tg C per year. This decline has been caused principally by a weakening of the sink on a per-hectare basis, and less so by the decline in forest area per se. Thus, the net gain in carbon in above-ground forest biomass declined more than 30%, from 0.37 Mg C ha^{-1} yr^{-1} in the 1980's and 1990's, to 0.24 Mg C ha^{-1} yr^{-1} in the 2000's, while total forest area declined less than 10% from an estimated 639×10^6 ha in 1985 to 590×10^6 ha by 2005.

The sink has been widely distributed and not driven by forests in one particular region (see Additional file 1: Table S1a). When divided into five regions based on large-scale geographic and biogeochemical divisions, individual plots in all five regions (Brazilian Shield, Guiana Shield, Upper Amazonia, South-West Amazonia, and East-Central Amazonia) have gained significantly. Among regions the long-term mean estimated gain varied relatively little, from a low of 58 Tg C per year in East-Central Amazonia, to a high of 123 Tg C per year in Brazilian shield (Additional file 1: Table S1b).

When results are broken down into biogeographic regions (see Additional file 1: Fig S1) and decade, the smaller sample sizes available imply reduced confidence in each individual combination of region by time period. Nevertheless, for each of the five regions in each of the three decades (i.e., for all 15 possible space–time combinations) the estimated mean rate of biomass change has been positive (Additional file 1: Table S1b). In 11 of these 15 possible combinations the lower confidence interval was also greater than zero, including for each of the five regions during the 1990's. While the results show how widespread and persistent the sink has been, the overall decline during the latest decade was not recorded everywhere. Rather, the decline has been sharp in Southwest Amazonia and the Brazilian Shield while in other regions it is not evident.

Over the whole period, the ground measurements suggest that for each of the nine Amazon nations that mature Amazon forests have provided a net carbon sink, ranging from 4 Tg C per year in the smallest country (French Guiana) to 243 Tg C per year in the largest (Brazil) (see Additional file 1: Table S2). The estimated Amazon-wide forest biomass carbon sink between 1980 and 2010 (430, [213, 669] Tg C yr^{-1}) has greatly exceeded the combined emissions from fossil fuel combustion (149 [131, 167] Tg C yr^{-1}) for the nine Amazon nations (Fig. 2). This holds also on a national basis for every country except Venezuela. Since the turn of the millennium, the carbon sink has declined while fossil fuel emissions have increased in most South American nations, but the

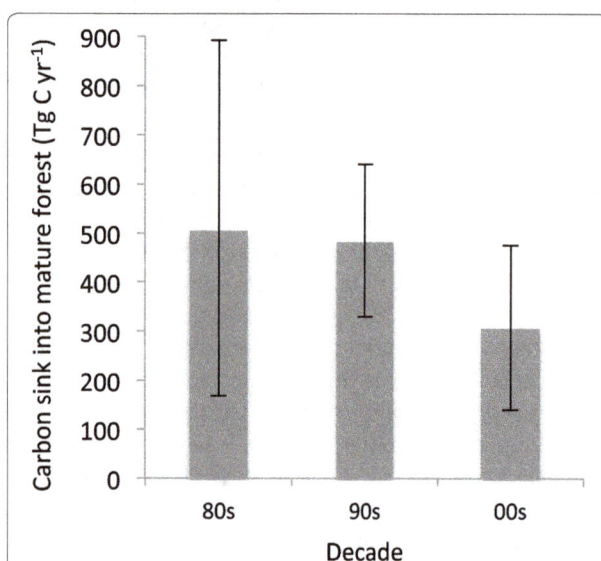

Fig. 1 Estimated carbon sink into mature forest biomass in the Amazon basin for each of the three decades since 1980. *Error bars* show 95% confidence intervals

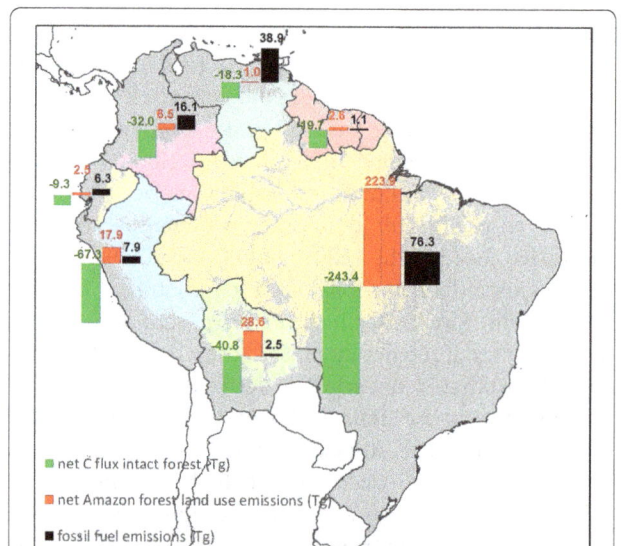

Fig. 2 Estimated Amazon carbon fluxes 1980–2010. For each nation three fluxes are represented: the net C flux mature forests (*green* and negative), the net fluxes from deforestation, i.e., losses from deforestation and degradation minus gains from regrowth (*red* and positive), and fossil fuel emissions (*black* and positive). Units are in Tg carbon per year (=10^{12} g C yr^{-1})

former is still likely to have exceeded the latter (306 (140, 476) vs. 180 Tg C (167, 193)).

As well as fossil fuel combustion, land-use changes in Amazonia have been substantial sources of carbon to the atmosphere. The 1980–2010 combined estimated flux from fossil fuel combustion, and Amazon deforestation, degradation, and fragmentation averaged 431 (326, 538) Tg C, a value which has been remarkably steady, composed of a generally declining land-use component and a generally increasing fossil field component (Table 1). Overall across the three decades, the mature forest sink has approximately mitigated these sources. Note that we estimate a net flux of just 1 Tg, remarkably close to zero, but in the latest decade the combined sources exceeded the mature forest sink for the first time in the record. Alternatively, if we assume conservatively that only the 'intact forest landscapes' have contributed sinks and that other mature forests were carbon–neutral, we estimate a somewhat smaller total, with the intact forest sink declining from 342 Tg C in the 1980s to 236 Tg C in the 2000s (Additional file 1: Table S3). Even under this conservative scenario, the forest sink considerably outweighs the fossil fuel emissions of the Amazon nations (Additional file 1: Fig. S2). Finally, for the directly comparable period (2001–2010), Amazon deforestation emissions as estimated from the online Global Forest Watch source average a total of 161 Tg C per year, while our PRODES-based estimate suggests total emissions of 201 Tg C per year in this decade.

Discussion

This is the first attempt to estimate the ecosystem service of carbon sequestration in mature forests in Amazonia on a long-term regional and national basis. The results suggest that, at least since 1980, the average annual carbon sink into mature forests of the Amazon nations has been at least twice the magnitude of carbon emissions from the same nations' burning of fossil fuels. Moreover, for every country except for Venezuela the net carbon uptake into mature Amazon forests has exceeded Amazon nations' total fossil fuel emissions. For most nations

the uptake has also exceeded the combined emissions due to fossil fuels and Amazon deforestation, degradation, and fragmentation. Despite lack of knowledge on forest area uncertainties the only comparison with an independent product for land cover class indicates that the GLC product is conservative for forest area in Colombia [22], suggesting that our mature forest sink estimate may be conservative. Further, since our PRODES-based deforestation-related carbon emission estimate exceeded by one fifth a comparable estimate derived from Global Forest Watch, it is possible our anthropogenic CO_2 emissions estimation methodology may over-estimate the deforestation source, further supporting the conclusion that natural forest sinks in Amazon have compensated for anthropogenic emissions. The mature forest sink is about 30% smaller if we alternatively assume that the only sinks are located in unbroken and expanses of natural forests of at least 500 km^2 ('intact forest landscapes'). This represents an extremely conservative and unlikely scenario. In fact, at least half the mature forest plots assembled are located outside these IFLs, including the longest-monitored plots (in Venezuela) and clusters with large, net sinks in Colombia, Ecuador, Peru, and Venezuela.

Thus, not only are the stocks of carbon in Amazon forests very large (exceeding 100 Pg in above- and below-ground biomass, e.g., [40]), but Amazon nations have also contributed to mitigating climate change via net carbon sequestration. The strength of this ecosystem service and its spatial and temporal pattern have implications both for understanding its possible ecological drivers, and for the effective management and conservation of tropical forests in the era of anthropogenic climate change. We first discuss the ecological implications, before addressing the wider implications.

Our analysis shows that the net sink for atmospheric carbon into mature Amazon forests has been an ecologically and geographically ubiquitous pattern. Thus, in all five regions defined a priori on biogeographic and biogeochemical criteria, the sink has been sustained for decades. The ecology and physical geography of these regions differ greatly. For example, while the forests in

Table 1 Net C fluxes for the Amazon basin 1980–2009.9, displayed decade by decade

Period	Mature forest Sink	Land use change	Fossil fuel emissions	Net flux
1980–1989.9	−504.4	317.9	105.2	−81.3
1990–1999.9	−482.1	271.7	139.5	−70.8
2000–2009.9	−305.9	275.4	180.0	149.5
1980–2009.9	−430.8	282.9	149.0	1.1

Fluxes are divided into carbon uptake by mature forests, the fossil fuel emissions, fluxes due to land use change and the resulting net flux. Land use change fluxes include emissions resulting from deforestation and forest degradation, and estimate for regrowth. Negative signs indicate removal of carbon from atmosphere, and positive signs indicate net C fluxes from land to the atmosphere. Units are in Tg carbon per year (=10^{12} g C yr^{-1})

Southern and Southwestern Amazonia have similar rates of wood productivity as those in the Guiana Shield [34], they typically contain just half the biomass [39], have almost completely different species and phylogenetic composition [32, 55] and greater diversity [54]. Trees in the south and southwest also die at twice the rate of those in the north-east [38], due largely to the strongly divergent geomorphology and soil physical and nutritional conditions [47, 48]. The consistency and long-term persistence of a carbon sink across such different forests indicates that the main driving mechanism is also ubiquitous and long-term. Our findings that the Amazon sink has been geographically widespread and persistent are also consistent with the larger tropical and global picture. Thus, there is compelling evidence from several measurement streams to show that the terrestrial ecosystem sink is persistent and large [e.g., 7, 37] and that most of this has been into forests including in the tropics [e.g., 40]. Together with basic expectations from theory and observations about the ecophysiological impact of increasing atmospheric CO_2 [e.g., 21], this spatial and temporal persistence implies that stimulation of tree growth by increasing carbon dioxide is at least partly responsible (cf. [52]. The fact that the sink has recently weakened only in the south and south-west, which are *also* the only regions which have experienced an increase in dry season intensity [29], is also instructive. This suggests that the recent Amazon droughts have exerted large-scale but not basin-wide influence. And, so far at least, while these droughts have reversed the carbon sink during individual drought years such as 2005 and 2010 [20, 24, 41], they have not yet done so on a sustained basis.

The findings also have several implications for Amazon forest management and policy. First and most obviously, from a historical perspective, if all of Amazonia has provided a carbon sink environmental service to the global climate, then it follows that the net carbon emissions of the Amazon nations—Brazil, Bolivia, Colombia, Ecuador, French Guyana, Guyana, Peru, Suriname and Venezuela—must have been seriously over-estimated in all assessments that omit to consider the carbon balance of mature forest ecosystems. While many northern countries include the carbon balance of their intact forest lands (which also tend to be a net sink e.g., [40]) in their reporting to the UNFCCC, Amazon countries have simply excluded carbon dynamics in old growth forests in their reporting.

Second, while there is rightly increasing emphasis on managing secondary forests for their carbon sink potential [e.g., 12] our results suggest that at a national level tropical secondary forests may not in fact provide the largest forest sinks. While potential maximum rates of carbon sequestration per unit area are high in secondary

forests for several decades following clearance [44], in Amazonia landscapes characterized by a mosaic of cropland, degraded and secondary forests are also at greatly enhanced risk of fire [5] or other degradation and deforestation processes [e.g., 2]. This, together with the large area that remains of structurally mature forest in Amazonia, means that the total carbon sequestration provided by mature forests has almost certainly been much greater than the net sequestration from secondary systems. Whether it continues to be so or not is of course unknown, but our estimates here are that in the decade since 2000 mature Amazon forests contributed 306 (140, 476) Tg C every year, while secondary forest recovery contributed 60 (34, 84) Mg C. The latter estimate is less than 30% of the potential estimated total annual sink for secondary forests in the neotropics if all were left to regrow (ca. 8 Pg over 40 years, [12]), but it is based on one high-resolution analysis [6]. Clearly, a research priority for the future must be to better understand the dynamics of forest carbon emissions in landscapes undergoing rapid land use change, including fragmentation, regrowth, deforestation, and degradation processes.

Third, and consequent on both points above, it remains feasible that in most Amazon nations the land remaining as forest can still provide net carbon sinks well into the future. Via a combination of protection of old-growth forests and some enhanced secondary forest recovery, the potential carbon sequestration benefits of Amazonia for mitigating climate change are strong. The extent to which these climate services are actually realised depends on many factors. While only some of these lie within the control of Amazon nations themselves, the protection of old-growth forests is a matter of national policy. The increased emphasis on national reporting of carbon fluxes following the Paris 2015 climate agreement means that tropical forest nations which protect remaining mature forests and carefully monitor and report the behaviour and subtle changes occurring within them may stand to benefit materially.

Conclusions

Results from standardised, ground-based monitoring of the growth and death of individual trees have been used to build a picture of the behaviour of mature forests across the Amazon basin since the 1980's. The picture that emerges is one of forests far from equilibrium, with both growth and mortality rates having risen and with a persistent and geographically very widespread difference between the two that implies a carbon sink into mature forests across the whole region. The net sink has substantially affected the long-term carbon budgets of all nine Amazon nations, exceeding the fossil fuel emissions in eight of them. While fossil fuel emissions have been increasing and the sink has recently weakened in some parts of the basin, mature forests in all nine nations continued to contribute substantial net sequestration of carbon over the most recent decade. Overall, in most tropical countries emissions and removals by forests dominate national net C flux profiles. If these developing countries are to contribute to global climate change mitigation, it is forests that will need to be managed to both increase removals and reduce emissions.

Whether or not Amazon nations will in turn benefit from this global ecosystem service in coming years is unclear. To achieve such benefits requires a better understanding of how carbon dioxide, climate and other 'indirect' anthropogenic factors are actually affecting old-growth forests. This in turn requires a significant increase in the level of investment in tropical forest monitoring, combining ground-based and remotely-sensing techniques, especially so in protected areas. At both national and global levels, a step-change in the magnitude and coordination of such work is needed in order to track the behaviour of these uniquely valuable ecosystems.

Abbreviations
GLC2000: Global Land Cover 2000; IFLs: intact forest landscapes; GHGs: greenhouse gases; PRODES: Projeto Prodes Monitoramento da Floresta Amazônica Brasileira por Satélite; RAINFOR: Amazon Forest Inventory Network (Red Amazónica de Inventarios Forestales).

Authors' contributions
OLP and RJWB designed the study, carried out the data analysis, and wrote the paper, using data generated by the RAINFOR consortium. Both authors read and approved the final manuscript.

Acknowledgements
We thank Georgia Pickavance for GIS analysis of IFL and GLC2000 datasets, and Ted Feldpausch for major contributions in the field. This paper has been possible by many other contributors to the RAINFOR collaboration who have individually and collectively supported the task of monitoring Amazon forests in many important ways. For this substantial effort we are grateful to E. Gloor, T. R. Baker, J. Lloyd, G. Lopez-Gonzalez, A. Monteagudo-Mendoza, Y. Malhi, S. L. Lewis, R. Vásquez Martinez, M. Alexiades, E. Álvarez Dávila, P. Alvarez-Loayza, A. Andrade, L. E. O. C. Aragão, A. Araujo-Murakami, E. J. M. M. Arets, L. Arroyo, G. A. Aymard C., O. S. Bánki, C. Baraloto, J. Barroso, D. Bonal, R. G. A. Boot, J. L. C. Camargo, C. V. Castilho, V. Chama, K. J. Chao, J. Chave, J. A. Comiskey, F. Cornejo Valverde, L. da Costa, E. A. de Oliveira, A. Di Fiore, T. L. Erwin, S. Fauset, M. Forsthofer, D. R. Galbraith, E. S. Grahame, N. Groot, B. Hérault, N. Higuchi, E. N. Honorio Coronado, H. Keeling, T. J. Killeen, W. F. Laurance, S. Laurance, J. Licona, W. E. Magnusson, B. S. Marimon, B. H. Marimon-Junior, C. Mendoza, D. A. Neill, E. M. Nogueira, P. Núñez, N. C. Pallqui Camacho, A. Parada, G. Pardo-Molina, J. Peacock, M. Peña-Claros, G. C. Pickavance, N. C. A. Pitman, L. Poorter, A. Prieto, C. A. Quesada, F. Ramírez, H. Ramírez-Angulo, Z. Restrepo, A. Roopsind, A. Rudas, R. P. Salomão, M. Schwarz, N. Silva, J. E. Silva-Espejo, M. Silveira, J. Stropp, J. Talbot, H. ter Steege, J. Teran-Aguilar, J. Terborgh, R. Thomas-Caesar, M. Toledo, M. Torello-Raventos, R. K. Umetsu, G. M. F. van der Heijden, P. van der Hout, I. C. Guimarães Vieira, S. A. Vieira, E. Vilanova, V. A. Vos, & R. J. Zagt, A. Alarcon, I. Amaral, P. P. Barbosa Camargo, I. F. Brown, L. Blanc, B. Burban, N. Cardozo, J. Engel, M. A. de Freitas, A. de Oliveira, T. S. Fredericksen, L. Ferreira, N. T. Hinojosa, E. Jimenez, E. Lenza, C. Mendoza, I. Mendoza Polo, A. Peña Cruz, M. C. Peñuela, P. Petronelli, J. Singh, P. Maquirino, J. Serano, A. Sota, C. Oliveira dos

Santos, J. Ybarnegaray and J. Ricardo. We thank the editor and two anonymous reviewers for their helpful suggestions.

Competing interests
Both authors declare that they have no competing interests.

Funding
Funding for the work reported here came principally from the Gordon and Betty Moore Foundation and the UK Natural Environment Research Council (Grants NE/B503384/1, NE/D01025X/1, NE/I02982X/1, NE/F005806/1, NE/D005590/1 and NE/I028122/1), and the EU Seventh Framework Programme (GEOCARBON-283080). O.P. is supported by an ERC Advanced Grant (T-FORCES) and is a Royal Society-Wolfson Research Merit Award holder. The study design, data collection, analysis, interpretation of data, and writing were done by the authors independently of the sources of funding.

Use of plants
The research involves working with plants in long-term permanent plots across Amazonia. All local, national or international guidelines and legislation has been followed, including where appropriate obtaining any required permissions for the fieldwork.

References
1. Aalde H, Gonzalez P, Gytarsky M, Krug T, Kurz WA, Ogle S, Raison J, Schoene D, Ravindranath NH, Elhassan NG, Heath LS, Higuchi N, Kainja S, Matsumoto M, Sanz Sánchez MJ, Somogyi Z. IPCC guidelines for national greenhouse gas inventories, chapter 4: forest land.
2. Alves DS, Escada MI, Pereira JL, De Albuquerque Linhares C. Land use intensification and abandonment in Rondônia, Brazilian Amazônia. Int J Remote Sens. 2003;24(4):899–903.
3. Aragão LE, Malhi Y, Metcalfe DB, Silva-Espejo JE, Jiménez E, Navarrete D, Almeida S, Costa AC, Salinas N, Phillips OL, Anderson LO. Above-and below-ground net primary productivity across ten Amazonian forests on contrasting soils. Biogeosciences. 2009;6(12):2759–78.
4. Aragao LE, Poulter B, Barlow JB, Anderson LO, Malhi Y, Saatchi S, Phillips OL, Gloor E. Environmental change and the carbon balance of Amazonian forests. Biol Rev. 2014;89(4):913–31.
5. Aragão LE, Shimabukuro YE. The incidence of fire in Amazonian forests with implications for REDD. Science. 2010;328(5983):1275–8.
6. Asner GP, Powell GV, Mascaro J, Knapp DE, Clark JK, Jacobson J, Kennedy-Bowdoin T, Balaji A, Paez-Acosta G, Victoria E, Secada L. High-resolution forest carbon stocks and emissions in the Amazon. Proc Natl Acad Sci. 2010;107(38):16738–42.
7. Ballantyne AP, Alden CB, Miller JB, Tans PP, White JW. Increase in observed net carbon dioxide uptake by land and oceans during the past 50 years. Nature. 2012;488(7409):70–2.
8. Bartholomé E, Belward AS. GLC2000: a new approach to global land cover mapping from Earth observation data. Int J Remote Sens. 2005;26(9):1959–77.
9. Boden TA, Marland G, Andres RJ. Global, regional, and national fossil-fuel CO_2 emissions. Oak Ridge, TN: Carbon Dioxide Information Analysis Center, Oak Ridge National Laboratory, U.S. Department of Energy; 2015. doi:10.3334/CDIAC/00001_V2013 [downloaded 8 Sept 2015].
10. Brienen RJ, Phillips OL, Feldpausch TR, Gloor E, Baker TR, Lloyd J, Lopez-Gonzalez G, Monteagudo-Mendoza A, Malhi Y, Lewis SL, Martinez RV, et al. Long-term decline of the Amazon carbon sink. Nature. 2015;519(7543):344–8.
11. Bugmann H, Bigler C. Will the CO_2 fertilization effect in forests be offset by reduced tree longevity? Oecologia. 2011;165(2):533–44.
12. Chazdon RL, Broadbent EN, Rozendaal DM, Bongers F, Zambrano AM, Aide TM, Balvanera P, Becknell JM, Boukili V, Brancalion PH, Craven D. Carbon sequestration potential of second-growth forest regeneration in the Latin American tropics. Sci Adv. 2016;2(5):e1501639.
13. Ciais P, Sabine C, Bala G, Bopp L, Brovkin V, Canadell J, Chhabra A, DeFries

R, Galloway J, Heimann M, Jones C. Carbon and other biogeochemical cycles. In: Climate change 2013: the physical science basis. Contribution of working group I to the fifth assessment report of the intergovernmental panel on climate change. Cambridge University Press; 2014. p. 465–570.

14. DeFries R, Achard F, Brown S, Herold M, Murdiyarso D, Schlamadinger B, de Souza C. Earth observations for estimating greenhouse gas emissions from deforestation in developing countries. Environ Sci Policy. 2007;10(4):385–94.

15. Doughty CE, Metcalfe DB, Girardin CA, Amézquita FF, Cabrera DG, Huasco WH, Silva-Espejo JE, Araujo-Murakami A, da Costa MC, Rocha W, Feldpausch TR. Drought impact on forest carbon dynamics and fluxes in Amazonia. Nature. 2015;519(7541):78–82.

16. Espírito-Santo FDB, Keller M, Braswell B, Nelson BW, Frolking S, Vicente G. Storm intensity and old-growth forest disturbances in the Amazon region. Geophys Res Lett. 2010;37(11):L11403. doi:10.1029/2010GL043146.

17. Espírito-Santo FD, Gloor M, Keller M, Malhi Y, Saatchi S, Nelson B, Junior RC, Pereira C, Lloyd J, Frolking S, Palace M, Shimabukuro YE, Duarte V, Monteagudo Mendoza A, López-González G, Baker TR, Feldpausch TR, Brienen RJW, Asner GP, Boyd DS, Phillips OL. Size and frequency of natural forest disturbances and the Amazon forest carbon balance. Nat Commun. 2014;5:3434. doi:10.1038/ncomms4434.

18. FAO - Food and Agricultural Organization. Global forest resources assessment 2010. FAO forestry paper 163, Rome, Italy; 2010. 378 pp.

19. Feldpausch TR, Banin L, Phillips OL, Baker TR, Lewis SL, Quesada CA, Affum-Baffoe K, Arets EJ, Berry NJ, Bird M, Brondizio ES. Height-diameter allometry of tropical forest trees. Biogeosciences. 2011;8(5):1081–106.

20. Feldpausch TR, Phillips OL, Brienen RJ, Gloor E, Lloyd J, Lopez-Gonzalez G, Monteagudo-Mendoza A, Malhi Y, Alarcón A, Álvarez Dávila E, Alvarez-Loayza P, et al. Amazon forest response to repeated droughts. Glob Biogeochem Cycles. 2016;30:964–82.

21. Franks PJ, Adams MA, Amthor JS, Barbour MM, Berry JA, Ellsworth DS, Farquhar GD, Ghannoum O, Lloyd J, McDowell N, Norby RJ. Sensitivity of plants to changing atmospheric CO_2 concentration: from the geological past to the next century. New Phytol. 2013;197(4):1077–94.

22. Fritz S, See L. Identifying and quantifying uncertainty and spatial disagreement in the comparison of global land cover for different applications. Glob Change Biol. 2008;14:1057–75.

23. Galbraith D, Malhi Y, Affum-Baffoe K, Castanho AD, Doughty CE, Fisher RA, Lewis SL, Peh KS, Phillips OL, Quesada CA, Sonké B. Residence times of woody biomass in tropical forests. Plant Ecol Divers. 2013;6(1):139–57.

24. Gatti LV, Gloor M, Miller JB, Doughty CE, Malhi Y, Domingues LG, Basso LS, Martinewski A, Correia CS, Borges VF, Freitas S. Drought sensitivity of Amazonian carbon balance revealed by atmospheric measurements. Nature. 2014;506(7486):76–80.

25. Global Forest Watch. World Resources Institute. www.globalforestwatch.org (2014). Accessed on 22 November 2016.

26. Gloor M, Phillips OL, Lloyd JJ, Lewis SL, Malhi Y, Baker TR, Lopez-Gonzalez G, Peacock J, Almeida S, Oliveira D, et al. Does the disturbance hypothesis explain the biomass increase in basin-wide Amazon forest plot data? Glob Change Biol. 2009;15(10):2418–30.

27. Gloor M, Gatti L, Brienen R, Feldpausch TR, Phillips OL, Miller J, Ometto JP, Rocha H, Baker T, De Jong B, Houghton RA. The carbon balance of South America: a review of the status, decadal trends and main determinants. Biogeosciences. 2012;9(12):5407–30.

28. Gloor MR, Brienen RJ, Galbraith D, Feldpausch TR, Schöngart J, Guyot JL, Espinoza JC, Lloyd J, Phillips OL. Intensification of the Amazon hydrological cycle over the last two decades. Geophys Res Lett. 2013;40(9):1729–33.

29. Gloor M, Barichivich J, Ziv G, Brienen R, Schöngart J, Peylin P, Cintra L, Barcante B, Feldpausch T, Phillips O, Baker J. Recent Amazon climate as background for possible ongoing and future changes of Amazon humid forests. Global Biogeochem Cycles. 2015;29(9):1384–99.

30. Hansen MC, Potapov PV, Moore R, Hancher M, Turubanova SA, Tyukavina A, Thau D, Stehman SV, Goetz SJ, Loveland TR, Kommareddy A. High-resolution global maps of 21st-century forest cover change. Science. 2013;342(6160):850–3.

31. Harris NL, Brown S, Hagen SC, Saatchi SS, Petrova S, Salas W, Hansen MC, Potapov PV, Lotsch A. Baseline map of carbon emissions from deforestation in tropical regions. Science. 2012;336(6088):1573–6.

32. Honorio Coronado EN, Dexter KG, Pennington RT, Chave J, Lewis SL, Alexiades MN, Alvarez E, Alves de Oliveira A, Amaral IL, Araujo-Murakami A, Arets EJ, et al. Phylogenetic diversity of Amazonian tree communities. Divers Distrib. 2015;21(11):1295–307.

33. Huntingford C, Zelazowski P, Galbraith D, Mercado LM, Sitch S, Fisher R, Lomas M, Walker AP, Jones CD, Booth BB, Malhi Y, et al. Simulated resilience of tropical rainforests to CO_2-induced climate change. Nat Geosci. 2013;6(4):268–73.

34. Johnson MO, Galbraith D, Gloor M, De Deurwaerder H, Guimberteau M, Rammig A, Thonicke K, Verbeeck H, Randow C, Monteagudo A, Phillips OL. Variation in stem mortality rates determines patterns of aboveground biomass in Amazonian forests: implications for dynamic global vegetation models. Glob Change Biol. 2016;22:3996–4013.

35. Lewis SL, Phillips OL, Baker TR, Lloyd J, Malhi Y, Almeida S, Higuchi N, Laurance WF, Neill DA, Silva JN, Terborgh J. Concerted changes in tropical forest structure and dynamics: evidence from 50 South American long-term plots. Philos Trans R Soc Lond B Biol Sci. 2004;359(1443):421–36.

36. Lewis SL, Lopez-Gonzalez G, Sonké B, Affum-Baffoe K, Baker TR, Ojo LO, Phillips OL, Reitsma JM, White L, Comiskey JA, Ewango CE. Increasing carbon storage in intact African tropical forests. Nature. 2009;457(7232):1003–6.

37. Luyssaert S, Schulze ED, Börner A, Knohl A, Hessenmöller D, Law BE, Ciais P, Grace J. Old-growth forests as global carbon sinks. Nature. 2008;455(7210):213–5.

38. Marimon BS, Marimon-Junior BH, Feldpausch TR, Oliveira-Santos C, Mews HA, Lopez-Gonzalez G, Lloyd J, Franczak DD, de Oliveira EA, Maracahipes L, Miguel A, et al. Disequilibrium and hyperdynamic tree turnover at the forest–cerrado transition zone in southern Amazonia. Plant Ecol Divers. 2014;7(1–2):281–92.

39. Mitchard ET, Feldpausch TR, Brienen RJ, Lopez-Gonzalez G, Monteagudo A, Baker TR, Lewis SL, Lloyd J, Quesada CA, Gloor M, Steege H. Markedly divergent estimates of Amazon forest carbon density from ground plots and satellites. Glob Ecol Biogeogr. 2014;23:935–46.

40. Pan Y, Birdsey RA, Fang J, Houghton R, Kauppi PE, Kurz WA, Phillips OL, Shvidenko A, Lewis SL, Canadell JG, Ciais P, et al. A large and persistent carbon sink in the world's forests. Science. 2011;333(6045):988–93.

41. Phillips OL, Aragão LE, Lewis SL, Fisher JB, Lloyd J, López-González G, Malhi Y, Monteagudo A, Peacock J, Quesada CA, Van Der Heijden G, et al. Drought sensitivity of the Amazon rainforest. Science. 2009;323(5919):1344–7.

42. Phillips OL, Gentry AH. Increasing turnover through time in tropical forests. Science. 1994;263(5149):954–8.

43. Phillips OL, Malhi Y, Higuchi N, Laurance WF, Núñez PV, Vásquez RM, Laurance SG, Ferreira LV, Stern M, Brown S, Grace J. Changes in the carbon balance of tropical forests: evidence from long-term plots. Science. 1998;282(5388):439–42.

44. Poorter L, Bongers F, Aide TM, Zambrano AM, Balvanera P, Becknell JM, Boukili V, Brancalion PH, Broadbent EN, Chazdon RL, Craven D, et al. Biomass resilience of Neotropical secondary forests. Nature. 2016;530:211–4.

45. Potapov P, Yaroshenko A, Turubanova S, Dubinin M, Laestadius L, Thies C, Aksenov D, Egorov A, Yesipova Y, Glushkov I, Karpachevskiy M. Mapping the world's intact forest landscapes by remote sensing. Ecol Soc. 2008;13(2):51.

46. PRODES: Brazilian government Deforestation estimates based on remote sensing: http://www.obt.inpe.br/prodes (2015).

47. Quesada CA, Lloyd J, Anderson LO, Fyllas NM, Schwarz M, Czimczik CI. Soils of Amazonia with particular reference to the RAINFOR sites. Biogeosciences. 2011;8:1–26.

48. Quesada CA, Phillips OL, Schwarz M, Czimczik CI, Baker TR, Patiño S, Fyllas NM, Hodnett MG, Herrera R, Almeida S, Alvarez DÃ. Basin-wide variations in Amazon forest structure and function are mediated by both soils and climate. Biogeosciences. 2012;9:2203–46.

49. Restrepo-Coupe N, da Rocha HR, Hutyra LR, da Araujo AC, Borma LS, Christoffersen B, Cabral OM, de Camargo PB, Cardoso FL, da Costa AC, Fitzjarrald DR. What drives the seasonality of photosynthesis across the Amazon basin? A cross-site analysis of eddy flux tower measurements from the Brasil flux network. Agric For Meteorol. 2013;15(182):128–44.

50. Rowland L, Da Costa AC, Galbraith DR, Oliveira RS, Binks OJ, Oliveira AA, Pullen AM, Doughty CE, Metcalfe DB, Vasconcelos SS, Ferreira LV. Death from drought in tropical forests is triggered by hydraulics not carbon starvation. Nature. 2015;528(7580):119–22.

51. Saatchi S, Asefi-Najafabady S, Malhi Y, Aragão LE, Anderson LO, Myneni RB, Nemani R. Persistent effects of a severe drought on Amazonian forest canopy. Proc Natl Acad Sci. 2013;110(2):565–70.

52. Schimel D, Stephens BB, Fisher JB. Effect of increasing CO_2 on the terrestrial carbon cycle. Proc Natl Acad Sci. 2015;112(2):436–41.

53. Song XP, Huang C, Saatchi SS, Hansen MC, Townshend JR. Annual carbon emissions from deforestation in the Amazon Basin between 2000 and 2010. PLoS ONE. 2015;10(5):e0126754.

54. Ter Steege H, Pitman N, Sabatier D, Castellanos H, Van Der Hout P, Daly DC, Silveira M, Phillips O, Vasquez R, Van Andel T, Duivenvoorden J. A spatial model of tree α-diversity and tree density for the Amazon. Biodivers Conserv. 2003;12(11):2255–77.

55. Ter Steege H, Pitman NC, Phillips OL, Chave J, Sabatier D, Duque A, Molino JF, Prévost MF, Spichiger R, Castellanos H, Von Hildebrand P. Continental-scale patterns of canopy tree composition and function across Amazonia. Nature. 2006;443(7110):444–7.

56. Wang X, Piao S, Ciais P, Friedlingstein P, Myneni RB, Cox P, Heimann M, Miller J, Peng S, Wang T, Yang H. A two-fold increase of carbon cycle sensitivity to tropical temperature variations. Nature. 2014;506(7487):212–5.

57. Wang W, Ciais P, Nemani RR, Canadell JG, Piao S, Sitch S, White MA, Hashimoto H, Milesi C, Myneni RB. Variations in atmospheric CO_2 growth rates coupled with tropical temperature. Proc Natl Acad Sci. 2013;110(32):13061–6.

13

From berries to blocks: carbon stock quantification of a California vineyard

Jorge Andres Morandé[1], Christine M. Stockert[2], Garrett C. Liles[3], John N. Williams[4], David R. Smart[2] and Joshua H. Viers[1,5*]

Abstract

Background: Quantifying terrestrial carbon (C) stocks in vineyards represents an important opportunity for estimating C sequestration in perennial cropping systems. Considering 7.2 M ha are dedicated to winegrape production globally, the potential for annual C capture and storage in this crop is of interest to mitigate greenhouse gas emissions. In this study, we used destructive sampling to measure C stocks in the woody biomass of 15-year-old Cabernet Sauvignon vines from a vineyard in California's northern San Joaquin Valley. We characterize C stocks in terms of allometric variation between biomass fractions of roots, aboveground wood, canes, leaves and fruits, and then test correlations between easy-to-measure variables such as trunk diameter, pruning weights and harvest weight to vine biomass fractions. Carbon stocks at the vineyard block scale were validated from biomass mounds generated during vineyard removal.

Results: Total vine C was estimated at 12.3 Mg C ha^{-1}, of which 8.9 Mg C ha^{-1} came from perennial vine biomass. Annual biomass was estimated at 1.7 Mg C ha^{-1} from leaves and canes and 1.7 Mg C ha^{-1} from fruit. Strong, positive correlations were found between the diameter of the trunk and overall woody C stocks ($R^2 = 0.85$), pruning weights and leaf and fruit C stocks ($R^2 = 0.93$), and between fruit weight and annual C stocks ($R^2 = 0.96$).

Conclusions: Vineyard C partitioning obtained in this study provides detailed C storage estimations in order to understand the spatial and temporal distribution of winegrape C. Allometric equations based on simple and practical biomass and biometric measurements could enable winegrape growers to more easily estimate existing and future C stocks by scaling up from berries and vines to vineyard blocks.

Keywords: Vineyard, Winegrape, Grapevine carbon partitioning, Carbon accounting, Carbon stocks, Aboveground biomass, Allometrics, Carbon sequestration, California

Background

Agriculture is a key human activity in terms of food production, economic importance and impact on the global carbon cycle. As the human population heads toward 9 billion or beyond by 2050, there is an acute need to balance agricultural output with its impact on the environment, especially in terms of greenhouse gas (GHG) production [1]. An evolving set of tools, approaches and metrics are being employed under the term "climate smart agriculture" (CSA) to help—from small and industrial scale growers to local and national policy setters—develop techniques at all levels and find solutions that strike that production-environment balance and promote various ecosystem services [2, 3]. California epitomizes the agriculture-climate challenge, as well as its opportunities. As the United States' largest agricultural producing state (2012 farmgate production valued at $44.7 billion, or 11% of the US total) agriculture also accounted for approximately 8% of California's greenhouse gas (GHG) emissions statewide for the period 2000–2013 [4, 5].

At the same time, California is at the forefront of innovative approaches to CSA [e.g., 6, 7]. Given the state's Mediterranean climate, part of an integrated CSA strategy will likely include perennial crops, such

*Correspondence: jviers@ucmerced.edu
[1] Environmental Systems, University of California, Merced, Merced, CA, USA
Full list of author information is available at the end of the article

as winegrapes, that have a high market value and store C long term in woody biomass [8]. Economically, wine production and retail represents an important contribution to California's economy, generating $61.5 billion in annual economic impact [9]. In terms of land use, 230,000 ha in California are managed for wine production, with 4.2 million tons of winegrapes harvested annually with an approximate $3.2 billion farm gate value [9]. This high level of production has come with some environmental costs, however, with degradation of native habitats, impacts to wildlife, and over abstraction of water resources [see 10].

Although many economic and environmental impacts of wine production systems are actively being quantified, and while there is increasing scientific interest in the carbon footprint of vineyard management activities [e.g., 11], efforts to quantify C capture and storage in annual and perennial biomass remain less well-examined [12, 13]. Studies from Mediterranean climates have focused mostly on C cycle processes in annual agroecosystems or natural systems [14, 15]. Related studies have investigated sources of GHGs [16, 17], on-site energy balance [18], water use [19] and potential impacts of climate change on productivity and the distribution of grape production [20].

The perennial nature and extent of vineyard agroecosystems have brought increasing interest from growers and the public sector to reduce the GHG footprint associated with wine production. The ongoing development of carbon accounting protocols within the international wine industry reflects the increased attention that industry and consumers are putting on GHG emissions and offsets. In principle, an easy-to-use, wine industry specific, GHG protocol would measure the carbon footprints of winery and vineyard operations of all sizes [21]. However, such footprint assessment protocols remain poorly parameterized, especially those requiring time-consuming empirical methods [22]. Data collected from the field, such as vine biomass, cover crop biomass, and soil carbon storage capacity are difficult to obtain and remain sparse, and thus limit the further development of carbon accounting in the wine sector [23]. Simple yet accurate methods are needed to allow vineyard managers to measure C stocks in situ and thereby better parameterize carbon accounting protocols. Not only would removing this data bottleneck encourage broader participation in such activities, it would also provide a reliable means to reward climate smart agriculture.

Empirical carbon estimation

Building on research that has used empirical data to compare soil and aboveground C stocks in vineyards and adjacent oak woodlands in California [12], this study

sought to estimate the C composition of a vine, including the relative contributions of its component parts (root, trunk and cordons, canes, leaves and fruit). By identifying the allometric relationships among trunk diameter, plant height, and other vine dimensions, growers could utilize a reliable mechanism for translating vine architecture and biomass into C estimates [24].

In both natural and agricultural ecosystems, several studies have been performed using allometric equations in order to estimate aboveground biomass to assess potential for C sequestration. For example, functional relationships between the ground-measured Lorey's height (basal area weighted height of all trees >10 cm in diameter) and aboveground biomass were derived from allometric equations in forests throughout the tropics [25]. Similarly, functional relationships have been found in tropical agriculture for aboveground, belowground, and field margin biomass and C [26–30]. In the vineyard setting, however, horticultural intervention and annual pruning constrain the size and shape of vines making existing allometric relationships less meaningful, though it is likely that simple physical measurements could readily estimate aboveground biomass.

To date, most studies on C sequestration in vineyards have been focused on soil C as sinks [e.g., 13] and some attempts to quantify biomass C stocks have been carried out in both agricultural and natural systems. In vineyards, studies in California in the late 1990s have reported net primary productivity (NPP) or total biomass values between 550 g C m^{-2} (5.5 Mg C ha^{-1}) and 1100 g C m^{-2} (11 Mg C ha^{-1}) [31]. In terms of spatial distribution, some data of standing biomass collected by Kroodsma et al. [8] from companies that remove trees and vines in California (Noni Enterprises and Orchard Removal, Fresno, California, USA; Wilson Agriculture Company, Shafter, California, USA; Volks and Sons Orchard Removal, Fresno, California, USA) yielded values of 1.0–1.3 Mg C ha^{-1} year^{-1} woody C for nuts and stone fruit species, and 0.2–0.4 Mg C ha^{-1} year^{-1} for vineyards. It has been reported that mature California orchard crops allocate, on average, one third of their NPP to the harvested portion [32] and mature vines 35–50% of the current year's production to grape clusters [33]. Pruning weight has also been quantified by two direct measurements which estimated 2.5 Mg of pruned biomass per ha for both almonds [34] and vineyards [31].

The incorporation of trees or shrubs in agroforestry systems can increase the amount of carbon sequestered compared to a monoculture field of crop plants or pasture [35]. Additional forest planting would be needed to offset current net annual loss of aboveground C, representing an opportunity for viticulture to incorporate the surrounding woodlands into the system. A study assessing C storage in California vineyards found that on average, surrounding forested wildlands had 12 times more aboveground woody C than vineyards and even the largest vines had only about one-fourth of the woody biomass per ha of the adjacent wooded wildlands [12].

Study objectives

The objectives of this study were to: (1) measure standing vine biomass and calculate C stocks in Cabernet Sauvignon vines by field sampling the major biomass fractions (i.e., roots, wood, canes, leaves and fruit); (2) calculate C fractions in berry clusters to assess C mass that could be returned to the vineyard from the winery in the form of rachis and pomace; (3) determine proportion of perennially sequestered and annually produced C stocks using easy to measure physical vine properties (i.e., trunk diameter, pruning weights or fruit weights); and (4) develop allometric relationships to provide growers and land managers with a method to rapidly assess vineyard C stocks. Lastly, we validate block level estimates of C with volumetric measurements of vine biomass generated during vineyard removal.

Methods
Study site

The study site is located in southern Sacramento County, California, USA (121°22'33"W, 38°18'19"N; Fig. 1a), and the vineyard is part of a property annexed into a seasonal floodplain restoration program, which has since removed the levee preventing seasonal flooding. The ensuing vineyard removal allowed destructive sampling for biomass measurements and subsequent C quantification.

The vineyard is considered part of the Cosumnes River appellation within the Lodi American Viticultural Area, a region characterized by its Mediterranean climate—cool wet winters and warm dry summers—and by nearby Sacramento-San Joaquin Delta breezes that moderate peak summer temperatures compared to areas north and south of this location. The study site is characterized by a mean summer maximum air temperature of 32 °C, has an annual average precipitation of 90 mm, typically all received as rain from November to April [36]. During summer time, the daily high air temperatures average 24 °C, and daily lows average 10 °C. Winter temperatures range from an average low 5 °C to average high 15 °C [37]. Total heating degree days for the site are approximately 3420 [38] and the frost-free season is approximately 360 days annually [24].

Similar to other vineyards in the Lodi region, the site is situated on an extensive alluvial terrace landform formed by Sierra Nevada outwash with a San Joaquin Series soil (fine, mixed, active, thermic Abruptic Durixeralfs). This soil-landform relationship is extensive, covering approximately 160,000 ha across the eastern Central Valley and it is used extensively for winegrape production. The dominant soil texture is clay loam with some sandy clay loam sectors; mean soil C content, based on three characteristic grab samples processed by the UC Davis Analytical Lab, in the upper 8 cm was 1.35% (sd = 0.77%) and in the lower 8–15 cm was 1.1% (sd = 0.1%).

The vineyard plot consisted of 7.5 ha of Cabernet Sauvignon vines, planted in 1996 at a density of 1631 plants ha^{-1} (3.35 m by 1.83 m spacing) with flood irrigation during spring and summer seasons. The vines were trained using a quadrilateral trellis system with two parallel

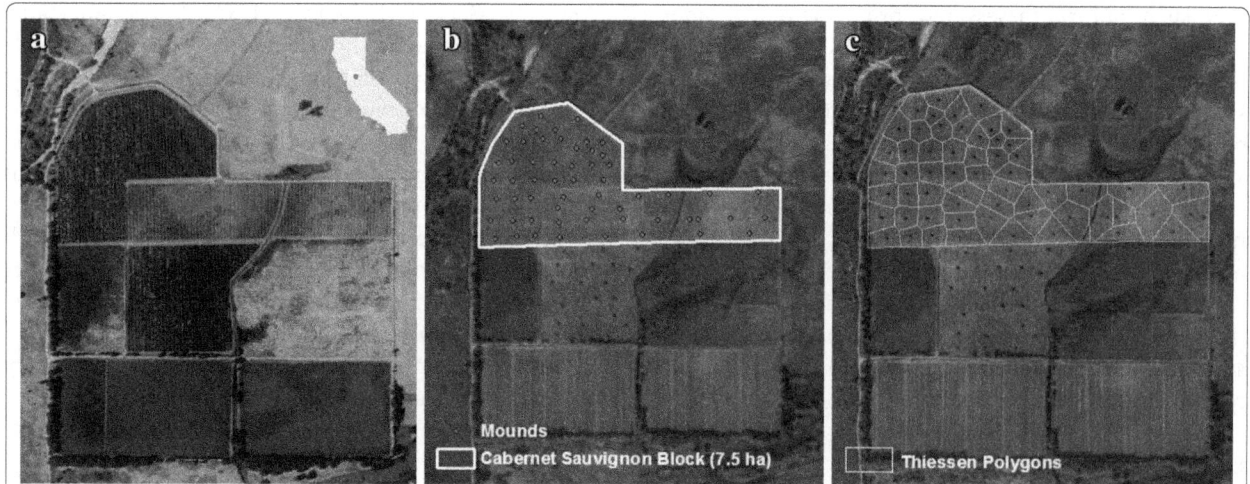

Fig. 1 Study vineyard in southern Sacramento County, California, USA. The vineyard prior to removal is shown in **a**. The boundaries of the Cabernet Sauvignon plot are outlined in **b** with *dots* representing woody debris mounds resulting from uprooted vines. The Thiessen polygons around mounds in **c** were used to estimate source area for C per hectare estimates

cordons and a modified Double Geneva Curtain structure attached to T-posts (Fig. 2). Atypically, these vines were not grafted to rootstock, which is used often in the region to modify vigor or limit disease (i.e., phylloxera).

Standing biomass quantification

In Sept.–Oct. of 2011, aboveground biomass was measured from 72 vines. The vineyard (7.5 ha) was divided equally in twelve randomly assigned blocks, and six individual vines from each block were processed into major biomass categories of leaf, fruit, cane and trunk plus cordon (Fig. 2). Grape berry clusters were collected in buckets, with fruit separated and weighed fresh in the field. Leaves and canes were collected separately in burlap sacks, and the trunks and cordons were tagged. Biomass was transported off site to partially air dry on wire racks and then fully dried in large ventilated ovens.

Plant tissues (i.e., leaves, canes, wood, roots, grape skins, rachis and seeds) were dried at 60 °C for 48 h and then ground to pass through a 250 µm mesh sieve using a Thomas Wiley® Mini-Mill (Thomas Scientific, Swedesboro, NJ). Total C (%) in plant tissues was analyzed using a PDZ Europa ANCA-GSL elemental analyzer at the UC Davis Stable Isotope Facility. For cluster and berry C estimations, grape clusters were randomly selected from all repetitions. Berries were removed from cluster rachis. While the berries were frozen, the seeds and skins were separated from the fruit flesh or "pulp", and combined with the juice (denoted juice + pulp herein). The rachis, skins and seeds were dried in oven and weighed. The pulp was separated from the juice + pulp with vacuum filtration using a pre-weighed Q2 filter paper (1–5 µm retention, Fisher Scientific). The filter paper with pulp was oven dried and weighed to get insoluble solid fraction (pulp). The largest portion of grape juice soluble solids are sugars. Sugars were measured at 25% using a Refractometer PAL-1 (Atago USA, Inc., Bellevue, WA). The C content of sugar was calculated at 42% using the formula of sucrose.

Belowground biomass was measured by pneumatically excavating the root system with compressed air applied at 0.7 Mpa (100 psi) for three of the 12 sampling blocks, exposing two vines each in 8 m³ pits. The soil was prewetted prior to excavation to facilitate removal and minimize root damage. A root restricting duripan, common in this soil, provided an effective rooting depth of about 40 cm at this site with only 5–10 fine and small roots (generally <20 mm diameter) able to penetrate below this depth in each plot. Roots were washed, cut into smaller segments and separated into four size classes (<2, 2–6, 6–20 and >20 mm), oven-dried at 60 °C for 48 h and weighed. Larger roots were left in the oven for 4 days. Stumps (i.e., fraction of the plant immediately above the

roots but belowground) were considered part of the root system for this analysis.

C estimates

In vineyard ecosystems, annual C is represented by fruit, leaves and canes, and is either removed from the system and/or incorporated into the soil C pools, which was not considered further. Structures whose tissues remain in the plant (i.e., trunk plus cordons and roots) were considered perennial C. Woody biomass volumes were measured and used for perennial C estimates. Cordon and trunk diameters were measured using a digital caliper at four locations per piece and averaged, and lengths were measured with a calibrated tape. Sixty vines were used for the analysis; twelve vines were omitted due to missing values in one or more vine fractions. All statistical estimates were conducted in R [39].

Mound volume and mass estimation

An earthmoving machine was used to uproot vines and gather them together to form mounds. Twenty-six mounds consisting of trunks plus cordons and canes were measured across this vineyard block (Fig. 1b). The mounds represented comparable spatial footprints within the vineyard area (Fig. 1c). Mound C stocks were estimated using their biomass contribution areas, physical size, density and either a semi-ovoid or hemispherical model.

Physical size

A real-time kinematic (RTK) global positioning system (Topcon HiperV) was used to map boundaries of each mound, with vertices placed every 1.5 m to measure circumference. Average mound height (m) was calculated using a stadia rod and laser inclinometer range finder. The circumsurficial distance (distance between two points measured across the pile surface) over the major axes (N–S and E–W) of each mound was measured with a calibrated cord. Combined, these measurements were used to estimate pile volume using semi-ovoid and hemispherical models (see below).

Semi-ovoid model

For the semi-ovoid model, length [*l*] and width [*w*] and mound height (*h*) were determined from direct field measurements. The mound volume was calculated as:

$$V_m = \frac{2}{3}\pi \cdot l \cdot w \cdot h$$

Hemispheric model

The hemispheric model used mapping data to resolve the geometric centroid in ArcGIS (v10 ESRI, Redlands) from each perimeter. A circular area from the average radius

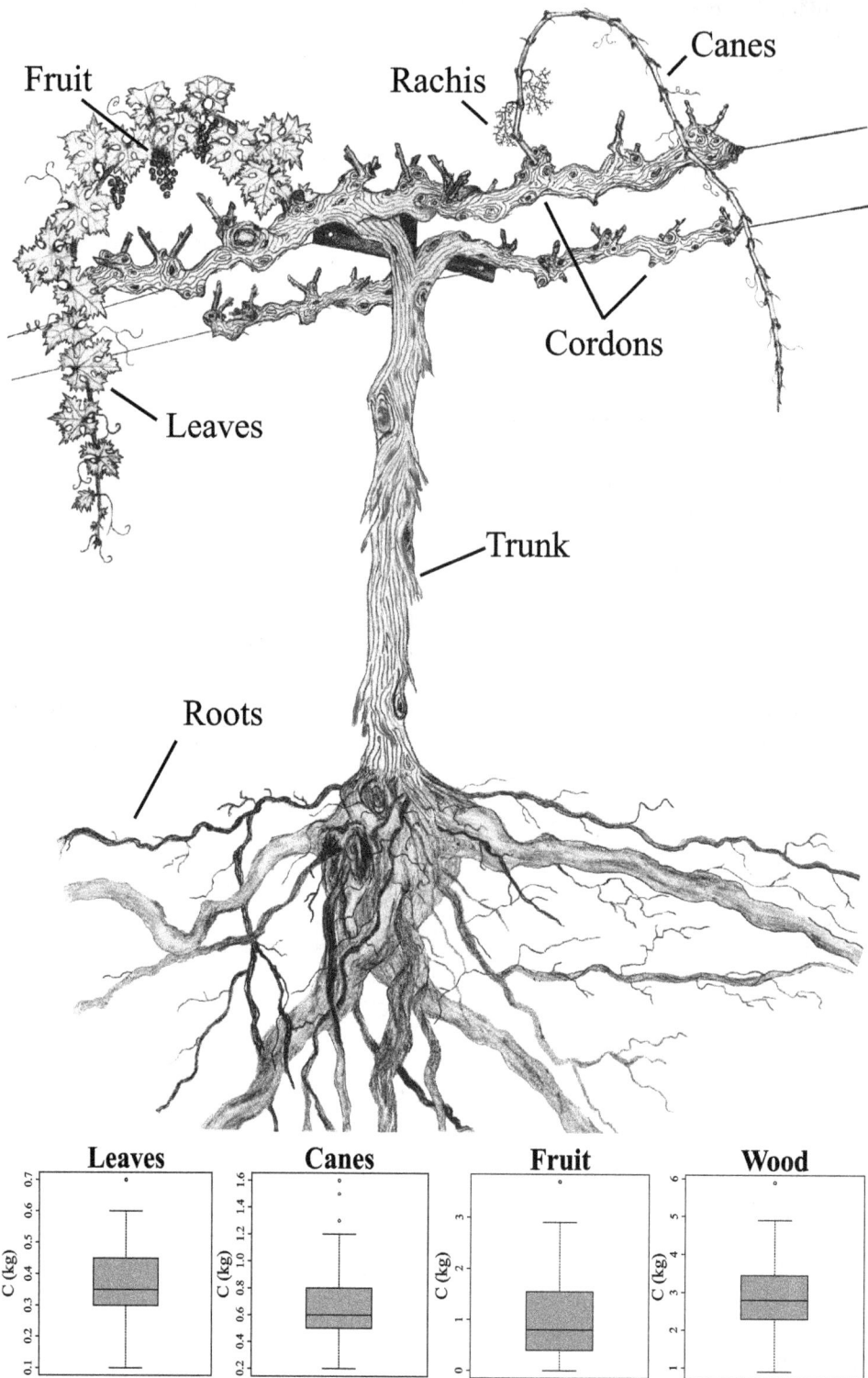

Fig. 2 Vine diagram plus *boxplots* describing major categories of C stocks measurements. Fruit was weighed separately in berries, seeds and rachis. Cordons represent the *horizontal arms* where the canes grow from. *Boxplots* show the median and range of C stocks for four categories in kg C vine^{-1}. Histogram-*boxplot* (*bottom*) shows the total C distribution per vine (kg C plant^{-1}) yielded by 72 samples (artwork credit C. M. Stockert)

of each mound centroid to perimeter vertex was then regressed against mapped mound area, resulting in an area-adjusted model. This area was back-transformed to arrive at best-fit radius [h] for each mound to represent one-half the volume of a sphere:

$$V_m = \frac{2}{3}\pi h^3$$

Mound C density
A standardized volume (~0.085 m^3) of biomass was collected by cutting out random sections of the same area from 12 mounds using a plastic container to insure size consistency. Plant material in the mounds included the fractions of trunk plus cordons, roots and canes, and the way the mound elements fill out the container simulated their spatial arrangement in the mound. Samples represent a range of biomass configurations (relative ratio of biomass volume:void) found across the vineyard block. Sample contents were divided into vine biomass classes (canes, wood, and roots) dried, and weighed. Relating sample mass with the collection volume supports the calculation of mound density (47.5 kg/m^3) and C mass. Vine category proportion data were compared to the measured vine proportion data to validate the basic assumption supporting these calculations. All biomass data were multiplied by a factor of 0.47 (average C calculated for the three fractions) to estimate C mass (kg). C data were scaled up to the individual mound, unit area, and vineyard totals (Mg ha^{-1}).

Results
Vine C stocks
Average C stocks per vine partition out to roughly equal one-third fractions of (i) roots, (ii) trunk plus cordons, and (iii) leaves, fruit and canes (Table 1; Fig. 3). The vine-based values when scaled to the spatial extent of the vineyard give an estimate of 12.3 Mg C ha^{-1} across the 7.5 ha site (92.3 Mg C total) based on 1631 vines ha^{-1} density. When partitioned into annual versus perennial contributions, 3.4 Mg C ha^{-1} (28%) was found to come from

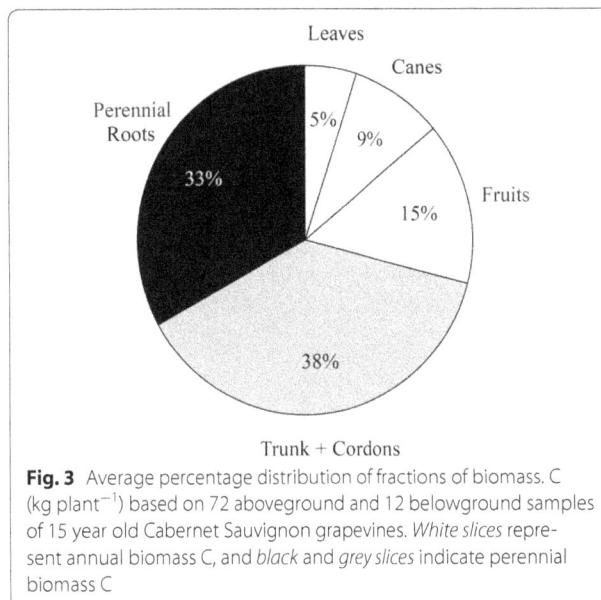

Fig. 3 Average percentage distribution of fractions of biomass. C (kg plant^{-1}) based on 72 aboveground and 12 belowground samples of 15 year old Cabernet Sauvignon grapevines. *White slices* represent annual biomass C, and *black* and *grey slices* indicate perennial biomass C

annual production in canes, leaves and fruit, and the remaining 8.9 C ha^{-1} (72%) was stored in the perennial woody fraction (trunk plus cordons and roots) (Table 2; Fig. 2). Boxplots in Fig. 2 show the variability of total C by biomass category.

Total C stocks per plant were variable (sd = 2.0 kg/26%). Fruit harvest accounted for approximately 10% of vine C by weight, about 26% of this C could be turned into soil C storage by returning the rachis and pomace (skin and seed) to the soil (Fig. 4), a source of biomass C contributing to long term C storage in agro-ecosystems and beneficial for GHG mitigation purposes. In this vineyard, the amount of return would average 0.44 Mg C ha^{-1} or 13% of the annual C fraction.

Per vine C was calculated to be 46% (7.7 kg; sd = 2.0 kg) of total dry biomass per vine (average = 16.8 kg), resulting in approximately one-third of each vine in annual (29%), aboveground wood (38%), and belowground root wood (33%) C (Fig. 3). Root C content was estimated

Table 1 Average biomass, and C for vine fractions and total plant are shown

Biomass fractions	n	% C	Dry biomass		C stocks		C stocks	
			kg vine^{-1}		kg C vine^{-1}		Mg C ha^{-1}	
Fruit	60	43	2.6	(2.0)	1.1	(0.9)	1.7	(1.4)
Leaves	60	45	0.9	(0.3)	0.4	(0.2)	0.6	(0.3)
Canes	60	48	1.4	(0.6)	0.7	(0.3)	1.1	(0.5)
Wood	60	48	6.2	(2.1)	3.0	(1.0)	4.8	(1.6)
Roots	3	44	5.7	(0.9)	2.5	(0.4)	4.1	(0.6)
Total			16.8	(4.4)	7.7	(2.0)	12.3	(2.5)

Standard deviations (sd) are shown in parentheses

Table 2 Fractions of C sequestration in vines by time and space

C distribution	Biomass fraction	Mg C ha^{-1}	Total	%
Temporal				
Annual	Fruit	1.7	3.4	28
	Leaves	0.6		
	Canes	1.1		
Perennial	Wood	4.8	8.9	72
	Roots	4.1		
Spatial				
Aboveground	Fruit	1.7	8.2	67
	Leaves	0.6		
	Canes	1.1		
	Wood	4.8		
Belowground	Roots	4.1	4.1	33

Annual growth represents the seasonal vegetative and reproductive development starting in spring and finishing in early fall. Most fruit is removed whereas leaves and canes return to soil. Wood constitutes the sum of trunk plus cordons

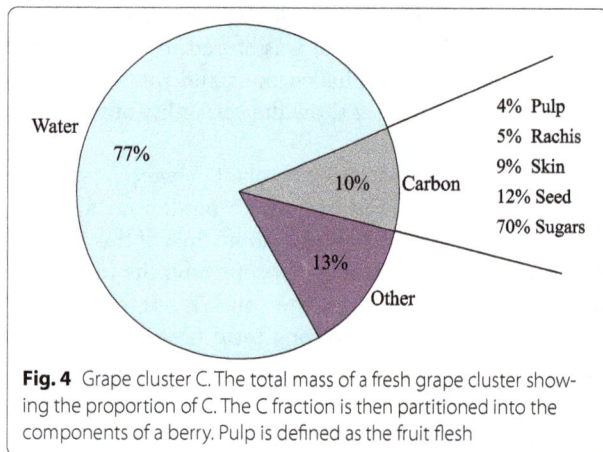

Fig. 4 Grape cluster C. The total mass of a fresh grape cluster showing the proportion of C. The C fraction is then partitioned into the components of a berry. Pulp is defined as the fruit flesh

as 44% of dry weight, of which 83.7% was stored in roots >6 mm diameter (including stump), and only ~4% was found in fine roots <2 mm diameter (Table 3).

Carbon stock estimates in woody tissue correlated positively with trunk diameter (r = 0.91) and more strongly than with other biomass categories of canes (r = 0.74), leaves (0.81), and fruit (r = 0.57) (Fig. 5). All three of the allometric relationships developed here—wet fruit weight, trunk diameter and pruning weight—showed relatively robust coefficients of determination when regressed against C content (Fig. 6) (i.e., $R^2 > 0.85$).

Mound C stocks

The secondary approach to estimate C stocks by fitting the regular hemispherical and semi-ovoid models produced comparable average biomass values. Our estimations of C stocks per ha quantifying mound biomass yielded an average of 9.93 ± 2.7 (semi-ovoid model) and 10.57 ± 3.6 Mg C ha^{-1} (hemispherical model) (Table 4). This compares favorably to 10.02 ± 1.9 Mg C ha^{-1} obtained by standing biomass considering C stocks estimations of trunk plus cordons, roots and canes (Table 1). Additionally, a paired T test was run to compare differences between the two mound methods finding no significant differences (95% CI; p = 0.2). A Welch Two Sample t test applied to check for possible significant differences between the Standing biomass and mound methods found no significant difference (95% CI p = 0.72).

Discussion

The present study provides results for an assessment of vineyard biomass that is comparable with data from previous studies, as well as estimates of belowground biomass that are more precise than previous reports. While most studies on C sequestration in vineyards have focused on soil C, some have quantified aboveground biomass and C stocks. For example, a study of grapevines in California found net primary productivity (NPP) values between 5.5 and 11 Mg C ha^{-1} [31]—figures that are comparable to our mean estimate of 12.4 Mg C ha^{-1}. For pruned biomass, our estimate of 1.1 Mg C ha^{-1} (2.3 Mg biomass ha^{-1}) were comparable to two assessments that estimated 2.5 Mg of pruned biomass ha^{-1} for both almonds [34] and vineyards [31]. Researchers

Table 3 Biomass and C fractions for five different root diameter classes including the stump

Root diameter classes (mm)	Root Biomass (Mg ha^{-1})	Root C (Mg C ha^{-1})	Fraction (%)
<2	0.4	0.16	3.8
2–6	1.2	0.51	12.5
6–20	2.9	1.27	31.3
>20	2.6	1.16	28.6
Stump	2.2	0.97	23.8
Total	9.3	4.1	100

Estimations per hectare are based on vine spacing = 1.83 × 3.35 (1631 vines ha^{-1})

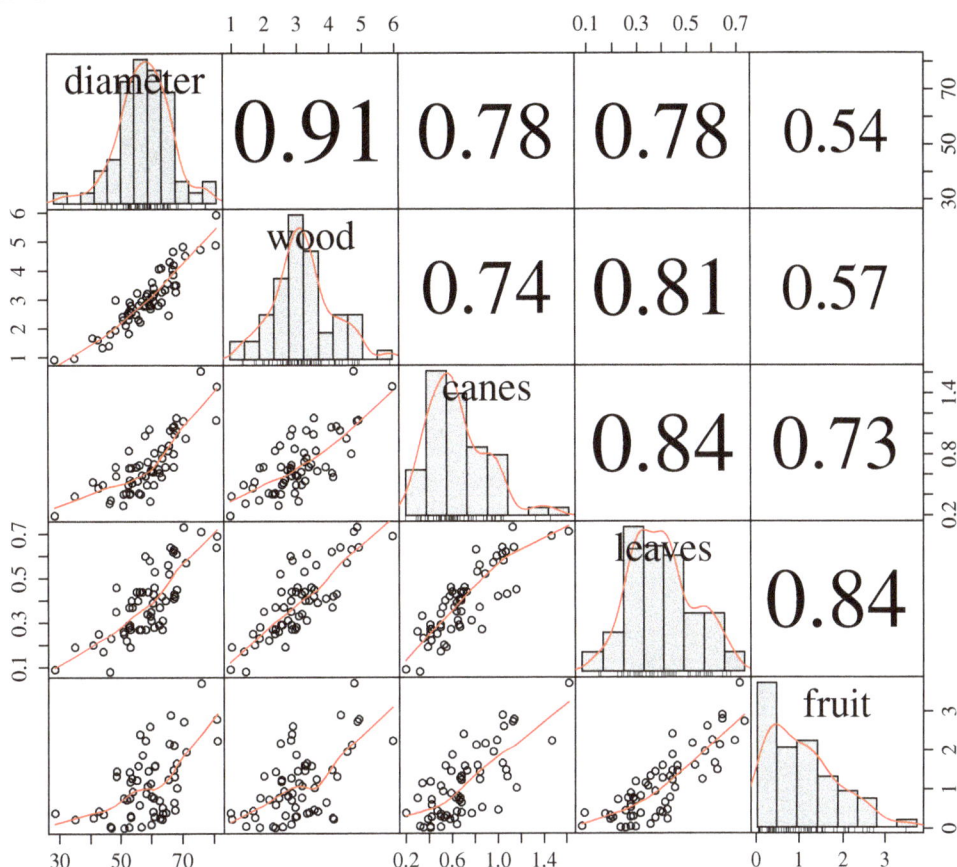

Fig. 5 *Scatterplot* matrix showing correlations between vine components. Vine trunk diameter (mm) and vine C stocks (kg C vine^{-1}) fractions in wood (trunk plus cordons, i.e. permanent scaffold), canes, leaves and fruit (grape clusters including rachis) are shown. The *lower left triangle* below the diagonal shows *scatterplots* with loess curves (*x* and *y axis* indicating values for each variable). The frequency histograms in the diagonal include kernel density estimation curves (only *x axis* indicating values for the variable). The *upper right triangle* shows Pearson correlation coefficients with all levels of significance p < 0.001, (n = 60)

reported that mature orchard crops in California allocated, on average, one third of their NPP to harvestable biomass [32], and mature vines allocated 35–50% of that year's production to grape clusters [33]. Our estimate of 50% of annual biomass C allocated to harvested clusters represent the fraction of the structures grown during the season (1.7 out of 3.4 Mg C ha^{-1}). Furthermore, if woody annual increments were considered (i.e., estimating differences between two seasons by applying our "trunk diameter" model) this proportion would be even lower. Likewise the observed 1.7 Mg ha^{-1} in fruit represents ~14% of total biomass (1.7 out of 12.3 Mg C ha^{-1}), which is within 10% of other studies in the region at similar vine densities [i.e., 40]. More importantly, this study reports the fraction of C that could be recovered from winemaking and returned to the soil for potential long term storage. However, this study is restricted to the agronomic and environmental conditions of the site, and the methodology would require validation and potential adjustment in other locations and conditions.

Few studies have conducted a thorough evaluation of belowground vine biomass in vineyards, although Elderfield [41] did estimate that fine roots contributed 20–30% of total NPP and that C was responsible for 45% of that dry matter. More recently, Brunori et al. [13] studied the capability of grapevines to efficiently store C throughout the growing season and found that root systems contributed to between 9 and 26% of the total vine C fixation in a model *Vitis vinifera sativa* L. cv Merlot/berlandieri rupestris vineyard.

The results of our study provide a utilitarian analysis of C storage in mature wine grape vines, including above- and belowground fractions and annual vs. perennial allocations. Such information constitutes the basic unit of measurement from which one can then estimate the contribution of wine grapes to C budgets at multiple scales—fruit, plant or vineyard level—and by region, sector, or in mixed crop analyses.

Our study builds on earlier research that focused on the basic physiology, development and allocation of biomass

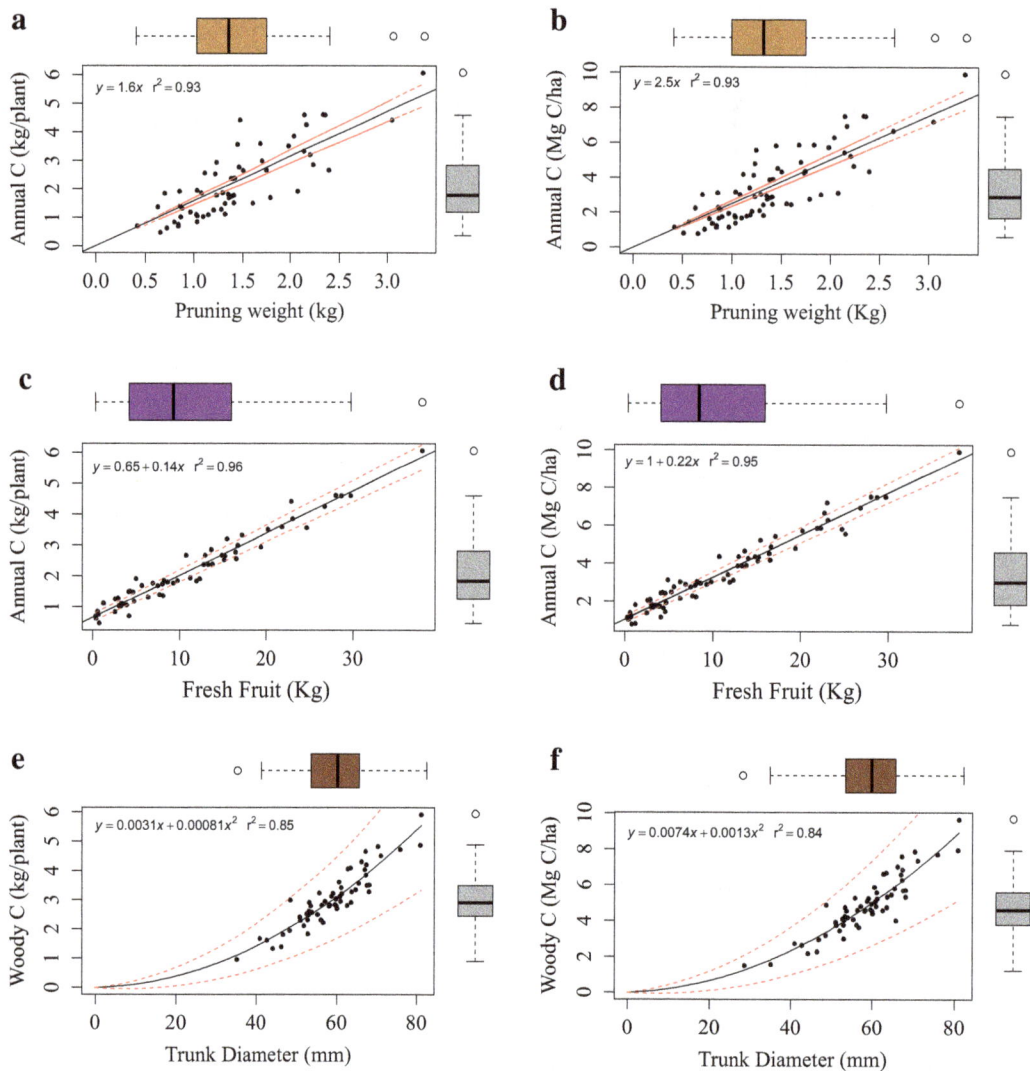

Fig. 6 Linear and quadratic vine allometrics with 95% confidence intervals. Allometrics for (**a**) pruning weight and annual C stocks ($R^2 = 0.93$, p < 0.001), **b** fresh fruit weight and annual C stocks ($R^2 = 0.96$, p < 0.001), and **c** trunk diameter and woody C stocks ($R^2 = 0.85$, p < 0.001) are shown. Annual C stocks represent the C content of canes, leaves and fruit together. Woody C stocks represent the C content of trunk plus cordons (perennial wood). Allometrics scaled to block level response (Mg C ha^{-1}) at planting density are also shown for pruning weight (**d**), wet fruit (**e**), and trunk diameter (**f**)

Table 4 Physical comparisons of biomass C measurements in vineyard block

Method	Model	Number of samples	Mean (Mg C ha^{-1})	Range (Mg C ha^{-1})	Sd (σ)
Standing biomass		60	10.02	5.9–16.2	1.9
Mounds	Semi-ovoid	26	9.93	6.2–17.0	2.7
	Hemispherical		10.57	5.9–23.4	3.6

Mounds included trunk plus cordons and canes. Estimations for standing biomass consider only the elements included in mounds to make them comparable

in vines [33, 42, 43]. Previous research has also examined vineyard-level carbon at the landscape level with coarser estimates of the absolute C storage capacity of vines of different ages, as well as the relative contribution of vines and woody biomass in natural vegetation in mixed vineyard-wildland landscapes [12]. The combination of findings from those studies, together with the more precise and complete (i.e. detailed, measurement-based above- and below-ground C estimates) carbon-by-vine structure assessment provided here, mean that managers now have access to methods and analytical tools that allow precise and detailed C estimates from the individual vine to whole-farm scales.

As carbon accounting (including offsets, credits and payments) in vineyard landscapes becomes more sophisticated, widespread and economically relevant, such vineyard-level analyses will become increasingly important for informing management decisions. The greater vine-level measuring precision that this study affords should also translate into improved scaled-up C assessments (e.g., county-, state- or sector-wide). In California alone, for example, there are more than 230,000 ha are planted in vines [21]. Given that for many, if not most of those hectares, the exact number of individual vines is known, it is easy to see how improvements in vine-level measuring accuracy can have benefits from the individual farmer to the entire sector.

Previous efforts to develop rough allometric woody biomass equations for vines notwithstanding [12], there is still a need to improve our precision in estimating of how biomass changes with different parameters. Because the present analysis was conducted for 15 year old Cabernet vines, there is now a need for calibrating how vine C varies with age, varietal and training system. There is also uncertainty around the influence of grafting onto rootstock on C accumulation in vines. As mentioned in the methods, the vines in this study were not grafted—an artifact of the root-limiting duripan approximately 50 cm below the soil surface. The site's location on the flat, valley bottom of a river floodplain also means that its topography, while typical of other vineyard sites per se, created conditions that limit soil depth, drainage and decomposition. As such, the physical conditions examined here may differ significantly from more hilly regions in California, such as Sonoma and Mendocino counties. Similarly, the lack of a surrounding natural vegetation buffer at this site compared to other vineyards (e.g., [12]) may mean that the ecological conditions of the soil communities may or may not have been broadly typical of those found in other vineyard sites. Thus, to the extent that future studies can document the degree to which such parameters influence C accumulation in vines or across sites, they will improve the accuracy and utility of C estimation methods and enable viticulturists to be among the first sectors in agriculture for which accurate C accounting is an industry-wide possibility.

The current study was also designed to complement a growing body of research focusing on soil-vine interactions [44–46]. Woody carbon reserves and sugar accumulation play a supportive role in grape quality, the main determinant of crop value in wine grapes. The extent to which biomass production, especially in belowground reservoirs, relates to soil carbon is of immediate interest for those focused on nutrient cycling, plant health and fruit production, as well as for those concerned with C storage [44, 47, 48].

The soil-vine interface may also be the area where management techniques can have the highest impact on C stocks and harvest potential [45, 46, 49, 50]. We expect the belowground estimates of root biomass and C provided here will be helpful in this regard and for developing a more thorough understanding of belowground C stores at the landscape level. For example, Williams et al. [12] estimated this component to be the largest reservoir of C in the vineyard landscape they examined, but they did not include root biomass in their calculations. Others have assumed root systems to be ~30% of vine biomass based on the reported biomass values for roots, trunk, and cordons [40]. With the contribution of this study, the magnitude of the belowground reservoir can now be updated.

Conclusions

Wine is a commodity of worldwide importance, and vineyards constitute a significant land use and contribution to economies across Mediterranean biome and beyond [10, 51]. Like orchards and tree plantations, grapevines are a perennial crop that stores C long-term in woody tissue, thereby helping to mitigate GHG emissions. Our study provides estimates of C in grape vines by vine component, as well as a simple measurement tool kit that growers can use to estimate the C in their vines and vineyard blocks. The equations presented here represent some of the first allometric models for estimating grapevine C from berries to blocks, with the hope that widespread use and refinement of these techniques may lead to recognition and credit for the C storage potential of vineyards and other perennial woody crops, such as orchards. The successful implementation of these methods, if applied widely to multiple cropping systems, could improve the precision of measurement and the understanding of C in agricultural systems relative to other human activities.

Authors' contributions
Study concept and design: DRS, JHV, JAM. Acquisition of data: CMS, GCL. Drafting of manuscript: JAM, CMS, JHV, JNW. Critical revision: JNW, GCL, DRS, JHV, CMS. All authors read and approved the final manuscript.

Author details
[1] Environmental Systems, University of California, Merced, Merced, CA, USA. [2] Department of Viticulture and Enology, University of California, Davis, Davis, CA, USA. [3] College of Agriculture, California State University, Chico, CA, USA. [4] Instituto Politécnico Nacional, CIIDIR-Unidad Oaxaca, Santa Cruz Xoxocotlán, Oaxaca, Mexico. [5] School of Engineering, University of California, Merced, Merced, CA, USA.

Acknowledgements

We would like to thank the Smart Lab for field work assistance and the Center for Watershed Sciences for administrative support. We also thank Teamrat Ghezzehei for comments on the analysis, and Judah Grossman and Rodd Kelsey for project coordination and collaboration.

Competing interests

The authors declare that they have no competing interests.

Funding

Funding for this research was provided by the California Department of Fish and Wildlife through the Ecosystem Restoration Program [Grant No. E1120001] as administered by The Nature Conservancy. All plant material research was conducted under the express written authorization of The Nature Conservancy, managing entity for the Cosumnes River Preserve where the work was performed. The Smart Lab at UC Davis and Viers Lab at UC Merced provided additional support as funded by the University of California system.

References

1. Godfray HCJ, Beddington JR, Crute IR, Haddad L, Lawrence D, Muir JF, et al. Food security: the challenge of feeding 9 billion people. Science. 2010;327(5967):812–8.
2. Lipper L, Thornton P, Campbell BM, Baedeker T, Braimoh A, Bwalya M, et al. Climate-smart agriculture for food security. Nat Clim Change. 2014;4(12):1068–72.
3. Palm C, Blanco-Canqui H, DeClerck F, Gatere L, Grace P. Conservation agriculture and ecosystem services: an overview. Agric Ecosyst Environ. 2014;187:87–105.
4. California environmental protection agency air resources board. California GHG Emission Inventory 2015. http://www.arb.ca.gov/cc/inventory/pubs/reports/ghg_inventory_trends_00-13%20_10sep2015.pdf. Accessed 18 Sep 2016.
5. United States Department of Agriculture. California Agricultural Statistics, 2012 Crop Year: National Agricultural Statistics Service, Pacific Regional Office; 2013. www.nass.usda.gov/Statistics_by_State/California/Publications/California_Ag_Statistics/Reports/2012cas-all.pdf. Accessed 15 Sep 2016.
6. Haden VR, Dempsey M, Wheeler S, Salas W, Jackson LE. Use of local greenhouse gas inventories to prioritise opportunities for climate action planning and voluntary mitigation by agricultural stakeholders in California. J Environ Plan Manag. 2013;56(4):553–71.
7. Jackson L, Haden VR, Wheeler SM, Hollander AD, Perlman J, O'Geen T, et al. Vulnerability and adaptation to climate change in California agriculture. California Energy Commission; 2012.
8. Kroodsma DA, Field CB. Carbon sequestration in California Agriculture, 1980–2000. Ecol Appl. 2006;16(5):1975–85.
9. Wine Institute G-F. Gomberg-Fredrikson Report. California Dept. of Food & Agriculture, US Tax & Trade Bureau, and US Dept. of Commerce; 2009.
10. Viers JH, Williams JN, Nicholas KA, Barbosa O, Kotze I, Spence L, et al. Vinecology: pairing wine with nature. Conserv Lett. 2013;6(5):287–99.
11. Marras S, Masia S, Duce P, Spano D, Sirca C. Carbon footprint assessment on a mature vineyard. Agric For Meteorol. 2015;214(215):350–6.
12. Williams J, Hollander A, O'Geen A, Thrupp L, Hanifin R, Steenwerth K, et al. Assessment of carbon in woody plants and soil across a vineyard-woodland landscape. Carbon Balance Manage. 2011;6(1):11.
13. Brunori E, Farina R, Biasi R. Sustainable viticulture: the carbon-sink function of the vineyard agro-ecosystem. Agric Ecosyst Environ. 2016;223:10–21.
14. Andrews SS, Mitchell JP, Mancinelli R, Karlen DL, Hartz TK, Horwath WR, et al. On-farm assessment of soil quality in California's central valley. Agron J. 2002;94(1):12–23.
15. Veenstra JJ, Horwath WR, Mitchell JP. Tillage and cover cropping effects on aggregate-protected carbon in cotton and tomato. Soil Sci Soc Am J. 2007;71(2):362–71.
16. Carlisle E, Smart DR, Williams LE, Summers M. California vineyard greenhouse gas emissions: Assessment of the available literature and determi-

17. Alsina MM, Fanton-Borges AC, Smart DR. Spatiotemporal variation of event related N2O and CH4 emissions during fertigation in a California almond orchard. Ecosphere. 2013;4(1):1.
18. Kavargiris SE, Mamolos AP, Tsatsarelis CA, Nikolaidou AE, Kalburtji KL. Energy resources' utilization in organic and conventional vineyards: energy flow, greenhouse gas emissions and biofuel production. Biomass Bioenergy. 2009;33(9):1239–50.
19. Herath I, Green S, Singh R, Horne D, van der Zijpp S, Clothier B. Water footprinting of agricultural products: a hydrological assessment for the water footprint of New Zealand's wines. J Clean Prod. 2013;41:232–43.
20. Hannah L, Roehrdanz PR, Ikegami M, Shepard AV, Shaw MR, Tabor G, et al. Climate change, wine, and conservation. Proc Natl Acad Sci. 2013;110(17):6907–12.
21. Wine Institute and International Partners to Release New Greenhouse Gas Protocol and Accounting Tool [press release]. San Francisco; 2008.
22. Forsyth K, Oemcke D. International Wine Carbon Calculator Protocol Version 1.2. Provisor Pty Ltd and Yalumba Wines, Hartley Grove, Urrbrae, SA 5064. Australia; 2008. p. 152.
23. Schultz HR. Climate change and viticulture: research needs for facing the future. J Wine Res. 2010;21(2–3):113–6.
24. Chave J, Andalo C, Brown S, Cairns M, Chambers J, Eamus D, et al. Tree allometry and improved estimation of carbon stocks and balance in tropical forests. Oecologia. 2005;145(1):87–99.
25. Saatchi SS, Harris NL, Brown S, Lefsky M, Mitchard ETA, Salas W, et al. Benchmark map of forest carbon stocks in tropical regions across three continents. Proc Natl Acad Sci. 2011;108(24):9899–904.
26. D'Acunto L, Semmartin M, Ghersa CM. Uncropped field margins to mitigate soil carbon losses in agricultural landscapes. Agric Ecosyst Environ. 2014;183:60–8.
27. Henry M, Tittonell P, Manlay R, Bernoux M, Albrecht A, Vanlauwe B. Biodiversity, carbon stocks and sequestration potential in aboveground biomass in smallholder farming systems of western Kenya. Agric Ecosyst Environ. 2009;129(1):238–52.
28. Kuyah S, Dietz J, Muthuri C, Jamnadass R, Mwangi P, Coe R, et al. Allometric equations for estimating biomass in agricultural landscapes: I. Aboveground biomass. Agric Ecosyst Environ. 2012;158:216–24.
29. Kuyah S, Dietz J, Muthuri C, Jamnadass R, Mwangi P, Coe R, et al. Allometric equations for estimating biomass in agricultural landscapes: II. Belowground biomass. Agric Ecosyst Environ. 2012;158:225–34.
30. Tadesse G, Zavaleta E, Shennan C. Effects of land-use changes on woody species distribution and above-ground carbon storage of forest-coffee systems. Agric Ecosyst Environ. 2014;197:21–30.
31. Christensen LP. Raisin production manual: UCANR Publications; 2000.
32. Rufat J, DeJong TM. Estimating seasonal nitrogen dynamics in peach trees in response to nitrogen availability. Tree Physiol. 2001;21(15):1133–40.
33. Williams LE. Growth and development of grapevines. In: Christensen LP, editor. Raisin production manual. Oakland: University of California Agriculture and Natural Resources; 2000. p. 17–23.
34. Holtz B, McKenry M, Caesar-TonThat T. Wood chipping almond brush and its effect on the almond rhizosphere, soil aggregation and soil nutrients. Acta Hortic. 2004;4:127–34.
35. Sharrow S, Ismail S. Carbon and nitrogen storage in agroforests, tree plantations, and pastures in western Oregon, USA. Agrofor Syst. 2004;60(2):123–30.
36. Kleinschmidt Associates. Cosumnes River preserve management plan—final. Report. Kleinschmidt Associates; 2008.
37. NOAA. Technical Memorandum NWS WR-272: Climate of Sacramento, CA. Technical memorandum. National Oceanic and Atmospheric Administration—Department of Commerce, USA; 2005. Contract No.: WR-272.
38. BizEE Software Unlimited. Custom Degree Day Data- Cosumnes River, Wilton, CA, US (121.23 W, 38.44 N) 2016. http://www.degreedays.net/. Accessed 12 Sept 2016.
39. R Core Team. R: a language and environment for statistical computing. In: R Foundation for Statistical Computing V, Austria. Vienna; 2015.
40. Keightley KE, Bawden GW. 3D volumetric modeling of grapevine biomass using Tripod LiDAR. Comput Electron Agric. 2010;74(2):305–12.
41. Elderfield H, Schlesinger WH. Biogeochemistry. An analysis of global

change, xiii + 588 pp. San Diego, London, Boston, New York, Sydney, Tokyo, Toronto: Academic Press. Price US $49.95 (paperback). ISBN 0 12 625155 X. CHAMEIDES, WL & PERDUE, EM 1997. Biogeochemical Cycles. A Computer-Interactive Study of Earth System Science and Global Change. xi + 224 pp. + disk. New York, Oxford: Oxford University Press. Price£ 37.50 (hard covers). ISBN 0 19 509279 1. Geological Magazine. 1998;135(06):819–42.

42. Williams LE. Grape. In: Zamski E, Schaffer AA, editors. Photoassimilate distribution in plants and crops: source–sink relationships. New York: Marcel Dekker; 1996. p. 851–81.

43. Williams LE, Biscay PJ. Partitioning of dry weight, nitrogen, and potassium in Cabernet Sauvignon grapevines from anthesis until harvest. Am J Enol Vitic. 1991;42(2):113–7.

44. Eldon J, Gershenson A. Effects of cultivation and alternative vineyard management practices on soil carbon storage in diverse Mediterranean landscapes: a review of the literature. Agroecol Sustain Food Syst. 2015;39(5):516–50.

45. Simansky V. Soil organic matter in water-stable aggregates under different soil management practices in a productive vineyard. Arch Agron Soil Sci. 2013;59(9):1207–14.

46. Steenwerth K, Belina KM. Cover crops enhance soil organic matter, carbon dynamics and microbiological function in a vineyard agroecosystem. Appl Soil Ecol. 2008;40(2):359–69.

47. Suddick EC, Ngugi MK, Paustian K, Six J. Monitoring soil carbon will prepare growers for a carbon trading system. Calif Agric. 2013;67(3):162–71.

48. Zarraonaindia I, Owens SM, Weisenhorn P, West K, Hampton-Marcell J, Lax S, et al. The soil microbiome influences grapevine-associated microbiota. Mbio. 2015;6(2):e02527.

49. Bosco S, Di Bene C, Galli M, Remorini D, Massai R, Bonari E. Soil organic matter accounting in the carbon footprint analysis of the wine chain. Int J Life Cycle Assess. 2013;18(5):973–89.

50. Agnelli A, Bol R, Trumbore SE, Dixon L, Cocco S, Corti G. Carbon and nitrogen in soil and vine roots in harrowed and grass-covered vineyards. Agric Ecosyst Environ. 2014;193:70–82.

51. Underwood EC, Viers JH, Klausmeyer KR, Cox RL, Shaw MR. Threats and biodiversity in the mediterranean biome. Divers Distrib. 2009;15(2):188–97.

14

Greenhouse gas emissions from tropical forest degradation: an underestimated source

Timothy R. H. Pearson[*][iD], Sandra Brown, Lara Murray and Gabriel Sidman

Abstract

Background: The degradation of forests in developing countries, particularly those within tropical and subtropical latitudes, is perceived to be an important contributor to global greenhouse gas emissions. However, the impacts of forest degradation are understudied and poorly understood, largely because international emission reduction programs have focused on deforestation, which is easier to detect and thus more readily monitored. To better understand and seize opportunities for addressing climate change it will be essential to improve knowledge of greenhouse gas emissions from forest degradation.

Results: Here we provide a consistent estimation of forest degradation emissions between 2005 and 2010 across 74 developing countries covering 2.2 billion hectares of forests. We estimated annual emissions of 2.1 billion tons of carbon dioxide, of which 53% were derived from timber harvest, 30% from woodfuel harvest and 17% from forest fire. These percentages differed by region: timber harvest was as high as 69% in South and Central America and just 31% in Africa; woodfuel harvest was 35% in Asia, and just 10% in South and Central America; and fire ranged from 33% in Africa to only 5% in Asia. Of the total emissions from deforestation and forest degradation, forest degradation accounted for 25%. In 28 of the 74 countries, emissions from forest degradation exceeded those from deforestation.

Conclusions: The results of this study clearly demonstrate the importance of accounting greenhouse gases from forest degradation by human activities. The scale of emissions presented indicates that the exclusion of forest degradation from national and international GHG accounting is distorting. This work helps identify where emissions are likely significant, but policy developments are needed to guide when and how accounting should be undertaken. Furthermore, ongoing research is needed to create and enhance cost-effective accounting approaches.

Keywords: Carbon stock, Deforestation, Forest fire, Woodfuel, REDD+, Timber harvest

Background

The degradation of forests in developing countries, particularly those within tropical and subtropical latitudes, is perceived to be an important contributor both to global greenhouse gas emissions and to development. Its impacts are understudied and poorly understood, and present a major challenge for national-level carbon inventories [7] and for addressing diminishing biodiversity [5]. International emission reduction programs (especially reducing emissions from deforestation and degradation, conservation of forest carbon stocks, sustainable management of forests and enhancement of forest carbon stocks—REDD+) have focused mostly on deforestation, which is easier to detect and thus more readily measured and monitored than forest degradation [13]. A key challenge for measuring and monitoring forest degradation is that it is difficult to detect using commonly-used remote sensing products, such as Landsat. Instead, much higher resolution imagery is needed to identify the more subtle changes in forest cover typical of forest degradation activity. The World Bank, a major REDD+ investor/donor, established a Carbon Fund [29] with a methodological framework that requires emissions from forest degradation to be accounted where 'significant', which is defined as more than 10% of 'forest-related emissions'. Yet it is unclear how to quantify and meaningfully

*Correspondence: tpearson@winrock.org
Winrock International, 2121 Crystal Drive, Suite 500, Arlington, VA 22101, USA

demonstrate "significance", or how to account for emissions cost-effectively when significant.

Forest degradation occurs when there is a direct, human-induced decrease in carbon stocks in forests resulting from a loss of canopy cover that is insufficient to be classed as deforestation [11, 17]. Moreover, the decrease in carbon stocks should be persistent, although the duration of this persistence has not been defined. Common drivers of forest degradation include timber harvesting (legal and illegal), fuel wood collection, non-stand replacing fires, and animal grazing in the forest (preventing forest regeneration) [11].

A handful of studies have attempted to assess and quantify emissions from human-driven forest degradation, including an assessment of the importance of drivers of forest degradation made by Hosonuma et al. [14]. This study was based on data only for the area of forest disturbed in 46 tropical and sub-tropical countries. Of the total area of disturbed forests in these countries, they found that 51% of the disturbed area was caused by timber harvesting, 31% by woodfuel harvest, 9% by fires, and 7% by grazing. While timber harvest was the most significant activity in South and Central America and Asia, woodfuel was the largest activity by proportion (48%) in Africa. These estimates only included a subset of tropical and subtropical countries; were not produced through an independent and consistent assessment; and offered no quantitative information on the magnitude of the greenhouse gas emissions and how they compare to those from deforestation.

Another assessment of the emissions from forest degradation in the tropics conducted by Houghton [15] was based on his bookkeeping model for the period 1990–2010. He estimated that the average annual *net* emissions from harvesting of timber and woodfuel (with the exclusion of the re-clearing of forest fallow within the shifting cultivation cycle) just 10% of the summed emissions from deforestation and degradation, with degradation emissions dominated by timber harvest with marginal emissions from woodfuel, and no emission from fires. Given the exclusion of other key causes of forest degradation, this study is incomplete and lacks consistency.

Recent work by Pearson et al. [24] focused on the perceived key cause of forest degradation: timber harvest and associated infrastructure (skid trails and logging roads). They showed that for nine major tropical timber producing countries, emissions from logging were on average equivalent to about 12% of those from deforestation. For those nine countries with relatively low emissions from deforestation, emissions from logging were found to be equivalent to half or more of those from deforestation, whereas for countries with the highest emissions from deforestation, emissions from logging were equivalent to <10% of those from deforestation.

These estimates are supported by the work of Asner and others in the Brazilian Amazon. Asner et al. [3] estimated logged areas ranged from 60 to 123% of previously reported deforestation areas. Huang and Asner [16] estimated that the inclusion of timber harvest elevated emissions by 15–19% over the emissions from deforestation alone.

Collection of traditional woodfuel (firewood and charcoal) for cooking and heating is common throughout the tropics, and can lead to forest degradation where removals exceed regrowth. Where annual harvest of woodfuel exceeds the forest's incremental growth in biomass, it is considered to be unsustainable, and leads to a decline of woody biomass and to net carbon emissions [4]. Bailis et al. [4] estimated that 27–34% of woodfuel harvest was unsustainable, particularly in East Africa and South Asia, and thus leads to significant forest degradation.

Fire is an important cause of forest disturbance and is commonly used to manage forest lands in the tropics and subtropics [28]. Fire is often used to transform forest, e.g. into croplands, but this is a land-use change and so is considered to be deforestation rather than forest degradation. When fires in forests are not associated with an intentional conversion for a land-use change, this is considered to be forest degradation. The work by van der Werf et al. [28] included an analysis for tropical latitudes that partitioned the forest fires into two classes: non-deforestation fires (i.e. forest degradation), and deforestation fires.

It is clear that no estimates of CO_2 emissions currently exist that incorporate all major forms of forest degradation. Thus, a systematic, consistent calculation approach is needed to allow for an estimation of all significant emissions across all tropical and subtropical developing countries. Such improved knowledge on emissions from forest degradation would allow decision makers to understand the extent of forest degradation and what opportunities there are to reduce associated emissions. As such, the goals of our work were to: (1) provide a consistent estimate of CO_2 emissions from the major causes of degradation in the tropical and subtropical forests of developing countries, and (2) compare the magnitude of the emissions caused by forest degradation and its sub-activities with those from deforestation in both absolute and relative terms. Results from such an analysis would provide guidance to national and international policy makers as to which forest lands to allocate resources so that national GHG emissions are reduced.

Methods

Our analysis of emissions from forest degradation covers 74 developing countries located mostly in tropical and subtropical latitudes. These countries contain 4.7 billion hectares of land area and 2.2 billion hectares of forest (Fig. 1). The three main causes of forest degradation included in this analysis are:

- Selective timber harvest in native forests.
- Woodfuel harvest—where removals exceed regrowth of forest C stocks.
- Fire—wildfires that do not cause a change in land-use.

All estimates of emissions for each activity are of gross emissions, and do not take into account how persistent the degradation might be or any regrowth and forest recovery. All estimates are derived from global databases and the scales, carbon pools, and time frame for each activity are given in Table 1.

Degradation
Selective timber harvest
The methodology described in Pearson et al. [24] was used to estimate total emissions from selective timber harvest. Emissions include those from (1) direct carbon loss of the extracted log (extracted log emissions—ELE); (2) the top and stump of the felled tree, plus trees incidentally killed or severely damaged surrounding the logging gap (logging damage factor—LDF); and (3) trees killed during the construction of logging infrastructure (logging infrastructure factor—LIF). This method combines these sources of emissions associated with selective timber harvesting to derive a single emission factor that is applied to the volume of timber extracted. The inverse of emissions is carbon stored therefore the calculation of emissions captures both all losses and the impact of carbon stored in long-term wood products (e.g., in furniture or buildings). Pearson et al. [24, Feldpausch et al. 9], and unpublished data from Ghana provide estimates of the emission factors for seven tropical forested countries based on field data collection as shown in Table 2. These factors were applied in this study based on what region the countries are best suited to represent, as shown in Table 2.

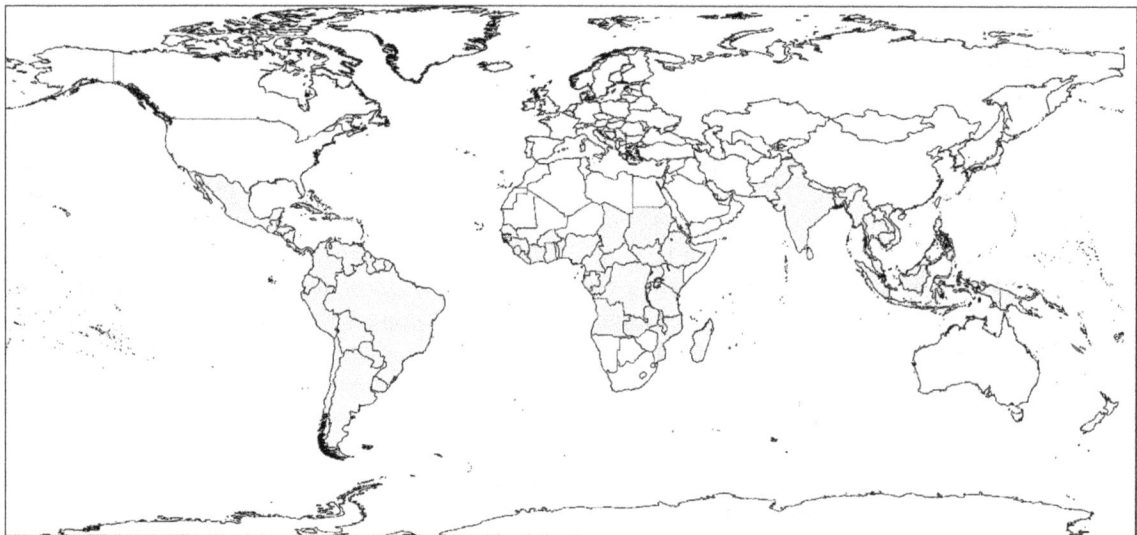

Fig. 1 Map of included countries (shaded in *blue*)

Table 1 Summary of activities, spatial scale, pools and time frame included in the analysis

Activity	Source	Spatial scale	Pools included	Time frame
Timber harvest	[24]	National	Above and belowground live biomass, harvested wood products	2005–2010
Wood fuel harvest	WISDOM Model	GADM Level 1	Above and belowground live biomass	2009
Fire	Global Fire Emissions Database [28]	50 km Summed to GADM Level 1	Above and belowground live biomass, dead wood, and litter	2005–2010
Deforestation	[12, 25]	Area—30 m Stocks—250 m Summed to GADM Level 1	Above and belowground live biomass, dead wood, litter, and soil carbon	2005–2010

GADM database of global administrative areas (http://www.gadm.org)

Table 2 Source of field data for development of timber harvesting emission factors (*ELE* extracted log emission, *LDF* logging damage factor, *LIF* logging infrastructure factor)

Region	ELE	LDF	LIF	Country
Central Africa	0.25	0.5	0.24	Republic of Congo
Rest of Africa	0.37	0.67	0.24	Ghana
Central America and Caribbean	0.28	1.26	0.27[a]	Belize
Andean countries[b] (Bolivia, Colombia, Ecuador, Paraguay, Peru, Venezuela)	0.30	1.23	0.27[a]	Bolivia
Brazil	0.38	0.71	0.27[a]	Brazil
Guyana, Suriname, French Guyana	0.36	0.99	0.98	Guyana
Asia	0.25	0.57	0.67	Indonesia

All factors are in units of Mg C m^{-3}

[a] The values for the LIF are from Feldpausch et al. [9]

[b] These countries are mostly Andean but grouped into once class

Average annual industrial roundwood production (IRP), a measure of the extracted volumes, for the period of 2005–2010 was obtained from the FAO Global Forest Resources Assessment database (FAOSTAT) [10], as well as the country reports submitted to the FAO as part of the Forest Resource Assessment (FRA) program. Because the reported IRP include volumes produced from native forests and forest plantations, the reported IRP was adjusted to ensure that only timber production from native forests was considered (to capture only emissions from selective logging). For the majority of timber-producing countries included in the analysis (representing 96% of the total IRP), country-specific harvest volumes from plantations for the 2005–2010 timeframe reported in Jürgensen et al. [20] were subtracted from the average total industrial roundwood production volume, as reported by FAOSTAT for the same time period. For countries not included in Jürgensen et al. [20], no adjustments were made, as we assumed that IRP from plantations (if they exist) were insignificant.

Woodfuel

Emissions from woodfuel were derived using the WISDOM model [4] that estimates the fraction of non-renewable biomass (NRB) in relation to supply and demand potential [4]. In the WISDOM model, woodfuel derived as a byproduct of deforestation activities was not included in order to avoid double-counting deforestation emissions. As the WISDOM model estimates only include the aboveground biomass pool, an expansion factor of 1.32 was applied to conservatively estimate the total biomass, based on the American Carbon Registry's Energy efficiency measures in thermal applications of nonrenewable biomass methodology [2], based on the CDM-approved methodology AMS-II.G, Version 05.0. This factor assumes that for every unit of biomass extracted from the forest, an additional 10% is left in the field from uncollected aboveground biomass. A further

20% is conservatively estimated to remain from root biomass.

Fire

The Global Fire Emissions Database (GFED; [28]) was used for estimates of emissions from forest fire. The GFED provides a global monthly layer with a cell size of 0.5 decimal degrees (approx. 50 × 50 km) of dry matter emissions that are classified into different sources and land cover types. Within the humid tropical forest biome, fire emissions from deforestation are decoupled from other emissions based on fire persistence (the length of time for which a fire burns in the same location). To avoid double-counting with deforestation emissions, only emissions from GFED-classified forest fires within latitudes 23° North and South (and not deforestation fires) were used in this degradation category. The GFED3 monthly layers from 2005–2010 were used for this study, and emissions estimates for only CO_2 are reported here in order to be consistent with other degradation activities.

Deforestation

Although there are several estimates of CO_2 emissions from tropical deforestation published fairly recently (e.g. [1, 6, 13, 15, 26, 30]), these estimates were not used because they were not consistent with respect to carbon pools included, area of study, definition of forest, inclusion of other land-use changes, gross versus net emissions, and years covered. As one of our goals was to compare estimates of degradation emissions with those of deforestation, we believed it was important to estimate the emissions from deforestation in a manner consistent with our analysis of forest degradation (Table 1).

Emissions were obtained by multiplying the average forest carbon stocks for each administrative unit by the area of forest loss. We used the Hansen et al. [12] dataset, derived from Landsat 7 ETM+ satellite images, to determine the area of deforestation. Deforestation data was based on a canopy closure of 20% to ensure that deforestation in countries with more open, drier forests were well captured. Areas shown as loss (between 2005 and 2010) were considered to be deforested, and were summed across level-one subnational administrative units as defined by the GADM (Database of Global Administrative Areas; political boundaries reflecting states or districts).

Tropical peatswamp forests under threat for deforestation are overwhelmingly located in Indonesia and Malaysia (more than 56% of area), with the remainder generally located in areas where pressure for deforestation is very low including at high altitudes in the mountains of Africa, South America and Papua New Guinea [23]. A spatial layer of peat forest areas in Indonesia and Malaysia was created using information from the Harmonized World Soil Database (HWSD; FAO/IIASA/ISRIC/ISS-CAS/JRC. Harmonized world soil database [8]. FAO, Rome, Italy and IIASA, Laxenburg, Austria 2012). All soil units classified as histosols (a soil consisting primarily of organic materials and defined as having 40 cm or more of organic soil material in the upper 80 cm) were assumed to be peat soil in these two countries. All areas of deforestation according to the Hansen et al. [12] layer that occurred on peat in Indonesia and Malaysia were assumed to be deforestation of peatswamp forests, and the method to estimate soil emissions is given in Table 3. The emissions for non-peat soils use the soil C stock to 30 cm deep given in the HWSD and the IPCC [18] land-use change factors (Table 3).

Carbon stocks of the non-soil pools were derived as detailed in Table 3. Biomass was averaged across the subnational administrative units and carbon stocks from all pools were assumed to be committed to the atmosphere immediately at the time of deforestation. Emissions were

obtained by multiplying the average forest carbon stocks for each administrative unit by the area of forest loss.

Results

We estimated that total emissions from forest degradation were 2.1 Gt CO_2e (Table 4) across the 74 countries assessed. Emissions associated with timber harvest accounted for more than half of the total degradation emissions (53%) followed by woodfuel (30%) and fire (17%). Emissions from forest degradation represented 25% of the estimated total emissions from deforestation plus forest degradation.

Although emissions from forest degradation for all countries included in this study accounted for just a quarter of the total emissions (deforestation and forest degradation combined), for 28 of the 74 countries (38%), more than half of the total emissions were derived from forest degradation. Estimates of emissions from all sources of forest degradation were less than 10% in only 11 countries (Fig. 2; recall that where forest degradation is less than 10% of emissions from all sources, it may be omitted from accounting under the World Bank methodological framework for REDD+). The highest proportion of degradation emissions relative to total emissions (>75%) were found to occur in the more arid countries of South Asia and north and east Africa (Fig. 2).

The magnitude of total degradation emissions was highest in the largest forested countries, led by Brazil and Indonesia (Fig. 3a, b). Timber production was the largest source of degradation emissions for these countries (Fig. 3c, d). Woodfuel emissions were highest in South Asia, Indonesia and in east Africa. Notable emissions from fire occurred in DRC and parts of the Brazilian Amazon (Fig. 3g, h). However, proportionally, fire

Table 3 Source of data for calculating emissions from deforestation

Pool	Source			
Aboveground live	Saatchi et al. biomass map ([25]; and unpublished update to 2011 increasing resolution from 500 to 250 m and adding additional ground data) Forest mask for year 2005 from Hansen et al. [12] to exclude non-forest biomass pixels			
Belowground live	Equations from Mokany et al. [22]			
Dead organic matter	Fraction of aboveground biomass [27]			
	Elevation (m)	Annual precipitation (mm year^{-1})	Deadwood fraction of AGB	Litter fraction of AGB
	<2000	<1000	0.02	0.04
	<2000	1000–1600	0.01	0.01
	<2000	>1600	0.06	0.01
	<2000	All	0.07	0.01
Soil organic matter	Peat soil emissions—annual emission factor for drained organic soil applied for 10 years (5.3 t CO_2 ha^{-1} year^{-1}; [19]) Non-peat soil emissions: C stock in top 30 cm from HWSD database Land use change soil factors from IPCC [18]			

Table 4 Estimated annual emissions from deforestation and forest degradation and relative proportions

Activity	Annual emission (Gt CO$_2$e year^{-1})	%
Degradation	2.06	25
Timber	1.09	53
Woodfuel	0.62	30
Fire	0.35	17
Deforestation	6.22	75

emissions were highest (about 75% or more of total emissions) for parts of Bolivia and Argentina in South America and for Central African countries (Fig. 3h).

The 35 countries with the greatest forest degradation emissions are divided into two groups—the top 10 with emissions >50 Mt CO$_2$ year^{-1} and the next 25 with emissions <50 Mt CO$_2$ year^{-1}—and are displayed with the varied proportion by source of emission (Fig. 4). The contribution of emissions by driver differs for these countries; timber harvest was the main cause for 5 of the top 10 countries, followed by woodfuel for three countries, and fire for the last two. For the countries in the second group, timber production was still the dominant cause for about half of them; and the dominant cause for remaining countries was equally divided into woodfuel and fire. Emissions from woodfuel are not correlated to the area of forest—several countries in Africa with relatively small areas of forest have high emissions from degradation due to woodfuel harvest. Emissions from woodfuel in general are relatively high in East Africa and South Asia, where it is a primary source of energy for cooking in not only in rural areas but also in urban areas (these two regions represent 71% of global woodfuel emissions; 439 Mt CO$_2$). For the relatively more developed countries of South and

Central America and the Caribbean, emissions caused by woodfuel harvests are insignificant. This is likely because alternate fuel sources are used and there is plenty of woodfuel available from timber harvesting offcuts.

This study reveals distinct patterns whereby dominant sources of emissions are split by region and continent (Fig. 5). South America and Southeast Asia contribute the most emissions from forest degradation (>51%), which can be attributed to their vast areas of high carbon stock forests. Forests in countries of Central America and the Caribbean as well as East Africa account for the least amount of degradation emissions (about 12%) due to their relatively small area of forests, many of which have low carbon stocks.

Discussion
Comparison of emissions from forest degradation
This study offers the first complete and consistent analysis of gross emissions from activities associated with the degradation of forest lands in developing countries in the tropical and subtropical latitudes. We estimated total forest degradation emissions of 2.1 Gt CO$_2$e year^{-1}, of which 53% was derived from timber harvest, 30% from woodfuel harvest, and 17% from forest fires.

Although Hosonuma et al. [14] did not quantify emissions, that study presented the proportion of total degradation resulting from each degradation activity (self-estimated by countries) for a subset of the countries included in our study area. Hosonuma et al. [14] estimated degradation emission sources as 51% from timber harvest, 31% from woodfuel and 9% from fires (compared to our results of 53, 30, 17). Breaking down by continent, Hosonuma et al. found that timber harvest exceeded 70% in South and Central America and Asia, but were just over 30% in Africa; woodfuel was 48% in

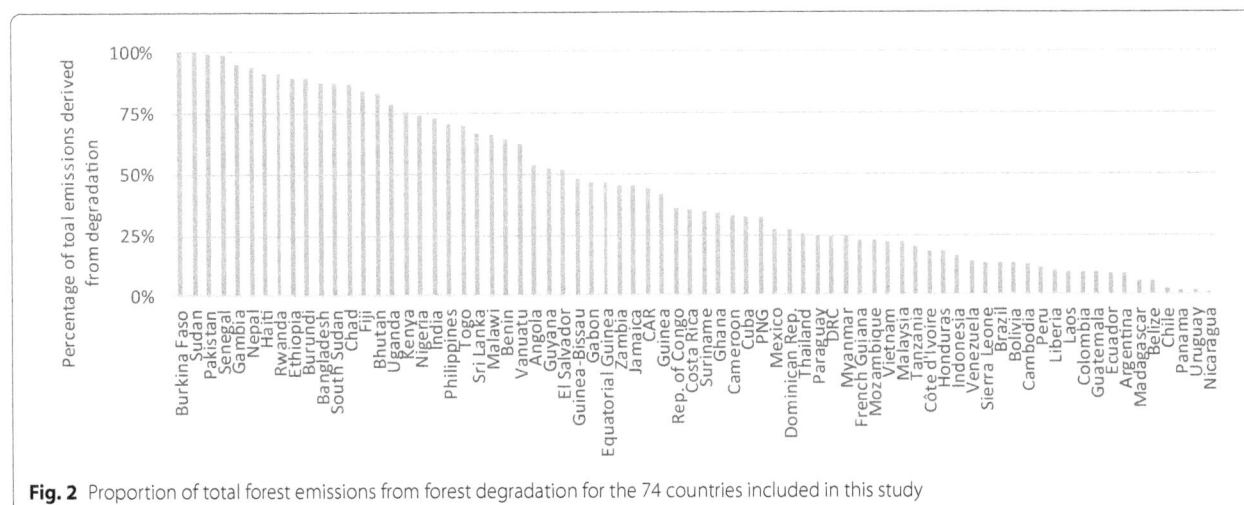

Fig. 2 Proportion of total forest emissions from forest degradation for the 74 countries included in this study

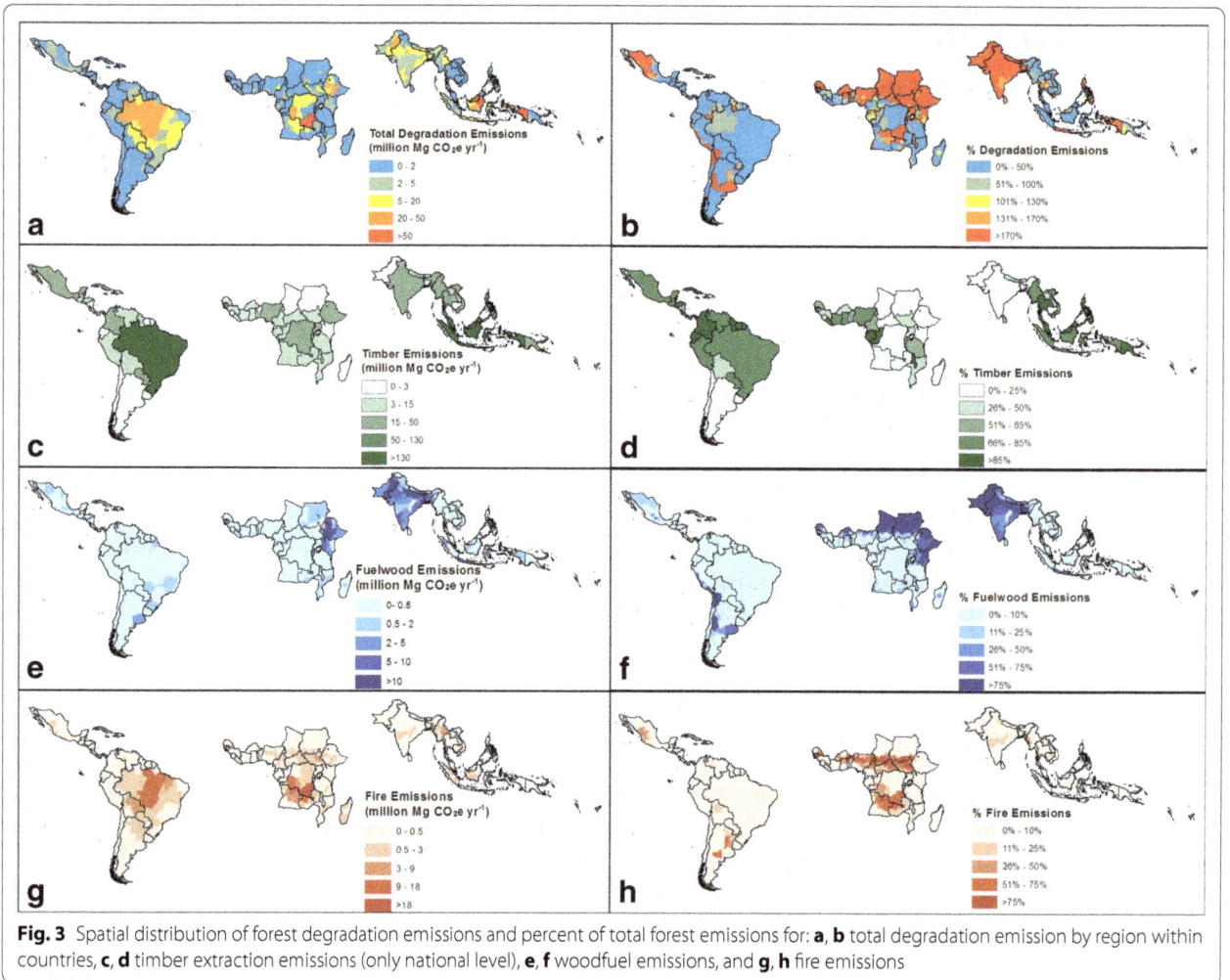

Fig. 3 Spatial distribution of forest degradation emissions and percent of total forest emissions for: **a**, **b** total degradation emission by region within countries, **c**, **d** timber extraction emissions (only national level), **e**, **f** woodfuel emissions, and **g**, **h** fire emissions

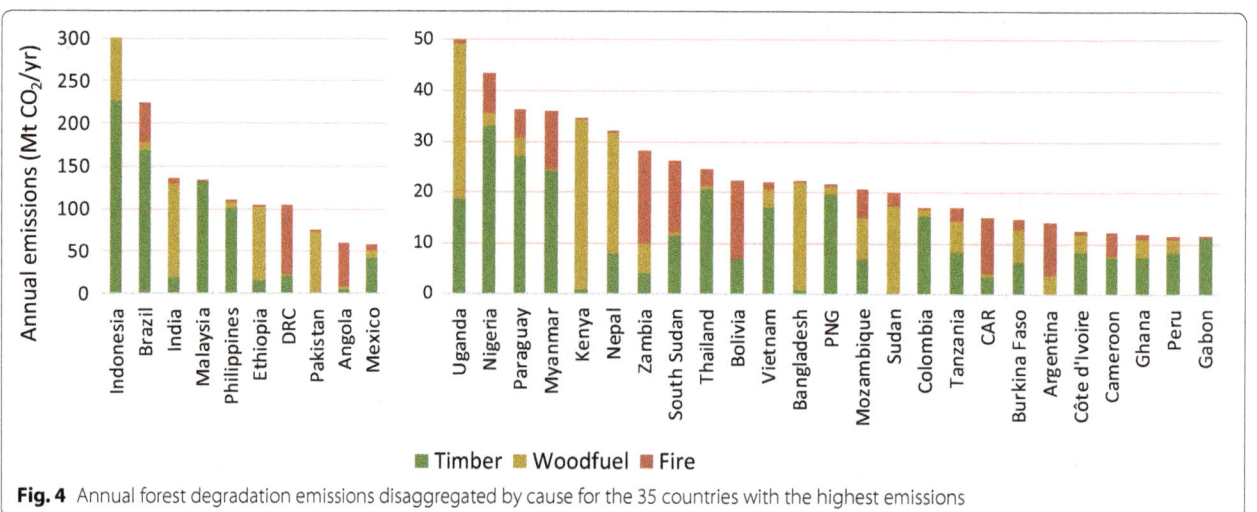

Fig. 4 Annual forest degradation emissions disaggregated by cause for the 35 countries with the highest emissions

Africa but less than 20% in Asia, and less than 10% in South and Central America; while fire was less than 20% in South and Central America, less than 10% in Africa and less than 5% in Asia. Thus the findings of Hosonuma et al. are largely in agreement with the findings of this study (Table 5) and highlight that the harvesting of

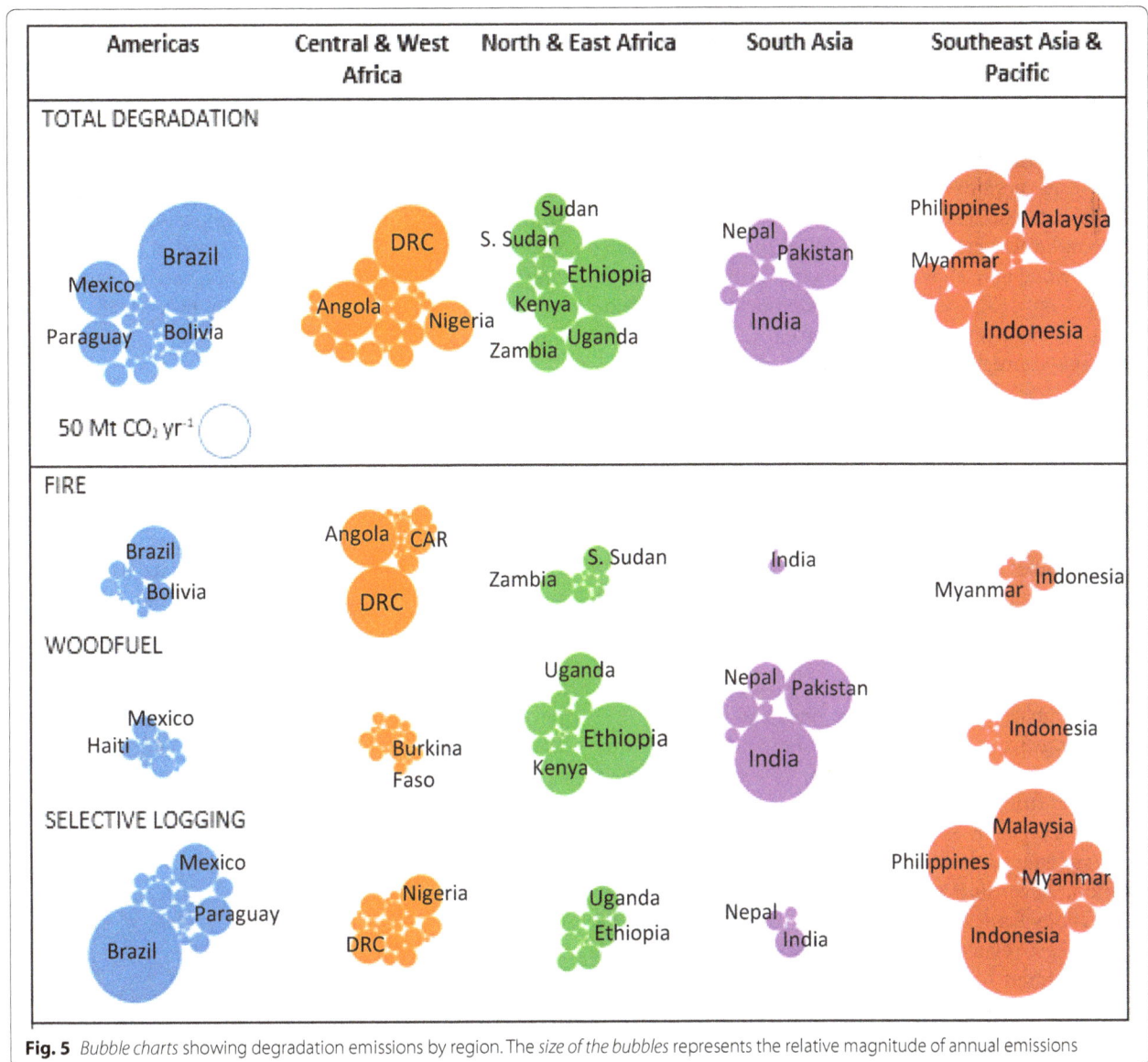

Fig. 5 *Bubble charts* showing degradation emissions by region. The *size of the bubbles* represents the relative magnitude of annual emissions

Table 5 Proportion of total forest degradation emissions by degrading activity by region

	Timber (%)	Woodfuel (%)	Fire (%)
America	69	10	21
Africa	31	36	33
Asia	61	35	5

timber and woodfuel are the largest contributor of emissions associated with forest degradation.

For another comparison we can specifically compare emissions from timber harvesting in the Brazilian Amazon. Huang and Asner [16] estimated annual gross emissions as 0.15–0.18 Gt CO_2e year^{-1}. Comparing just the nine Brazilian states that comprise the Amazon region, our study estimates emissions to be 0.28 Gt CO_2e year^{-1}, or more than 1.5 times higher than those reported by Huang and Asner. However, the Huang and Asner study explicitly stated that their estimate of gross annual emissions was likely to have substantially underestimated emissions due to the exclusion of areas that were deforested in subsequent years.

Emissions from deforestation versus forest degradation

The estimate of gross deforestation emissions presented in this study (average annual for 2005–2010 is 6.22 Gt CO_2) is included primarily to serve as a basis for consistent comparison with the estimates of degradation emissions. Recent published estimates of deforestation emissions [1, 6, 13, 15, 26, 30] have been smaller than our estimate, ranging from 2.3 to 4.2 Gt CO_2 year^{-1}. There are several reasons for the discrepancy between these estimates, including a focus on net rather than gross emissions, different time periods which will capture lower historical rates of deforestation—e.g. 2000–2005 [13] to 2001–2013 [30]—and different study areas. All of the estimates generally include only aboveground biomass carbon stocks in trees (except [13], which also included belowground biomass), yet our estimate includes all five IPCC carbon pools, including aboveground, belowground, dead wood, litter, soil, and peat. Belowground biomass of forests is about 20% or more of aboveground biomass and dead wood and litter will account for at least another 5% of aboveground biomass. Emissions from mineral soil due to cultivation generally account for another 20–25% of aboveground stocks. Taking all these factors into account, the emissions from the other studies could increase by as much as a factor of 1.5, or to a range of 3.5–6.3 Gt CO_2 year^{-1}, while still not including significant peat soil emissions in Indonesia and Malaysia. In light of all this, we conclude that our estimate of deforestation emissions is in line with other recently published estimates mentioned above.

Emissions from forest degradation are not an insignificant source of CO_2 and account for 25% of the summed emissions from deforestation and forest degradation of 8.28 Gt CO_2 year^{-1}. In other words, degradation emissions are equivalent to about a third of those from deforestation. According to the World Bank's Carbon Fund, if emissions from forest degradation are more than 10% of all forest-related emissions, they must be included and accounted for. As we have shown, emissions from all sources of forest degradation were less than 10% in only 11 out of the 74 countries, and thus all the remaining countries would need to include forest degradation in their accounting system. The guidelines, however, only give instructions on summed forest degradation but not on individual activities. For example, in Colombia summed degradation emissions were equal to 9% of total emissions, but all the emissions are from timber harvest and thus could be excluded under FCPF rules. In contrast, the summed degradation emissions in Peru were 11% but the timber harvest emissions comprised 8% of total degradation. While Peru's emissions from timber degradation are less significant than in Colombia, since total degradation emissions make up more than 10%, Peru would be required to also account for fire and woodfuel even though they sum to just 3% of emissions. Thus, there is a need for policies that better articulate the inclusion and exclusion of activities rather than the summed forest degradation level.

Significance of degradation emissions

The consistent estimates of emissions produced in this study allow us to consider the significance of total emissions resulting from forest degradation. To better illustrate this significance, we directly compared our estimates with emissions by country and emission sector as listed by the WRI CAIT database (http://cait.wri. org) for 2010. According to this comparison, degradation emissions are only significantly exceeded by the energy and agriculture sectors (Fig. 6). On a country basis, total emissions from forest degradation exceed all but the seven highest emitting countries (Fig. 6).

Uncertainties and omitted sources

The purpose of this analysis was to demonstrate the scale of emissions from forest degradation in a manner that is to the best of our knowledge consistent and accurate. This requires accurate information on extent of the type of forest degradation and the associated emissions. For selective logging, there was concern about the data used to estimate emissions, as it may have included timber volumes derived from plantations. However, steps were taken to ensure that our estimates of IRP capture extraction rates only from native forests. The logging emission factors were developed using data only from a limited number of countries yet have very small error bounds, and the emission

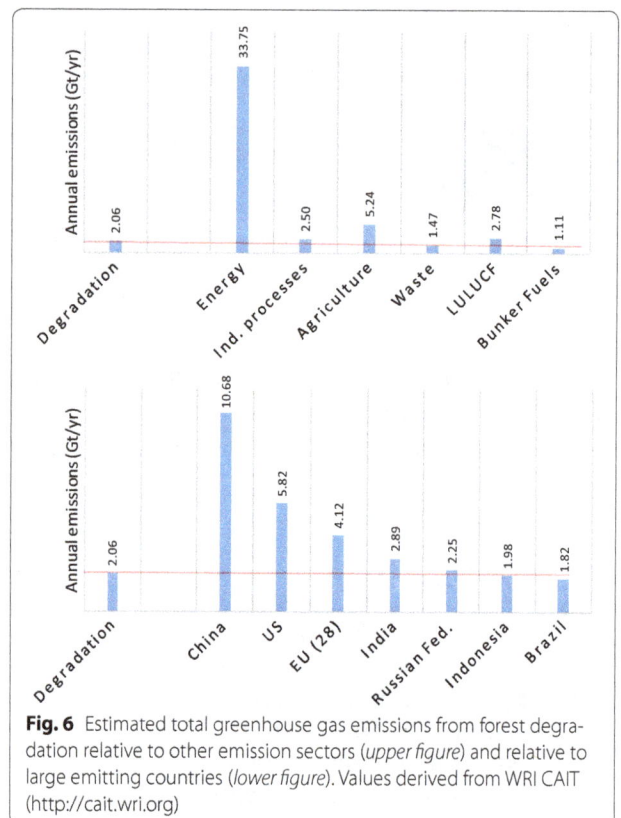

Fig. 6 Estimated total greenhouse gas emissions from forest degradation relative to other emission sectors (*upper figure*) and relative to large emitting countries (*lower figure*). Values derived from WRI CAIT (http://cait.wri.org)

sources considered have significant relationships with forest characteristics [24]. The fire analysis is spatially-specific and globally-consistent, and was designed to avoid double counting fire degradation emissions with fire emissions resulting from or associated with deforestation. The most uncertain emission source is woodfuel, given that the data are derived from a single year.

Estimates of emissions from timber harvest are likely to be underestimated due to the omission of illegal logging, assuming illegal logging is not included in national official statistics of IRP. It is important to acknowledge that research indicates that as much as 72% of logging is illegal in the Brazilian Amazon, 61% in Indonesia and 65% in Ghana [21].

Another omission is degradation through overgrazing. This source was included in Hosonuma et al. [14], who reported that this activity is responsible for 7% of the pantropical area of forest degradation (the least important form of degradation in the study). In addition, the impact of grazing is predominantly on regeneration, with damage to seedlings and saplings. The impact on forest carbon stocks is therefore small in the short term, though may be greater in later years as future generations of emergent trees are removed.

Conclusions

Our estimates show annual forest degradation emissions of 2.1 billion tons of carbon dioxide across 74 developing countries. To further illustrate the significance of this number: it exceeds both the total emission from highway vehicles (1.7 billion tons of carbon dioxide equivalents per year; fueleconomy.gov accessed 1/27/17), and the total emissions from power generation in the USA (1.9 billion tons of carbon dioxide equivalents per year; eia.gov accessed 1/27/17).

Our study demonstrates that, almost without exception, forest degradation emissions are significant. Indeed, by our estimates 85% of the countries studied surpass the defined minimum threshold and would be required to estimate forest degradation emissions under World Bank requirements for participation in the Carbon Fund REDD+ program.

Yet emissions from forest degradation are overlooked and not accounted in any complete or systematic way. It is imperative that this source of greenhouse gas emissions be better understood so that strategies that tap into the mitigation potential of addressing them may be developed. These strategies might in turn also offer significant economic and development opportunities.

This paper serves as a starting point to demonstrate the importance of forest degradation as a source of greenhouse gases, and to show where emissions are most significant—and thus where interventions may have the greatest impact.

Abbreviations
CO_2: carbon dioxide; ELE: extracted log emissions; FAO: Food and Agriculture Organization; GHG: greenhouse gases; HWP: harvested wood products; IPCC: Intergovernmental Panel on Climate Change; IRP: industrial roundwood production; LDF: logging damage factor; LIF: logging infrastructure factor; REDD+: reducing emissions from deforestation and degradation, conservation of forest carbon stocks, sustainable management of forests and enhancement of forest carbon stocks; NRB: non-renewable biomass.

Authors' contributions
The study and manuscript preparation was led and conceptualized by TRHP, SB provided scientific guidance on analysis and was the co-lead author, GS led the spatial analysis and LM conducted timber harvest analyses and data visualization. All authors read and approved the final manuscript.

Acknowledgements
We acknowledge helpful comments from anonymous reviewers. We thank Jeff Murray for copy editing of an advanced version of the manuscript.

Competing interests
The authors declare that they have no competing interests.

Funding
The database that forms the basis of this analysis was initially prepared under funding from the World Bank (contract 7167342). Subsequent support for analysis was derived from the Interamerican Development Bank (contract INE/CCS-RG-T2036-SN2/14).

References
1. Achard F, Ebeuchle R, Mayaux P, et al. Determination of tropical deforestation rates and related carbon losses from 1990 to 2010. Glob Change Biol. 2014;20:1–15.
2. American Carbon Registry. Switch from non-renewable biomass for thermal applications. Approved carbon accounting methodology under American Carbon Registry. http://americancarbonregistry.org/carbon-accounting/standards-methodologies/switch-from-non-renewable-biomass-for-thermal-applications. Accessed 25 Oct 2016.
3. Asner GP, Knapp DE, Broadbent EN, Oliveira PJC, Keller M, Silva JN. Selective logging in the Brazilian Amazon. Science. 2005;310:480–2.
4. Bailis R, Drigo R, Ghilardi A, Masera O. The carbon footprint of traditional woodfuels. Nat Clim Chang. 2015;5:266–72.
5. Barlow J, Lennox GD, Ferreira J, et al. Anthropogenic disturbance in tropical forests can double biodiversity loss from deforestation. Nature. 2016. doi:10.1038/nature18326.
6. Baccini A, Goetz SJ, Walker WS, et al. Estimated carbon dioxide emissions from tropical deforestation improved by carbon-density maps. Nat Clim Chang. 2012;2:182–5.
7. Bustamante MMC, Roitman I, Aide TM, Alencar A, et al. Toward an integrated monitoring framework to assess the effects of tropical forest degradation and recovery on carbon stocks and biodiversity. Glob Chang Biol. 2016;22:92–109.
8. FAO/IIASA/ISRIC/ISS-CAS/JRC. Harmonized world soil database (version 1.2). FAO, Rome, Italy and IIASA, Laxenburg, Austria. 2012.
9. Feldpausch TR, Jirka S, Passos CAM, Jasper F, Riha SJ. When big trees fall: Damage and carbon export by reduced impact logging in southern Amazonia. For Ecol Manag. 2005;219(2–3):199–215
10. Food and Agriculture Organization of the United Nations. FAOSTAT database on agriculture. http://faostat3.fao.org/browse/F/*/E. Accessed 25 Oct 2016.
11. GOFC-GOLD. A sourcebook of methods and procedures for monitoring and reporting anthropogenic greenhouse gas emissions and removals associated with deforestation, gains and losses of carbon stocks in forests

remaining forests, and forestation. GOFC-GOLD Report version COP19-2, (GOFC-GOLD Land Cover Project Office, Wageningen University, The Netherlands); 2013.

12. Hansen MC, Potapov PV, Moore R, et al. High-resolution global maps of 21st-century forest cover change. Science. 2013;342:850–3.

13. Harris NL, Brown S, Hagen SC, et al. Baseline map of carbon emissions from deforestation in tropical regions. Science. 2012;336:1573–6.

14. Hosonuma N, Herold M, De Sy V, De Fries RS, et al. An assessment of deforestation and forest degradation drivers in developing countries. Environ Res Lett. 2012;7:1–12.

15. Houghton RA. Carbon emissions and the drivers of deforestation and forest degradation in the tropics. Curr Opin Environ Sustain. 2012;4:597–603.

16. Huang M, Asner GP. Long-term carbon loss and recovery following selective logging in Amazon forests. Global Biogeochem Cycles. 2010. doi:10.1 029/2009GB003727.

17. IPCC. Definitions and methodological options to inventory emissions from direct human-induced degradation of forests and devegetation of other vegetation types. In: Penman J, Gytarsky M, Hiraishi T, Krug T, Kruger D, Pipatti R, Buendia L, Miwa K, Ngara T, Tanabe K, Wagner F, editors. Miura: Institute for Global Environmental Strategies (IGES); 2003.

18. IPCC. 2006 IPCC guidelines for national greenhouse gas inventories. Prepared by the National Greenhouse Gas Inventories Programme. In: Eggleston HS, Buendia L, Miwa K, Ngara T, Tanabe K, editors. IGES, Japan. Volume 4 Agriculture, forestry and other land use. Paustian K, Ravindranath NH, Van Amstel A (coordinating lead authors); 2006.

19. IPCC. 2013 Supplement to the 2006 guidelines: wetlands. Prepared by the task force on national greenhouse gas inventories of the IPCC. Chapter 2: Drained inland organic soils. Drosler M, Verchot LV, Freibauer A, Pan G (coordinating lead authors); 2013.

20. Jürgensen C, Kollert W, Lebedys A. Assessment of industrial roundwood production from planted forests. Planted Forests and Trees Working Paper Series No. 48. Rome: FAO; 2014.

21. Lawson S, MacFaul L. Illegal logging and related trade; indicators of the global response. London: Chatham House; 2010.

22. Mokany K, Raison JR, Prokushkin AS. Critical analysis of root: shoot ratios in terrestrial biomes. Glob Chang Biol. 2006;12:84–96.

23. Page SE, Rieley JO, Banks CJ. Global and regional importance of the tropical peatland carbon pool. Glob Chang Biol. 2011;17:798–818.

24. Pearson TRH, Brown S, Casarim FM. Carbon emissions from tropical forest degradation caused by logging. Environ Res Lett. 2014. doi:10.1088/1748-9326/9/3/034017.

25. Saatchi SS, Harris NL, Brown S, et al. Benchmark map of forest carbon stocks in tropical regions across three continents. Proc Natl Acad Sci. 2011;108:9899–904.

26. Tyukavina A, Baccini A, Hansen MC, et al. Aboveground carbon loss in natural and managed tropical forests from 2000 to 2012. Environ Res Lett. 2015;10:1–14.

27. UNFCCC. Estimation of carbon stocks and change in carbon stocks in dead wood and litter in A/R CDM project activities Version 2.0.0. EB 67 Report Annex 23. 2012.

28. van der Werf GR, Randerson JT, Giglio L, et al. Global fire emissions and the contribution of deforestation, savannah, forest, agriculture, and peat fires (1997–2009). Atmos Chem Phys. 2010;10:11707–35.

29. World Bank. Carbon Fund Methodological Framework. 2013. https://www.forestcarbonpartnership.org/carbon-fund-methodological-framework. Accessed 25 Oct 2016.

30. Zarin DJ, Harris NL, Baccini A, et al. Can carbon emissions from tropical deforestation drop by 50% in five years? Glob Chang Biol. 2015. doi:10.1111/gcb.13153.

15

Combining airborne laser scanning and Landsat data for statistical modeling of soil carbon and tree biomass in Tanzanian Miombo woodlands

Mikael Egberth[1]* , Gert Nyberg[2,6], Erik Næsset[3], Terje Gobakken[3], Ernest Mauya[3], Rogers Malimbwi[4], Josiah Katani[4], Nurudin Chamuya[5], George Bulenga[4] and Håkan Olsson[1]

Abstract

Background: Soil carbon and biomass depletion can be used to identify and quantify degraded soils, and by using remote sensing, there is potential to map soil conditions over large areas. Landsat 8 Operational Land Imager satellite data and airborne laser scanning data were evaluated separately and in combination for modeling soil organic carbon, above ground tree biomass and below ground tree biomass. The test site is situated in the Liwale district in southeastern Tanzania and is dominated by Miombo woodlands. Tree data from 15 m radius field-surveyed plots and samples of soil carbon down to a depth of 30 cm were used as reference data for tree biomass and soil carbon estimations.

Results: Cross-validated plot level error (RMSE) for predicting soil organic carbon was 28% using only Landsat 8, 26% using laser only, and 23% for the combination of the two. The plot level error for above ground tree biomass was 66% when using only Landsat 8, 50% for laser and 49% for the combination of Landsat 8 and laser data. Results for below ground tree biomass were similar to above ground biomass. Additionally it was found that an early dry season satellite image was preferable for modelling biomass while images from later in the dry season were better for modelling soil carbon.

Conclusion: The results show that laser data is superior to Landsat 8 when predicting both soil carbon and biomass above and below ground in landscapes dominated by Miombo woodlands. Furthermore, the combination of laser data and Landsat data were marginally better than using laser data only.

Keywords: Soil carbon, Biomass, Landsat 8 OLI, Airborne laser, Miombo woodlands

Background

The Miombo woodlands of Tanzania are under pressure for several reasons, among them a general population increase which brings a need for subsistence agriculture as well as small scale charcoal production [1, 2]. The loss of natural ecosystems is a common pattern which occurs when subsistence agriculture increases in the transition towards an urbanized society with more intensive agriculture [3]. In the case of Tanzania, the National Forest Resources Monitoring and Assessment of Tanzania (NAFORMA; [4]) estimates that the annual consumption of forest exceeds the available resources by 19.5 million m^3 and Hansen et al. [5] estimated a net loss of 17,000 km^2 of forests and woodlands above 5 m height in Tanzania between the years 2000 and 2012. This deficit is currently met by overharvesting inaccessible forests and illegal harvesting in protected areas, thus diminishing the overall forest and woodland area.

*Correspondence: mikael.egberth@slu.se
[1] Department of Forest Resource Management, Swedish University of Agricultural Sciences, Umeå, Sweden
Full list of author information is available at the end of the article

Miombo woodlands are a mosaic of areas with different tree densities, often with a varying degree of degradation. The woodlands are also often mixed with agricultural fields that are covered by crops or have open soil, depending on season. There is a large number of criteria used for defining Miombo degradation [6] of which the United Nations Framework Convention on Climate Change (UNFCCC) definition is related to loss of carbon stock during a certain time period [7].

Soil organic carbon (SOC) is an important part of the soil ecosystem; the disturbance of natural forests in tropical areas, as well as the conversion of forests and woodlands to agricultural land is known to generally reduce SOC [8–12]. Traditionally soil maps have been created where soil types have been classified into taxonomic units. Land degradation is however a continuous process and it is of interest to investigate to what degree remote sensing in combination with field plot data can be an aid for following this process over time [13].

The first attempts to use remote sensing for estimation of SOC were based on the fact that soils with a higher organic matter content, i.e., higher proportion SOC, generally appear darker. This led to studies relating data from electro-optical sensing with organic matter [14–16]. Recent research covering large areas in East Africa confirms that optical satellite imagery could be used to predict SOC as well as other soil properties. Vågen et al. [10] obtained a R^2 of 0.79 when modeling SOC in Ethiopia and Vågen and Winowiecki [17] obtained a R^2 of 0.65 when their study material was extended to also include test sites in Kenya and Tanzania. In both studies, SOC were modeled from Landsat ETM+ data for plots of 1000 m^2, whereas soil data were averaged from one sample from each of four subplots. Winowiecki et al. [11] subsequently obtained an R^2 of 0.85 when modeling SOC on 166 of these plots near Lushoto, Tanzania, and using Rapid Eye optical satellite data instead of Landsat data.

There is also a need for the development of accurate methods for estimation of above ground tree biomass (AGB) and below ground tree biomass (BGB) for carbon accounting, including the measuring, reporting and verification (MRV) needed within countries' efforts to reduce emissions from deforestation and forest degradation (REDD+) ([18–20]; http://www.un-redd.org), as well as for national and regional planning of forest resources. In the case of Tanzania, the sample based NAFORMA inventory is a key source for national level data about forests and woodlands [4, 21], but remote sensing methods used in combination with the field plots will allow estimates both for smaller areas or estimates with lower error [22].

Optical satellite data have been used for estimation of AGB since the launch of the first Landsat satellite 1972

[23]. The use of regression is one of the standard methods for modelling biomass using remote sensing data as independent variables and data from ground reference plots as dependent variables [24]. Landsat multispectral satellite data are a natural first hand choice among the remote sensing data sources, since the data are freely available, have a suitable pixel size of 30 m and wavelength bands suitable for forest monitoring, are regularly provided and offer a data continuity since the 1980s. In particular, the new Operational Land Imager (OLI) sensor onboard Landsat 8 also offers improved performance, such as better signal to noise ratios [25]. Additionally, the new European Sentinel 2 satellite system provides free optical satellite images but with more wavelength bands than Landsat [26].

As an example of early Landsat studies in dry tropical forests, Roy and Ravan [27] used Landsat TM for regression modelling of AGB in dry forest areas in India and obtained an $R^2(adj)$ value of 0.70 on a sample plot level. Gizachew et al. [28] modeled total tree biomass (defined as AGB + BGB) from Landsat 8 OLI data in a recent study in the Liwale district in Tanzania. The field data consisted of 500 plots from the NAFORMA inventory, distributed within an area of 15,700 km^2. They obtained a RMSE of 49% (63% after cross validation) for plot level modeling of total tree biomass using only the Normalized Difference Vegetation Index (NDVI) from one Landsat 8 OLI image as the independent variable.

Airborne laser scanning (ALS) produces a point cloud with three dimensional coordinates for laser returns from the ground and vegetation. ALS data therefore often provide more information about tree canopies than "two dimensional" spectral data from optical satellite data and will generally provide the best data for modeling above ground tree biomass and other tree-size related variables. However, the discrete return ALS systems that are commonly used are not very reliable for estimation of vegetation near the ground. Mauya et al. [29] modeled AGB with ALS using Linear Mixed Modeling (LMM) and similar plot sizes to Gizachew et al. [28]. The obtained RMSE after cross validation was 28.4% for forests, 47.7% for woodlands, and 80.2% for other land cover types. Næsset et al. [22] investigated the use of different remote sensing data sources for a sampling study based on a subset of the plots previously used in the Liwale study area. In addition to the results related to precision for sampling based estimates, their results showed that ALS data provided the best plot level models for AGB ($R^2 = 0.64$) followed by high resolution satellite images from RapidEye ($R^2 = 0.53$). Use of interferometric radar data (InSAR), as well as a global Landsat product and PALSAR L-band satellite data performed less well with an R^2 of 0.25, 0.11 and 0.05, respectively.

Important features of the three dimensional canopy structure can be derived from ALS data, while Landsat, or other similar sensors, measure reflected light in several wavelengths. Since these data sources provide complementary information, the best results can be expected to be obtained when ALS and Landsat data are used in combination. Ediriweera et al. [30] estimated AGB by combining Landsat 5 TM data and ALS data for two study areas in Australia: one subtropical rainforest area, and one Eucalyptus forest. They found that the ALS data performed better than the Landsat data for both sites. The combination of Landsat TM data and ALS data improved R^2 for the Eucalyptus forest by 3%, but did not improve the model for the tropical rainforest.

The purpose of the current study was to compare the usefulness of Landsat 8 OLI data and ALS data, separately and in combination, for modeling of SOC, AGB and BGB in the Miombo woodlands of Tanzania.

Methods
Study area
The study area is located in Liwale District, one of six districts of the Lindi region of southeastern Tanzania. The area is part of the Eastern Miombo Woodland ecoregion, which covers a relatively unbroken area in the interior regions of southeastern Tanzania and the northern half of Mozambique, as well as parts of southeastern Malawi. The study area is a rectangular block of 11.25 km × 32.50 km (total area 36,562 ha), and is the same area as used by Naesset et al. [22] (Figs. 1, 2).

The study area consists of Miombo woodlands, mixed with shifting cultivation and permanent fields of cashew trees together with food crops (Fig. 3). In the upper left

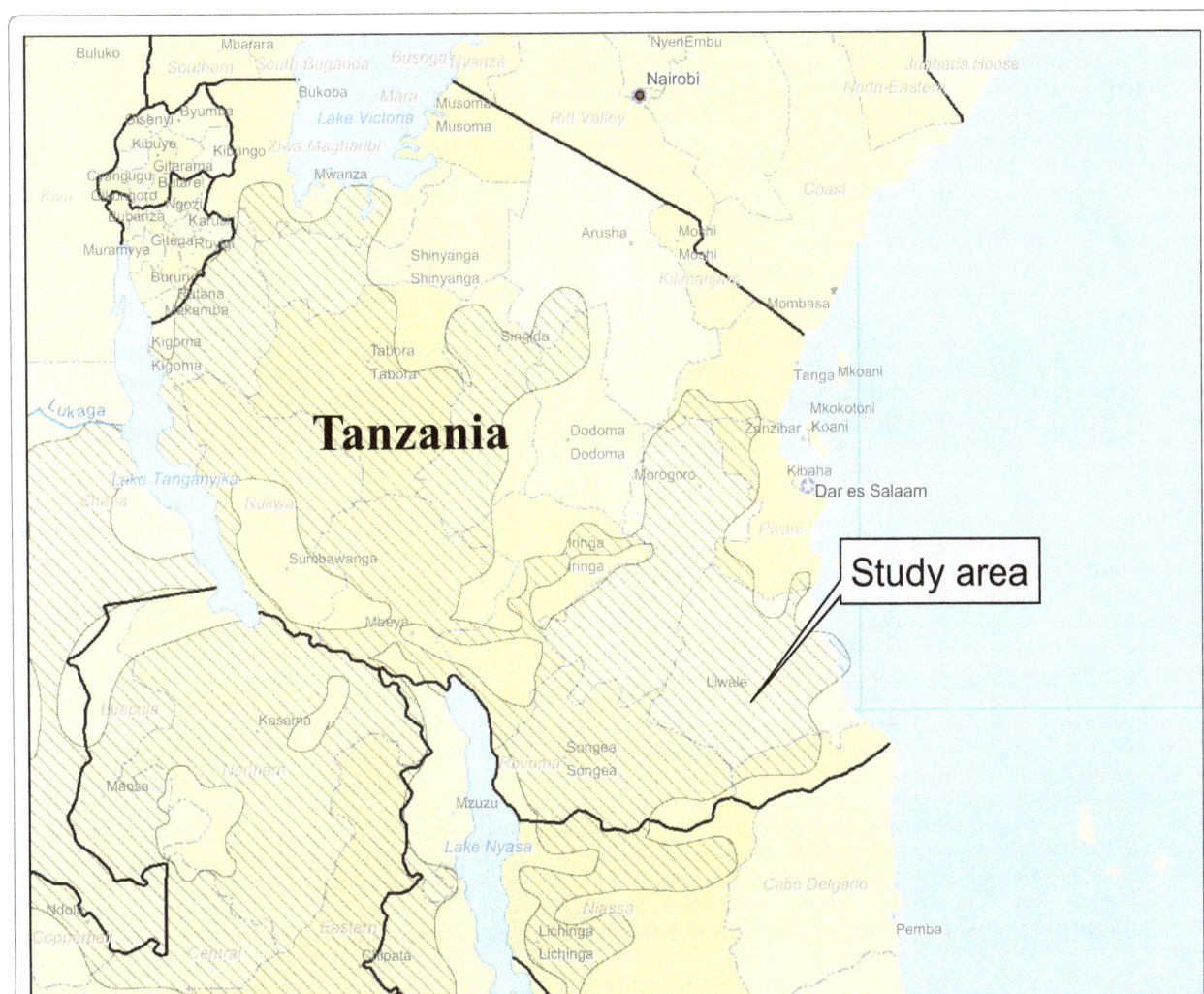

Fig. 1 Location of the study area, the striped pattern roughly indicates Miombo woodland distribution in the area

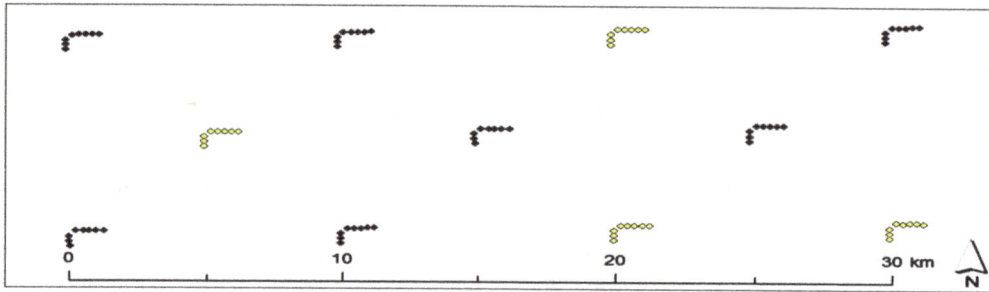

Fig. 2 Location within the study area of the 11 clusters containing eight plots each that were used for collection of field data. The locations of four of these clusters, marked as *yellow*, are identical to the clusters used in the NAFORMA program

corner of the study area a forest protection area occupies approximately 4000 ha, i.e., 11% of the total study area.

The climate in the Liwale area is characterized by two rain periods a year and a main dry season. The shorter period of rain is from late November to January and the longer period is from March to May. The main dry season is between July and October. The annual precipitation is in the range 600–1000 mm. The soils in the Eastern Miombo Woodlands are relatively nutrient poor which limits the agricultural potential. There is

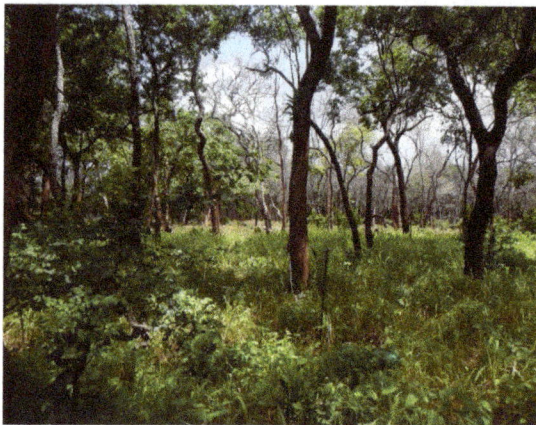

Production forest, AGB =99 Mg/ha.

Agricultural land or shifting cultivation. AGB= 99 Mg/ha. Note piles of branches and leaves around some of the stems. These will be set on fire to remove the trees.

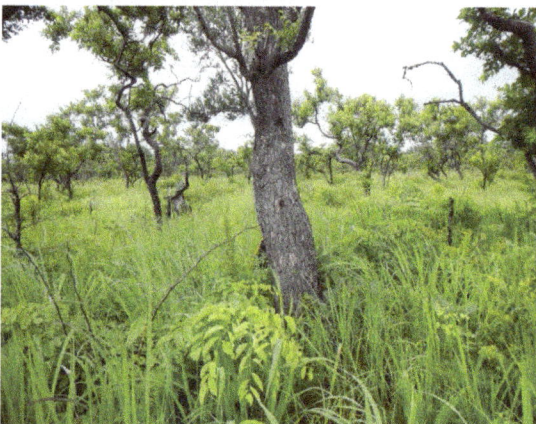

Protection forest, AGB 33= Mg/ha.

Agricultural land, AGB= 0 Mg/ha.

Fig. 3 Photos from different types of land taken on measured field plots within the study area, production forest, agricultural land and shifting cultivation are land use classes defined in NAFORMA [31]

also a widespread presence of tsetse fly (*Glossina* spp.) and vectors of *trypanosomiasis,* which affect the possibilities for settlement of both humans and livestock. However, population growth has increased demand for arable land, thus soils that in earlier years were not profitable enough are now to an increasing degree being utilized.

The Miombo woodlands of Liwale are characterized by high tree species diversity including highly valuable timber species such as *Brachystegia* spp., *Julbernardia* spp. and *Pterocarpus angolensis.* According to the field sample survey conducted within the project, the study area consists of 61% forest or woodland, 14% grassland and 25% cultivated land. The definitions used are according to NAFORMA [31]. In the wooded areas human disturbances occur in the form of harvesting for timber, charcoal burning, honey collection and game hunting. Fire is also an important factor in the Miombo woodlands, underlined by the seasonality in precipitation which leaves the vegetation dry for several months. In the study area, the lack of cattle also leaves large amounts of grasses that dry and therefore are easily set alight (Fig. 4).

Field data measurements

The field plots used as reference data were located in eleven clusters that were systematically sampled (Fig. 1). The locations of four of these clusters are identical to the clusters used in the NAFORMA program [21]. The additional clusters were located in order to obtain a denser systematic cluster design. Each cluster consisted of eight field measured sample plots with a radius of 15 m and a distance of 250 m between the plots. The cluster design as well as the plot size and field protocol were adopted from the NAFORMA program [31] apart from the fact that two plots in each cluster were removed. The removal of these plots was the result of the width of one flight line which could not cover the whole cluster.

The field measurements on the 88 circular sample plots were conducted during January–February 2014. On each

plot, information such as land use, land cover, and disturbance history were recorded, and the plots were photographed. Handheld GPS receivers were used to navigate to the predefined plot centers. For the 32 previously established plots, the plot centers were found and identified according to marks placed at the first measurement [31]. On all plots, the plot center coordinates were determined by means of combined differential global positioning system (GPS) and global navigation satellite system (GLONASS) using a 40-channel dual frequency survey grade receiver as field unit. The field unit was placed in the center of each plot on a 2.9 m rod and data recording lasted for 19–55 min (mean 30 min) with a 1-s logging rate. A second receiver was used as a base station located in Liwale town. Before the positioning of the plots started, the coordinates of the base station antenna were determined with precise point positioning with GPS and GLONASS data collected continuously for 24 h, following recommendations from Kouba [32]. The distances between the plots and the base station were <76 km. Pinnacle version 1.00 post-processing software was used to compute coordinates with the base station as reference. The standard errors of the planimetric plot coordinates reported by Pinnacle ranged from 0.01 to 0.28 m with an average of 0.05 m (Additional file 1).

Soil carbon

Soil carbon measurements were taken from NAFORMA's original plot numbers 4, 7 and 10 from each cluster, using the method described in NAFORMA's biophysical field manual [31]. On the border of each soil sampling plot, four minipits were located in the four cardinal directions. At each vertical minipit wall, starting from the top, a volumetric soil sample was collected from three depths, 0–10, 10–20 and 20–30 cm. Soil samples from the respective depths were bulked into one per plot. Soils were analyzed for carbon content according to Walkley and Black [33] and bulk density [34] and then converted to ton C hectare^{-1}. After removal of five outliers, 28 plots having a valid soil carbon measurement remained

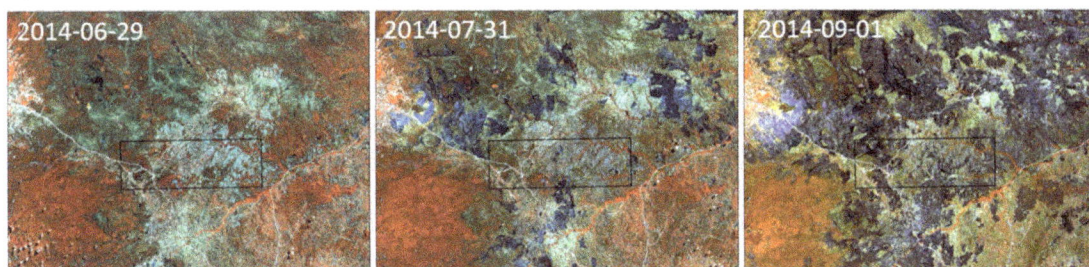
Fig. 4 Around June, bush fires start to appear in the study area. Here illustrated in three Landsat 8 OLI images where black burnt areas clearly spread during July to September

for modeling. The removal of measurements from five plots is unfortunate when considering the small sample, however, a boxplot analyses revealed three extreme outliers. These three samples showed SOC values over 4% and >130 ton SOC ha^{-1} in the top 30 cm of soil which is unrealistic on these predominately sandy soils. Two additional samples had Bulk Density values of <1, i.e. the density of water. These five samples were excluded as lab or sampling errors.

Above and below ground tree biomass

The tree measurements were acquired using concentric circular plots to define the diameter limits of trees to be included in the measurements on each part of a plot. The radii of the concentric circles were 2, 5, 10 and 15 m [21], and trees with diameter at breast height (dbh) greater than 1, 5, 10, and 20 cm, respectively, for the concentric plots of increasing size were measured. A botanist determined and recorded tree species for every tree. Every fifth tree on a plot was selected as a sample tree for height measurement using Suunto hypsometers. For trees without height measurements, tree height was predicted according to diameter-height models constructed from the sample trees. Ground reference AGB and BGB on a plot was calculated by summing individual tree biomass predictions using single-tree allometric tree-species independent models of total AGB and BGB [35] with dbh and tree height as independent variables. AGB on the plots ranged from 0 to 133.5 Mg ha^{-1} with a mean and standard deviation of 51.3 and 45.9 Mg ha^{-1}, respectively. BGB on the plots ranged from 0 to 56.5 Mg ha^{-1} with a mean and standard deviation of 18.6 and 14.5 Mg ha^{-1}, respectively.

Three plots with unusually high biomass were analyzed and removed after confirming that single large trees close to the plot boundary were influencing the measurements to an unproportional degree given that about half the canopy were outside the plots.

Remotely sensed data
Airborne laser scanning data

The ALS data were acquired on 1 March 2014 using a Leica ALS70 laser scanner mounted on a Cessna 404 two-engine fixed-wing aircraft. Twenty-two parallel flight-lines were flown as a block with three additional flight lines perpendicular to the main direction of the block. The maximum half scan-angle was 20 degrees. The flying speed was 77 m s^{-1} and the altitude was 1200 m above ground level. The data were acquired at a pulse repetition frequency of 193.2 kHz and the resulting average pulse density on the ground was 11.9 pulses m^{-2}. The data were processed and every echo was classified as "ground" or "non-ground" by the contractor (TerraTec

AS, Norway) using TerraScan software and the progressive TIN densification algorithm [36]. Heights relative to the TIN surface were computed for every echo.

The software FUSION/LDV [37] was used for computation of metrics from the laser returns and a total of 66 variables were used as candidates for being included as independent variables in the regression models. The ALS metrics were computed using point elevations above ground within the 15 m radius field surveyed plots. A height threshold of 1.5 m was used for most variables to exclude sub-canopy vegetation from the tree canopies. We also calculated various ratios of all returns above 3, 5, 7.5 and 10 m. If two variables had a Pearson correlation above 0.99 or below −0.99 one of the variables was removed before modeling.

Landsat data

Six Landsat 8 OLI images acquired during the period 12 May and 1 September 2014 were almost cloud free over the study area. The reason for this specific time period is that the ground measurements where obtained during spring 2014 and during that year cloud free images before and after the specified dates did not exist. These six images were downloaded from USGS (http://earthexplorer.usgs.gov/). Both Standard Terrain Correction (Level 1T) OLI data and Provisional Landsat 8 Surface reflectance product images (LaSRC, version 2.2) were downloaded for further testing.

The ground control points used for Level 1T correction are derived from the GLS2000 data set (http://landsat.usgs.gov/science_GLS.php). The bands analyzed to determine which OLI scene to use were OLI bands 2–7 and NDVI which had a pixel size of 30 m × 30 m. In the final modelling three OLI band ratios (Band5/Band4, Band6/Band4, Band7/Band4) were added and therefore the statistics obtained from this initial screening of suitable image acquisitions might differ slightly from those obtained for the final models. Based on studies of the images, including analysis of correlations, scatter plots and best subset regressions with the data to be modeled as dependent variables, one of the Landsat images was selected for further modelling of SOC and another image for modeling of AGB and BGB. Image data corresponding to the field plots were extracted using bilinear interpolation.

Regression analysis

Final models for prediction of SOC, AGB and BGB were developed using three sets of sensor data: only Landsat 8 OLI, only ALS, and the combination of Landsat 8 OLI and ALS. Of the total 79 variables used, 10 were obtained from the Landsat 8 OLI data including NDVI and three ratios that by experience is known to be of importance for biomass assessments, 59 from the ALS dataset as derived using FUSION/LDV [37], 7 from combinations of two or more of the FUSION/LDV generated ALS variables and three from the combination of ALS and Landsat 8 OLI data. When modeling forest biomass from only Landsat data, intensity data are, in particular from the mid infra-red bands, often important since much of the tree-size related signal is driven by shadows [38]. Compared to spectral data, ALS data models the tree size related information better. What mainly remain to be modeled with the spectral data is thus the difference between vegetation types, which indirectly also influence the biomass. A combination of different ALS metrics, Landsat 8 OLI bands and ratios between these bands, as well as ALS metrics and spectral data, might therefore improve the regression models.

A best subset routine [39] using stepwise exhaustive search was used to select a number of models with two to six variables. Selection of the final models then depended on studies of model statistics such as Mallow's Cp and Akaike information criterion (AIC), residual analysis and correlation analysis. We applied different log models, multiplicative models and models with square root transformed variables but in the end multiple linear regression performed as well as more complicated models.

Results

Selection of Landsat 8 OLI image

The usefulness of the six available Landsat 8 OLI images as well as the processing levels 1T or surface reflectance calibration (SRC) were compared using visual inspection, scatter plots, correlations and best subset regressions. Table 1 shows the results in terms of R^2 from the best subset regressions using three explanatory variables [39].

Based on studies of the scenes, scatter plots, and the coefficients of determination presented in Table 1, we selected the Landsat 8 OLI scene from 12 May 2014, (scene id LC81660672014132LGNlarge 00), for the further modelling of AGB and BGB. This image was cloud free for 86 of the 88 plots.

We selected the Landsat 8 OLI scene from 31 July 2014 (scene id LC81660672014212LGN00) for modeling of SOC. This image was cloud free for 87 of the 88 plots. We also decided to use the surface reflectance calibrated product (SRC) only, even though the differences between the two levels of radiometric calibration did not influence the final result to a large extent.

Final models

The final models for prediction of SOC, AGB and BGB, using only spectral information from Landsat 8 OLI SRC data, only ALS and the combination of both data sources are presented in Table 2.

Scatter plots of observed versus predicted SOC on the measured plots are shown in Fig. 5.

Discussion

A first observation from Table 1 is that the Landsat 8 OLI data from the end of July, which is about three months into the dry season, were best for modeling SOC. On the other hand, the R^2(adj) values for modeling AGB and BGB from OLI decreased steadily from the May 12th to the July 31st images. The Landsat 8 OLI data from May 12th were therefore best for modeling AGB and BGB. The area disturbed by recent fires also increased between July and September (Fig. 4). In the July 31st image one to four plots were disturbed by fire as seen by fire scars, and in the September 1st image the amount was between 40 and 50 plots. The number of burnt plots was extracted using visual interpretation of the satellite images and is therefore in some cases difficult to classify with certainty. The studies of MODIS and AVHRR satellite data time series have also confirmed a seasonal pattern with high values for tasseled cap greenness [40] and NDVI [41] during the rainy season. Given the increased availability of free optical satellite data, there are therefore good reasons for carefully selecting the optimal image for the given task. It is also evident from Table 1 that the standard Landsat 8 OLI level 1T data performed similarly to the ground reflectance calibrated Landsat data for the task of modeling with field plot data as independent variables.

Table 1 Adjusted coefficients of determination in percent [R²(adj), %] for best subsets regressions (italics) with three explanatory variables for six different Landsat 8 OLI images and two different processing levels

Modeled variable	Processing level	12 May 2014	13 June 2014	29 June 2014	15 July 2014	31 July 2014	1 Sept 2014
SOC	1T	8.9	19.5	20.9	21.8	30.0	30.0
SOC	SRC	11.1	19.8	19.9	20.4	*32.4*	27.9
AGB	1T	30.4	24.1	18.4	15.2	4.5	8.8
AGB	SRC	*30.5*	23.9	17.9	14.6	8.0	9.5
BGB	1T	31.8	22.4	17.7	13.7	4.0	4.6
BGB	SRC	*31.6*	22.6	17.5	13.9	6.8	5.1

Table 2 Results from plot level regression analysis of soil carbon (SOC), above ground tree biomass (AGB) and below ground tree biomass (BGB) using Landsat 8 OLI, ALS and the combination of these data sources

Data source	Model[a,b]	R^2(adj) %	RMSE Mg ha^{-1}	RMSE %
OLI 140731	SOC = −113.8 + 0.0637 B7 + 22.93 B5/B4	34.6	16.2	27.9
ALS	SOC = 74.97 − 0.000500 XL1 + 0.425 XL2 + 0.02500 XL3	42.4	15.2	26.2
ALS + OLI 140731	SOC = −4.2 + 36.93 B7/B4 − 0.000429 XL1 + 0.00433 XL4	56.0	13.3	22.9
OLI 140512	AGB = 35.3 − 0.0661 B6 − 18.41 B5/B4 + 55.20 B6/B4	38.1	30.6	66.2
ALS	AGB = 5.92 − 5.05 P60 + 1.248 PFR50 + 0.576 PFR75	64.4	23.3	50.3
ALS + OLI 140512	AGB = 80.8 − 0.02178 B5 − 3.38 P60 + 1.499 PFR75	66.0	22.7	49.1
OLI 140512	BGB = 39.4 − 0.02009 B5 + 7.12 B6/B4	40.1	11.6	62.2
ALS	BGB = 1.99 − 1.960 MAD − 0.943 P70 + 0.6644 PFR500	71.5	8.1	43.4
ALS + OLI 140512	BGB = 19.43 − 0.00549 B5 − 0.03186 XLS1 + 0.6417 PFR500	71.8	8.1	43.3

The R^2_{adj} statistic is for the model and RMSE values from 'leave one out cross validation' (LOOCV)

XL and XLS = combination of different variables. For full variable explanation see [37]

XL1 = "return 1 count above −1.00" + "total return count above −1.00"

XL2 = "percentage first returns above 1.50"/"P80"

XL3 = "P90" * "percentage first returns above 1.50" * "return 1 count above −1.00"/"total return count above −1.00"

XL4 = "P90" * "percentage first returns above 1.50" * "return 1 count above −1.00"/"return 2 count above −1.00"

XLS1 = "percentage first returns above 1.50" * "P80"/("NDVI + 1"); MAD = elev MAD median

[a] All regression coefficients were statistically significant at 5% level

[b] B = Landsat 8 OLI band (1, 2,…,8); P = Height percentiles of lidar vegetation echoes (0, 10,…,90); PFR = Percentage first lidar returns above heightbreak in dm (50, 75, 100)

The issue addressed in this article is to which degree the combination of Landsat 8 OLI and ALS data can improve models of AGB, BGB and SOC, compared to only using data from one of these sensors separately. It was found that modeling of AGB performed substantially better with ALS than with OLI data (Table 2). This is expected and in accordance with other studies from forest covered landscapes where ALS and optical satellite data have been compared [42, 43]. It was also found that the model for predicting AGB based on ALS was only marginally improved when adding optical satellite data as additional independent variables. This is also in accordance with earlier studies, for example Ediriweera et al. [30] who found improvements in one forest type but not in another, when Landsat TM data were added to an ALS based biomass model. The results for BGB followed the results for AGB, which is as expected since they are modeled from the same field survey of tree stems.

A unique finding is that ALS data that describe the forest and woodland canopy could also be used for modeling of soil carbon. In this study, ALS data were superior to Landsat 8 OLI for modeling SOC. The combination of ALS and Landsat 8 OLI data further improved the models obtained in comparison with either of these sensors separately. A reason for this is that ALS is superior to two-dimensional optical satellite data for the purpose of modeling tree biomass, and SOC generally is positively correlated with tree cover [11]. It is also logical that the

spectral data provide additional information when combined with ALS, since the spectral data will contribute with both information about soil colour and field layer vegetation that is not captured by the ALS point cloud. There are only a few studies where SOC has been modeled based on tree cover data from ALS. Kristensen et al. [44] tested this at a site in a boreal forest in southern Norway and found a weak correlation between tree canopy density and height obtained from ALS and the organic layer C stock. They found a stronger correlation with organic layer C stock and topographical wetness index obtained from the ALS based elevation model. The use of terrain variables for modeling SOC was however not tested in the present study.

Vågen and Winowiecki [17], Vagen et al. [45] and Winowiecki et al. [11] obtained even higher R^2 values when modeling SOC with optical satellite data than obtained in the present study. The reasons contributing to this might be that they used larger plots with more soil samples per plot, and that they had many more field plots which allowed development of more complex models as well as enabling the inclusion of a greater span of data, which tends to improve R^2. The importance of a large plot size was also noted by Mauya et al. [29] who showed that when using ALS data for predicting AGB in a tropical rain forest of Tanzania R2 increased from around 0.4 for 700 m^2 plots to around 0.75 for 2000 m^2 plots. The plot radius 15 m used in this study was mainly because

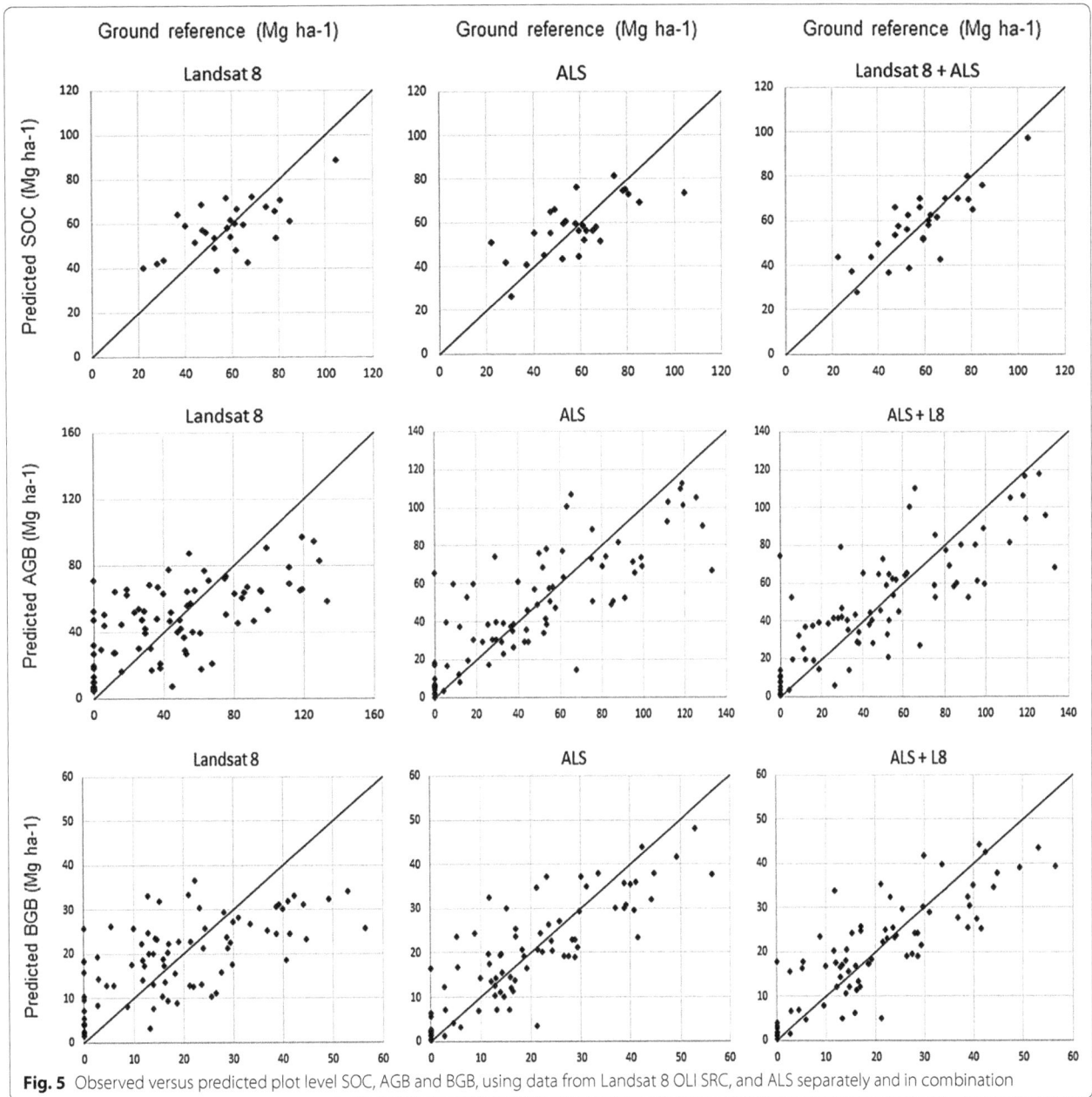

Fig. 5 Observed versus predicted plot level SOC, AGB and BGB, using data from Landsat 8 OLI SRC, and ALS separately and in combination

we used the same field instructions as in the NAFORMA inventory. The fact that three plots had to be omitted because of single large trees near the plot borders indicate however that much larger plots might be needed in the woodlands of Africa than in the boreal where plot radius of about 10 m most often are used.

When modeling AGB with OLI, we obtained plot level R^2(adj) of 38% and RMSE 63% using three explanatory variables. The results for modeling of BGB were similar. As a comparison, Gizachew et al. [28] obtained a plot level RMSE of 49% for AGB + BGB by using only NDVI

from an OLI image acquired over Liwale on 31 July 2014. Their study area was 15,700 km^2 in size and covered a substantial part of the Liwale district and their model was trained with 500 plots from the original NAFORMA inventory, which is stratified for tree biomass [21]. Two of the three explanatory variables we used for modeling AGB with OLI contained short wave infrared (SWIR) bands. These bands are missing on some remote sensing sensors, and their importance for the modeling of forest biomass was already noted when Landsat TM was new [38]. One reason for their importance for forest biomass assessment is probably that the shadows from the trees are more evident in these bands [38]. It should however be observed that the high solar angles in the tropics reduces the effect of tree shadows, and care should therefore be taken when transferring research results about optical forest remote sensing from other latitudes.

Næsset et al. [22] used the same field plots as in the current study, as part of a study regarding remote sensing data as an aid in large area sampling. They obtained the same RMSE, 63%, for modeling of AGB with RapidEye data as we obtained with OLI in this study, while their results with InSAR, Global tree cover maps from Landsat, and PALSAR products were slightly less good. Their model of AGB using ALS has about the same R^2, but the RMSE is slightly lower in the present study, probably since outliers were removed in this study. Our model for AGB with ALS is also very similar to the accuracy and R^2 obtained by Mauya et al. [29] who modeled AGB with ALS over the larger 15,700 km^2 area in Liwale, using plots from the NAFORMA inventory.

There are several sources of error that should be noted. The number of available field plots was limited, especially since only 28 plots were used for modeling SOC. Of the 88 plots with tree biomass measurements, three were not used because large trees near the plot boundary considerably disturbed the relationship with the remote sensing data. Two additional plots were cloud covered in the satellite image used for the biomass modeling. We tried to avoid overfitting by using simple regression functions with few explanatory variables.

Another source of uncertainty that may have affected the model fit is the sub-sampling of trees within a plot as described by the field protocol [31]. The smaller trees were only recorded in the center of each plot, and for a radius of >10 m (the outer 393 m^2 of each plot) only trees with dbh >20 cm were recorded. The amount of biomass for the smaller trees was estimated from the recordings in the inner parts of the plots. When inspecting the plots visually using the ALS point clouds, we noticed some plots in which smaller trees were present in the outer part of some plots for which ALS echoes were included in the AGB prediction for the plot, while no or only a few trees had been recorded in field. To take full advantage of ALS data to improve forest parameter estimates, field protocols should reflect the utility of measuring the same trees on the ground as observed by the remote sensor.

The overall conclusions from this study are that SOC, AGB and BGB can be modeled in Miombo woodlands and forests with Landsat 8 OLI and similar satellite data such as from Sentinel 2 or SPOT, but that even better results are obtained when using ALS data. However, the best results were obtained by combining Landsat 8 OLI data and ALS, in particular when modeling SOC.

Abbreviations

AGB: above ground tree biomass; ALS: airborne laser scanning; BGB: below ground tree biomass; GLONASS: global navigation satellite system; GPS: global positioning system; MRV: measuring, reporting and verification; NAFORMA: National Forest Resources Monitoring and Assessment of Tanzania; NDVI: Normalized Difference Vegetation Index; OLI: Operational Land Imager; REDD+: reduce emissions from deforestation and forest degradation; SOC: soil organic carbon; SRC: surface reflectance calibrated; UNFCCC: United Nations Framework Convention on Climate Change.

Authors' contributions

All authors have made substantial contribution to the successful completion of this manuscript. HO, ME, GN and EN designed the study. ME did most of the data processing as well as the statistical analysis. He also did most of the writing together with HO. GN together with JK and GB did most of the analysis concerning Soil carbon. All authors contributed with the planning of the field campaign, collection of, and analysis of field data. EN and TG planned the survey with airborne laser data and also performed the biomass estimations that were used in later analysis. All authors also took part in, read and approved the final manuscript.

Author details

[1] Department of Forest Resource Management, Swedish University of Agricultural Sciences, Umeå, Sweden. [2] Department of Forest Ecology and Management, Swedish University of Agricultural Sciences, Umeå, Sweden. [3] Department of Ecology and Natural Resource Management, Norwegian University of Life Sciences, Ås, Norway. [4] Department of Forest Mensuration and Management, Sokoine University of Agriculture, Morogoro, United Republic of Tanzania. [5] Tanzania Forest Services Agency, Ministry of Natural Resources and Tourism, Morogoro, United Republic of Tanzania. [6] Department of Business Administration, Technology and Social Sciences, Luleå University of Technology, Luleå, Sweden.

Acknowledgements

This study was financed by the Swedish Research Council, Grant No. SWE-2012-125. The project "Enhancing the measuring, reporting and verification (MRV) of forests in Tanzania through the application of advanced remote sensing techniques" funded by the Royal Norwegian Embassy in Tanzania as part of the Norwegian International Climate and Forest Initiative, also contributed to the data collection. Terratec AS, Norway is acknowledged for collection and processing of the airborne laser data.

Competing interests

The authors declare that they have no competing interests.

Funding

Swedish Research Council; Award Number SWE-2012-125.

References

1. Campbell B, Center for International Forestry Research. The Miombo in transition: Woodlands and welfare in Africa. Jakarta: CIFOR; 1996.
2. Schaafsma M, Morse-Jones S, Posen P, Swetnam RD, Balmford A, Bateman IJ, et al. Towards transferable functions for extraction of non-timber forest products: a case study on charcoal production in Tanzania. Ecol Econ. 2012;80:48–62.
3. Foley JA, DeFries R, Asner GP, Barford C, Bonan G, Carpenter SR, et al. Global consequences of land use. Science. 2005;309(5734):570–4.
4. Anon. National Forest Resource Monitoring and Assessment of Tanzania. Main results: Ministry of Natural Resources and Tourism; 2015.
5. Hansen MC, Potapov PV, Moore R, Hancher M, Turubanova SA, Tyukavina A, et al. High-resolution global maps of 21st-century forest cover change. Science. 2013;342(6160):850–3.

6. Simula M. Towards defining forest degradation: comparative analysis of existing definitions. Rome: FAO; 2009.

7. Eggleston S, Buendia L, Miwa K, Ngara T, Tanabe K. In: IPCC Guidelines for National Greenhouse Gas Inventories, editor. Hayama: IPCC (Intergovernmental Panel on Climate Change); 2006.

8. Ogle S, Breidt F, Paustian K. Agricultural management impacts on soil organic carbon storage under moist and dry climatic conditions of temperate and tropical regions. Biogeochemistry. 2005;72(1):87–121.

9. Schlesinger WH. Biogeochemistry: an analysis of global change. 2nd ed. San Diego: Academic Press; 1997.

10. Vågen TG, Lal R, Singh B. Soil carbon sequestration in sub-Saharan Africa: a review. Land Degrad Dev. 2005;15(1):53–71.

11. Winowiecki L, Vågen T-G, Massawe B, Jelinski NA, Lyamchai C, Sayula G, Msoka E. Landscape-scale variability of soil health indicators: effects of cultivation on soil organic carbon in the Usambara Mountains of Tanzania. Nutr Cycl Agroecosyst. 2016;105(3):263–74.

12. Don A, Schumacher J, Freibauer A. Impact of tropical land-use change on soil organic carbon stocks—a meta-analysis. Glob Change Biol. 2011;17(4):1658–70.

13. Post WM. Monitoring and verifying changes of organic carbon in soil. Clim Change. 2001;51(1):73–99.

14. Alexander JD. A color chart for organic matter. Crops Soils. 1969;21:15–7.

15. Jarmer T, Hill J, Lavee H, Sarah P. Mapping topsoil organic carbon in non-agricultural semi-arid and arid ecosystems of Israel. Photogramm Eng Remote Sens. 2010;76(1):85–94.

16. Steinhardt GC, Franzmeier DP. Comparison of organic-matter content with soil color for silt loam soils of Indiana. Commun Soil Sci Plant Anal. 1979;10(10):1271–7.

17. Vågen TG, Winowiecki LA. Mapping of soil organic carbon stocks for spatially explicit assessments of climate change mitigation potential. Environ Res Lett. 2013;8(1):015011.

18. Burgess ND, Bahane B, Clairs T, Danielsen F, Dalsgaard S, Funder M, et al. Getting ready for REDD plus in Tanzania: a case study of progress and challenges. Oryx. 2010;44(3):339–51.

19. Herold M, Roman-Cuesta RM, Mollicone D, Hirata Y, Pv Laake, Asner GP, et al. Options for monitoring and estimating historical carbon emissions from forest degradation in the context of REDD+. Carbon Balance Manag. 2011;6:13.

20. Joseph S, Herold M, Sunderlin WD, Verchot LV. REDD plus readiness: early insights on monitoring, reporting and verification systems of project developers. Environ Res Lett. 2013;8(3):034038.

21. Tomppo E, Malimbwi R, Katila M, Makisara K, Henttonen HM, Chamuya N, et al. A sampling design for a large area forest inventory: case Tanzania. Can J Forest Res. 2014;44(8):931–48.

22. Næsset E, Ørka HO, Solberg S, Bollandsås OM, Hansen EH, Mauya E, et al. Mapping and estimating forest area and aboveground biomass in Miombo woodlands in Tanzania using data from airborne laser scanning, TanDEM-X, RapidEye, and global forest maps: a comparison of estimated precision. Remote Sens Environ. 2016;175:282–300.

23. Boyd DS, Danson FM. Satellite remote sensing of forest resources: three decades of research development. Prog Phys Geogr. 2005;29(1):1–26.

24. Rahman MM, Csaplovics E, Koch B. Satellite estimation of forest carbon using regression models. Int J Remote Sens. 2008;29(23):6917–36.

25. Roy DP, Wulder MA, Loveland TR, Woodcock CE, Allen RG, Anderson MC, et al. Landsat-8: science and product vision for terrestrial global change research. Remote Sens Environ. 2014;145:154–72.

26. Drusch M, Del Bello U, Carlier S, Colin O, Fernandez V, Gascon F, et al. Sentinel-2: ESA's optical high-resolution mission for GMES operational services. Remote Sens Environ. 2012;120:25–36.

27. Roy P, Ravan S. Biomass estimation using satellite remote sensing data—an investigation on possible approaches for natural forest. J Biosci. 1996;21(4):535–61.

28. Gizachew B, Solberg S, Nasset E, Gobakken T, Bollandsas OM, Breidenbach J, et al. Mapping and estimating the total living biomass and carbon in low-biomass woodlands using Landsat 8 CDR data. Carbon Balance Manag. 2016;11:13.

29. Mauya E, Ene L, Bollandsås O, Gobakken T, Næsset E, Malimbwi R, et al. Modelling aboveground forest biomass using airborne laser scanner data in the Miombo woodlands of Tanzania. Carbon Balance Manag. 2015;10(1):1–16.

30. Ediriweera S, Pathirana S, Danaher T, Nichols D. Estimating above-ground biomass by fusion of LiDAR and multispectral data in subtropical woody plant communities in topographically complex terrain in North-eastern Australia. J For Res. 2014;25(4):761–71.

31. Anon. National Forestry Resource Monitoring and Assessment of Tanzania (NAFORMA). Field Manual. Biophysical survey. In: Tourism MoNR, editor. Dar es Salaam: Forestry and Beekeeping Division; 2010.

32. Kouba J. A simplified yaw-attitude model for eclipsing GPS satellites. GPS Solutions. 2009;13(1):1–12.

33. Walkley A, Black IA. An examination of the Degtjareff method for determining soil organic matter, and a proposed modification of the chromic acid titration method. Soil Sci. 1934;37(1):29–38.

34. Anderson JM, Ingram JSI. Tropical soil biology and fertility: a handbook of methods. 2nd ed. Wallingford: CAB; 1993.

35. Mugasha WA, Eid T, Bollandsas OM, Malimbwi RE, Chamshama SAO, Zahabu E, et al. Allometric models for prediction of above- and belowground biomass of trees in the Miombo woodlands of Tanzania. For Ecol Manag. 2013;310:87–101.

36. Axelsson PE. Processing of laser scanner data—algorithms and applications. ISPRS J Photogramm Remote Sens. 1999;54(2–3):138–47.

37. McGaughey RJ. FUSION/LDV: software for LIDAR data analysis and visualization. 3.42 ed. USDA, United States Department of Agriculture; 2015.

38. Horler DNH, Ahern FJ. Forestry information-content of thematic mapper data. Int J Remote Sens. 1986;7(3):405–28.

39. Lumley T, Miller A. Regression subset selection, package leaps, 2.9 ed. Regression subset selection including exhaustive search. 2009.

40. Mayes MT, Mustard JF, Melillo JM. Forest cover change in Miombo woodlands: modeling land cover of African dry tropical forests with linear spectral mixture analysis. Remote Sens Environ. 2015;165:203–15.

41. Pelkey NW, Stoner CJ, Caro TM. Vegetation in Tanzania: assessing long term trends and effects of protection using satellite imagery. Biol Conserv. 2000;94(3):297–309.

42. Hudak AT, Crookston NL, Evans JS, Falkowski MJ, Smith AMS, Gessler PE, et al. Regression modeling and mapping of coniferous forest basal area and tree density from discrete-return lidar and multispectral satellite data. Can J Remote Sens. 2006;32(2):126–38.

43. Tonolli S, Dalponte M, Neteler M, Rodeghiero M, Vescovo L, Gianelle D. Fusion of airborne LiDAR and satellite multispectral data for the estimation of timber volume in the Southern Alps. Remote Sens Environ. 2011;115:2486–98.

44. Kristensen T, Naesset E, Ohlson M, Bolstad PV, Kolka R. Mapping above- and below-ground carbon pools in boreal forests: the case for airborne lidar. PLoS ONE. 2015;10(10):e0138450.

45. Vågen TG, Winowiecki LA, Abegaz A, Hadgu KM. Landsat-based approaches for mapping of land degradation prevalence and soil functional properties in Ethiopia. Remote Sens Environ. 2013;134:266–75.

PERMISSIONS

The contributors of this book come from diverse backgrounds, making this book a truly international effort. This book will bring forth new frontiers with its revolutionizing research information and detailed analysis of the nascent developments around the world.

We would like to thank all the contributing authors for lending their expertise to make the book truly unique. They have played a crucial role in the development of this book. Without their invaluable contributions this book wouldn't have been possible. They have made vital efforts to compile up to date information on the varied aspects of this subject to make this book a valuable addition to the collection of many professionals and students.

This book was conceptualized with the vision of imparting up-to-date information and advanced data in this field. To ensure the same, a matchless editorial board was set up. Every individual on the board went through rigorous rounds of assessment to prove their worth. After which they invested a large part of their time researching and compiling the most relevant data for our readers.

The editorial board has been involved in producing this book since its inception. They have spent rigorous hours researching and exploring the diverse topics which have resulted in the successful publishing of this book. They have passed on their knowledge of decades through this book. To expedite this challenging task, the publisher supported the team at every step. A small team of assistant editors was also appointed to further simplify the editing procedure and attain best results for the readers.

Apart from the editorial board, the designing team has also invested a significant amount of their time in understanding the subject and creating the most relevant covers. They scrutinized every image to scout for the most suitable representation of the subject and create an appropriate cover for the book.

The publishing team has been an ardent support to the editorial, designing and production team. Their endless efforts to recruit the best for this project, has resulted in the accomplishment of this book. They are a veteran in the field of academics and their pool of knowledge is as vast as their experience in printing. Their expertise and guidance has proved useful at every step. Their uncompromising quality standards have made this book an exceptional effort. Their encouragement from time to time has been an inspiration for everyone.

The publisher and the editorial board hope that this book will prove to be a valuable piece of knowledge for researchers, students, practitioners and scholars across the globe.

LIST OF CONTRIBUTORS

Da B Tran and Tho V Hoang
The Vietnam Forestry University, Hanoi, Vietnam

Paul Dargusch
School of Geography, Planning and Environmental Management, The University of Queensland, Brisbane, QLD, Australia

Ida Theilade
Faculty of Science, Institute of Food and Resource Economics, University of Copenhagen, Rolighedsvej 25, 1958 Frederiksberg C, Denmark

Ervan Rutishauser
Carbo-ForExpert, 1248 Hermance, Switzerland

Michael K Poulsen
Nordic Agency for Development and Ecology (NORDECO), Skindergade 23, 1159 Copenhagen K, Denmark

Beatrice Tarimo
Department of Ecology and Natural Resource Management, Norwegian
University of Life Sciences, 1432 Ås, Norway.
Department of Geoinformatics, School of Geospatial Sciences and Technology, Ardhi University, Dar es Salaam, Tanzania

Øystein B Dick
Department of Mathematical Sciences and Technology, Norwegian University of Life Sciences, 1432 Ås, Norway

Terje Gobakken and Ørjan Totland
Department of Ecology and Natural Resource Management, Norwegian
University of Life Sciences, 1432 Ås, Norway

Wenli Huang, Anu Swatantran, Laura Duncanson, Hao Tang, George Hurtt and Ralph Dubayah
Department of Geographical Sciences, University of Maryland, College Park, USA

Kristofer Johnson
USDA Forest Service, Northern Research Station, Newtown Square, PA, USA

Jarlath O'Neil Dunne
Rubenstein School of the Environment and Natural Resources, University of Vermont, Burlington, USA

David V. D'Amore and Paul E. Hennon
U.S. Department of Agriculture, Forest Service, Pacific Northwest Research Station, Juneau Forestry Sciences Laboratory, 11175 Auke Lake Way, Juneau, AK 99801, USA

Kiva L. Oken
Quantitative Ecology and Resource Management, University of Washington, Seattle, WA 98195, USA.

Paul A. Herendeen
Graduate Degree Program in Ecology, Colorado State University, Fort Collins, CO 80523, USA

E. Ashley Steel
U.S. Department of Agriculture, Forest Service, Pacific Northwest Research Station, 400 N 34th Street, Suite 201, Seattle, WA 98103, USA

Coeli M. Hoover and James E. Smith
USDA Forest Service, Northern Research Station, Durham, NH, USA

Gregory P. Asner, Sinan Sousan, David E. Knapp and Roberta E. Martin
Department of Global Ecology, Carnegie Institution for Science, 260 Panama St, Stanford, CA 94305, USA

Paul C. Selmants
Department of Natural Resources and Environmental Management, University of Hawaii at Manoa, 1910 East–West Rd., Honolulu, HI 96822, USA

R. Flint Hughes and Christian P. Giardina
USDA Forest Service, Pacific Southwest Research Station, Institute of Pacific Islands Forestry, 60 Nowelo Street, Hilo, HI 96720, USA

Yuanming Ni
Department of Business and Management Science, Norwegian School of Economics, Helleveien 30, 5045 Bergen, Norway

Gunnar S. Eskeland
Department of Business and Management Science, Norwegian School of Economics, Helleveien 30, 5045 Bergen, Norway
Centre for Applied Research (SNF), Helleveien 30, 5045 Bergen, Norway

Jarl Giske
Department of Biology, University of Bergen, 5020 Bergen, Norway

Jan-Petter Hansen
Department of Business and Management Science, Norwegian School of Economics, Helleveien 30, 5045 Bergen, Norway
Department of Physics and Technology, University of Bergen, Allegt. 55, 5007 Bergen, Norway

Askia M. Mohammed
CSIR-Savanna Agricultural Research Institute, Nyankpala, Tamale, Ghana

James S. Robinson, David Midmore and Anne Verhoef
School of Archaeology, Geography and Environmental Science, University of Reading, Reading

Roberto Pilli, Giacomo Grassi, Raúl Abad Viñas and Nuria Hue Guerrero
European Commission, Joint Research Centre, Institute for Environment and Sustainability, Via E. Fermi 2749, 21027 Ispra, VA, Italy

Werner A. Kurz
Natural Resources Canada, Canadian Forest Service, Victoria, BC V8Z 1M5, Canada

N. Neeti
Department of Natural Resources, TERI University, New Delhi, India

R. Kennedy
College of Earth, Ocean, Atmospheric Sciences, Oregon State University, Corvallis, OR, USA

Oliver L. Phillips and Roel J. W. Brienen
School of Geography, University of Leeds, Leeds LS2 9JT, UK

Timothy R. H. Pearson, Sandra Brown, Lara Murray and Gabriel Sidman
Winrock International, 2121 Crystal Drive, Suite 500, Arlington, VA 22101, USA

Mikael Egberth and Håkan Olsson
Department of Forest Resource Management, Swedish University of Agricultural Sciences, Umeå, Sweden

Gert Nyberg
Department of Forest Ecology and Management, Swedish University of Agricultural Sciences, Umeå, Sweden
Department of Business Administration, Technology and Social Sciences, Luleå University of Technology, Luleå, Sweden

Erik Næsset, Terje Gobakken and Ernest Mauya
Department of Ecology and Natural Resource Management, Norwegian University of Life Sciences, Ås, Norway

Rogers Malimbwi, Josiah Katani and George Bulenga
Department of Forest Mensuration and Management, Sokoine University of Agriculture, Morogoro, United Republic of Tanzania

Nurudin Chamuya
Tanzania Forest Services Agency, Ministry of Natural Resources and Tourism, Morogoro, United Republic of Tanzania

Index

www.ingramcontent.com/pod-product-compliance
Lightning Source LLC
Chambersburg PA
CBHW082018190326
41458CB00010B/3218